MAIN CURRENTS IN WESTERN ENVIRONMENTAL THOUGHT

MAIN CURRENTS IN WESTERN ENVIRONMENTAL THOUGHT

Peter Hay

Indiana University Press
Bloomington and Indianapolis

This book is dedicated
to past and present students of
KGE515/KGE815 Environmental Values.

They have been my best teachers.

This book is a publication of

Indiana University Press
601 North Morton Street
Bloomington, IN 47404-3797 USA

http://iupress.indiana.edu

Telephone orders 1-800-842-6796
Fax orders 812-855-7931
Orders by e-mail iuporder@indiana.edu

The paper used in this publication meets the minimum requirements of American National Standard for Information Sciences—Permanence of Paper for Printed Library Materials, ANSI Z39.48-1984.

Cataloging information is available from the Library of Congress.

ISBN 0-253-34053-5 (cloth)
ISBN 0-253-21511-0 (paper)

1 2 3 4 5 07 06 05 04 03 02

CONTENTS

PREFACE

Introducing his book, *Green Political Theory*, Robert Goodin writes:

> my aim is not simply to set out, fully and faithfully, what self-styled greens happen for better or worse to believe. Considering the diversity of positions ... that project would quickly degenerate into taxonomic tedium even if it did not prove altogether impossible (1992: vii).

It is just such a book — a book to 'set out, fully and faithfully, what self-styled greens happen for better or worse to believe' — that, some years ago, I determined to write. Whether I have merely demonstrated the validity of Goodin's observation is not for me to say. But I hope not.

I have tried to give a faithful account of the main feeder-streams that debouch into the swift-running river of 'western' environmental thought. I specify 'western' because I have no wish to be arrogantly ethnocentric. There are forms of environmentalism emerging within indigenous and other non-western contexts that draw little upon developments in the industrial west (and as a citizen of an over-developed nation within the *southern* hemisphere I continue to prefer 'west' to 'north'). I cannot write of these other environmentalisms; hence the hedge within the title. Given this acknowledged limitation, however, I hope to have provided an all but comprehensive overview of the complex, burgeoning, endlessly moving swirl that is western environmentalism.

I should also state that where strands of thought have frayed into a myriad sub-strands, as many have in recent years, I have adopted one of two strategies. I have either focused upon seminal contributions, or I have presented the thought of one or more contributors who can be held to be emblematic of recently occurring developments. Where it has seemed feasible, though, I have striven to be comprehensive.

In seeking breadth I have also been mindful of that other axis, the temporal one. Since the 1970s environmental thought has evolved at an extraordinary pace, yesterday's orthodoxy rapidly becoming today's history. Many of the people contributing to this dynamism have evolved commensurately. Where an individual's thought has shifted markedly I have endeavoured to acknowledge this. Where evolution has been more subtle, less readily apparent, I mostly have not. It will be the case, then, that some people will have ideas attributed to them from which they have since, at least in part, moved on. In trying to account for the rills and runnels of environmental thought, even those through which a lessened volume now flows, this has been unavoidable. Nevertheless, to any parties aggrieved on this account I unreservedly apologise.

Each chapter concludes with a 'problems of' sub-section. The ideas and beliefs of the environment movement have been, and continue to be hotly contested. The terms in which outsider criticisms have been couched are beyond the scope of this book. But each strand of environmental thought is also contested, often as vigorously, from within the family of environmentalist positions. It is these, the terms in which *insider* critiques have been constructed, that are the focus of the concluding section of each chapter.

I also pondered long over the vexed question of referencing. I have written with an intelligent, non-specialist reader in mind, and I am aware that for such people textual references are irksome. I *have* managed to dispense with notes, but not with quotes: on the contrary, I have felt it important to allow those who have contributed significantly to environmental thought to express directly their ideas as much as possible, and so I have had recourse to quotation beyond what may be considered usual. I have attempted to integrate quotes seamlessly, so that it should be possible to read straight into, and through them without a sense of dislocation. Such an approach to reading is, indeed, strongly advised. As for sources for citation, I have given prominence to all writings historically important to the development of environmental thought, but I have also made extensive use of less well-known writings when it has seemed to me that they suit a particular need. The book, thus, contains a very large undifferentiated bibliography. I considered several possible schemes before deciding that the standard, single comprehensive list would probably still serve the reader's needs best. The book's text will make evident which bibliographic items constitute the environmentalist canon.

I must acknowledge the unbegrudged assistance of the many people who have helped in the course of writing this book. Discussions at various times with Wolfgang Rudig, John Cameron, Vicki King, Warwick Fox, Robyn Eckersley, Peter Christoff, Michael Jacobs, Janis

Birkeland, Tim Doyle, Tim Bonyhady, Barry Lopez, Jim Russell, John Todd, Julie Davidson, Peter Wilde, Peter Horsley, Richard Flanagan, Tim Boston, Derek Verrall, Jan Howard, David Hogan, Ken Walker, Jan Pakulski, Dallas Hanson, Elaine Stratford, Kate Crowley, Jerry de Gryse, Andy Dobson, Richard Herr, Marcus Haward, Jamie Kirkpatrick, Tony McCall, Christine Milne, Mary Jenkins, Adam Simpson, Paul Sinclair, John Young, Peter Patmore, Geoff Holloway and the late Angus McIntyre constitute the mere iceberg-tip of debate that provided insight and energy. Some of these good and generous folk will not have known how much I valued their intellectual comradeship, and how much, in a practical sense, they have contributed to the book's realisation (for this latter acknowledgement they may not thank me, of course!). To Wolfgang Rudig, who set me on the path in the first place, and to Warwick Fox, who tendered enthusiasm at a time of crisis, I owe especial debts. Thanks are also due to Cheryl Vertigan and the staff on the reference desk at the Morris Miller Library, University of Tasmania, and to my friend and trusty aide in the office of the Centre for Environmental Studies, Glenda Fahey.

And I must finally register my gratitude to two special collectivities. Firstly, to my family — Anna, Tom, Maddy, Murphy, Dougal, Duffy and Pushkin. Were it not for them I would not write; there would be no point. To Anna, Tom and Maddy, thanks for the mettlesome vitality of a family going about its daily, dynamic business. And secondly, my thanks to the students to whom this book is dedicated. They are now too numerous to mention. But I will say this: never could a teacher have wanted for more inspiring, noble and open-hearted company with which to share his working hours.

1 THE ECOLOGICAL IMPULSE

WHAT INSPIRES A 'GREEN' SYMPATHY?

It is difficult to plot the boundaries and contours of the environment movement with precision. Is the environment movement to be equated with the green movement? Are 'environmentalism', 'ecologism' and 'green' synonymous? Are these — all or any of them — political philosophies, ethical systems or mode-of-living designations?

These shades of difference are mightily disputed. Attempts have been made to differentiate 'environmentalism' from 'ecology', 'regarding the former as reformist and the latter as revolutionary' (Giddens 1994: 203). Thus, Murray Bookchin unfavourably contrasts 'environmentalism' with 'ecology', the former being 'more based on tinkering with existing institutions, social relations, technologies and values than on changing them' (1980: 77; see also 1990: 160–62). Similarly, Andrew Dobson uses the term 'ecologism' for application to the green *movement*; this is clearly distinguished from 'environmentalism', which 'would argue for a "managerial" approach to environmental problems, secure in the belief that they can be solved without fundamental changes in present values or patterns of production and consumption' (1990:13). This book will displease those who wish to demarcate these terms, because it treats them as synonyms, for two related reasons.

Firstly, the people who constitute the movement are themselves not fussed about such shades of difference, as they tend to use the terms 'environmental', 'ecological' and 'green' interchangeably. They presumably do so because to them the terms connote common meanings, and thus it seems hardly necessary to drive wedges of pedantry betwixt terms that, in any case, shimmer with political resonance. (The important exception occurs in economic thought, where 'environmental' economics and 'ecological' economics are clearly distinct schools of thought.) And secondly, if the terminology oscillates, so does the amorphous entity it seeks to designate. Commentators have been at a loss to define the precise boundaries of the environment (or 'green') movement; to declaim authoritatively that this group (a birdwatching society, for example) is within, whilst this other group (a save-our-playground action group, for example) is not. The 'real thing', in other words, wavers at the edges.

This raises a crucial issue: the nature of the dialectic between the activist and intellectual components of the green/environment movement. Not all the values and theoretical preoccupations that are much debated within green thought have a commensurate currency within green activism, just as dilemmas that occupy activists are not necessarily given the same attention within the literature.

The obvious case in point concerns structure and organisation. Activist and non- or seldom-active supporters tend to concentrate more on issues and outcomes than on matters of structure and organisation, though they are not, of course, oblivious to such considerations — witness the debilitating controversies within Die Grünen and the British Green Party on these and related matters. Green theory, on the other hand, is increasingly focused upon questions of structure and process: the nature of green democracy, to bureaucratise or not to bureaucratise, and so on. There is a recognisable tendency, for example, to see the construction of a new democratic polity appropriate to the emerging times as the core green project. This will come as something of a surprise to the activist fighting to stop inner-city freeway construction, or for wetlands preservation, or for the adoption of a carbon tax, or for an end to the clear-felling of virgin forests.

I do not mean to belittle the theoretical project of establishing the conditions and processes of green democracy. On the contrary, I think it to be of surpassing importance and have myself contributed to this growing debate. The point I wish to make is that *the wellsprings of a green commitment — at both the activist and more passive levels of identification — are not, in the first instance, theoretical; nor even intellectual. They are, rather, pre-rational.* Though such a commitment may be subsequently justified via recourse to an intellectually generated system of ideas, most people come to a position of green identification *in the first instance* via some trigger of impulse.

This is a generalisation, of course. It does not hold for all green-identifying people. Many people, for instance, come to the green movement from one or other of the socialist traditions. They do so via a front-of-brain process and bring much of their previous ideological baggage with them; for some, indeed, the earlier commitment remains primary. But the generalisation does, I think, hold up.

And what is the 'pre-rational impulse' that, for most people, establishes identification with the green movement? I submit that it is a deep-felt consternation at the scale of the destruction wrought, in the second half of the twentieth century, and in the name of a transcendent human progression, upon the increasingly embattled lifeforms with which we share the planet. It is an instinctual and deep-felt horror at what Holmes Rolston has called 'the maelstrom of killing and insensitivity to forms of life' (1985: 720) that characterises our times.

This is not all that startling a position. Green literature is spiced with observations that rub shoulders with my own. Paul Shepard's thesis that western culture's widespread psychological dislocation (with its attendant planetary harm) can be sourced to our 'heroic' project of disembedding ourselves from nature (Shepard 1982) nudges in the same direction. And Alan Drengson seems to have something similar in mind when he writes, in *The Trumpeter*, of a latent 'ecological unconscious' (1992: 1–2). So do Devall and Sessions when they write of 'the biocentric *intuition* that species have a right to exist and follow their own evolutionary destinies' (1985: 126); so does Skolimowski when he talks of a commitment to ecophilosophy as a non-objective '*disposition* of the human mind' (1981: 32); so does John Hay when he writes of 'the people who know that water, trees and soil measure the terms of their own existence, who have seen them and their wild creatures disappear and have found the world more empty as a result' (1985: 167–68); and so does Snyder when he writes that 'the fertility of the soil, the magic of animals, the power-vision in solitude ... are rooted in the belly' (1980: 3). Even Edward O. Wilson's well-known 'biophilia' thesis — that there is an 'innately emotional affiliation of human beings to other living organisms' (1993: 31; see also 1984; Kellert and Wilson 1993) — seems not too distant from my own observation.

But its most compelling expression is perhaps to be found in Canadian naturalist John Livingston's brilliant little book, *The Fallacy of Wildlife Conservation* (1981). Livingston examines the usual rationally put arguments for wildlife preservation and finds them all wanting. He concludes:

> there can be no 'rational' argument for wildlife preservation, just as there can be no logical explanation of quality experience. It now seems to me that argument *itself* — in the sense of reasonable dialogue — is not only inappropriate to our subject matter but also may be destructive of it. There is no 'logic' in feeling, in experiencing, in states of being. Yet, these same phenomena appear to be the prerequisite for wildlife preservation (1981: 117).

Again, it is stressed that we are dealing in generalisation here. There are people for whom the green trigger is dismay at the greatly enhanced potential of late-industrial technology for increased *human* danger, exploitation and alienation, and for the obliteration of meaningful *human* environments. It seems probable that this is more the case in Europe than it is in North America and Australasia. In Europe the green movement took sprout from seeds within pre-existing radicalisms, with issue concerns and political priorities attendant upon those prior radicalisms merging with newer concerns, whereas in North America and Australasia it can be shown that the green movement has a much weaker relationship with earlier movements and pre-existing systems of ideas (Hay and Haward 1988). Nevertheless, even within the more complex circumstances of green recruitment in Europe the obliteration of once-abundant life seems to be the most potent greening agent.

ROMANTICISM: THE FIRST BROADCAST EXPRESSION OF AN 'ECOLOGICAL IMPULSE'

Until the massive humankind/other nature dislocation brought on by the industrial revolution, the 'ecological impulse' stood in no need of coherent articulation. Not surprisingly, as the transformative impact of early industrialisation became apparent towards the end of the eighteenth century, for the first time an 'ecological impulse' found widespread, articulated expression. This was the so-called 'romantic' movement. As many commentators today — critics in the main, but not always so — portray the *present* environment movement as a form of neo-romanticism, it is worth looking in rather more detail at the ideas and values of the early nineteenth century romantics.

It is difficult to say with precision just what romanticism is — or rather was, for romanticism arose at a particular point of European history in response to certain specific historical conditions. But in one sense it was a nineteenth century movement of reaction against the values, tastes and ideas of the preceding century. That century, the so-called 'Age of Enlightenment', was the most intellectually dynamic period of European history.

The Enlightenment was the age when the secular and the human triumphed over the ecclesiastical and the divine, the nexus between political power and religious authority was broken, and into the void swept a ferment of scientific, philosophical and technological change. It was the great age of European expansion, of the technological innovation that launched the industrial revolution towards the end of the century, of radical new ideas of the greatness of 'man' that culminated in the French Revolution of 1789. The hallmarks of the new age were a faith in the capacity of human reason, particularly via the unfettered application of the scientific method, to know all things, to master nature, and to set humanity on a steady, upward path of eternal progress.

The legacy of the Enlightenment can hardly be overstated. Its key precepts are still the foundational assumptions of western society, and all the great political isms of the present have a common base in these tenacious core assumptions, with only the declining and increasingly marginalised conservative tradition evincing any ambiguity on the matter.

As the full force of the political, economic and social consequences of the Enlightenment became apparent in the crude, dislocative early years of the industrial revolution, romanticism came into being as appalled reaction. It reached back to an earlier, pre-industrial time that was not beset with the social and physical disruptions the romantics found so disturbing in their own day, and which allowed for human sensitivity and individual spiritual fulfilment in a way in which the new hurly-burly world of industrial and political ferment did not. But it *was* first and foremost a reaction; as such, it is easier to articulate what romanticism was against than what it stood for. Lacking a coherent philosophical core, romanticism unfolded differently in different parts of the western world. Indeed, one of its main features was a defence of cultural nationalism against what were seen to be the standardising, cosmopolitan values of the Enlightenment.

One consequence of romanticism's cultural nationalism is that it is hard to type politically. Romantics might be monarchical or republican; conservative, liberal, anarchist, or socialist. In religion they might be orthodox, pantheist or atheist. In England the socialists William Morris and John Ruskin are located within the romantic tradition; but so are the politically conservative poets William Wordsworth and Samuel Coleridge. Romanticism extolled the 'common man', who, against a backdrop of bewildering change, nevertheless retained his sense of having a place in a larger scheme of things and, against the Enlightenment's misplaced faith in its philosophers, was a source of *genuine* wisdom. This could underpin both a conservative and a radical politics. Thus, it is to be found in the conservative romanticism of Wordsworth and the socialist romanticism of Morris.

The relationship between romanticism and science is as confusing as that between romanticism and politics. The Enlightenment was the age in which science, as the application of human reason in its most rigorous form, came into its own. To react against the key values of the Enlightenment was to react against modern science.

The romantic reaction is, however, most properly seen as a reaction against a *particular* science — the technologically applied science in which the all-knowing, all-powerful human stood above and apart from nature, manipulating it in 'his' own interests. Romanticism thus *edged* towards the assertion that living nature is a unity from which humankind cannot be separated; this is not so much anti-science as demanding of a differently founded science, one based upon ecological insights. But, whilst romanticism may have moved in the direction of such a science, in my opinion it failed to quite get there (though the English and North American romantics did have a science, for they had a great interest in 'natural history'; on this see Worster 1985: 59–76, 88–95), its vision of science stopping short of a fully fledged ecological conception.

Donald Worster disagrees. He deems it entirely ecological: 'the Romantic view of nature was what later generations would come to call an ecological perspective; that is, a search for holistic or integrated perception, and an emphasis on interdependence and relatedness in nature' (Worster 1985: 82; a similar view is advanced by Bate 1991). For the romantics, Worster argues, life on earth was seamless. It might be broken up for convenient study, but that was an artificial exercise. Wordsworth used the metaphor of the tree of life, a metaphor that is still prominent within environmental philosophy today (for example, W. Fox 1990: 261–62; Skolimowski 1992). Here, then, is a view of the natural world, and an attendant science, that is biological and concerned with interconnectedness, rather than being, as was the case with the new technologically intensive science, mathematical and concerned with atomisation.

But the picture is, as I have said, far from simple, because not all romantic traditions elevated nature to centre stage; on the Continent, by contrast, 'nature' was only consistently prominent in the writings of the German *naturphilosophie* movement, in which Goethe was preeminent. Romantic perceptions of the natural world are perhaps best appreciated through the writings of their English and North American exponents, and it is to them we now turn.

The 'call of nature' was a fundamental impulse behind English romantic poetry. The so-called 'Lakes poets' (Wordsworth and Coleridge) continually juxtaposed civilisation and 'Nature', to the pronounced detriment of the former.

As a young man Wordsworth supported the French Revolution but, sickened by its excesses, he became its vehement critic, and his early radical romanticism became deeply conservative, supportive of elitist and narrowly based political power. His reaction against industrialism was ultimately more important than his reaction against the radical excesses of the French Revolution. But the failure of the French Revolution is important in one respect: as its disenchanted supporters turned away from politics, the media of romantic expression became poetry, music and art. Thus, Wordsworth applied his romantic vision to his verse. He was deeply influenced by nature, and it is he who is considered the quintessential voice of English romanticism.

Just as his politics evolved, so too did Wordsworth's attitude to 'Nature' change. Whilst the younger Wordsworth had seen it primarily as a source of inspiration, the older man took an apparently more ecological view, seeing nature as humankind's all-encompassing home.

But there are tensions here. There are tensions between 'wild nature' and ruralism. There is a pronounced backward-lookingness to Wordsworth, a yearning for a pre-industrial 'Golden Age'; but the idealised past is decidedly more garden-like than wild. It is the clean, quiet rhythms of the countryside to which Wordsworth pays homage. More importantly, there is the problem of whether Wordsworth's vision of nature ultimately celebrates an ecological worldview or the very opposite — the triumph of extreme individualism. I have argued (P.R. Hay 1988) that the similarity between Wordsworth's view of nature and a modern ecological sensibility can be overstated. Even the most ecologically informed of Wordsworth's writings respond to nature *primarily* as a source of heightened imaginative sensibility. Thus, we find, in 'Tintern Abbey' (the most enrapt philosophic expression of the poet's absorption in nature):

> ... Nature never did betray
> The heart that loved her; 'tis her privilege,
> Through all the years of this our life, to lead
> From joy to joy

His 'immersion in nature' was thus primarily 'a process of elevation of the *human* spirit' (P.R. Hay 1988: 46): as a vigorous assertion of human individuality it may almost be deemed *anti*-ecological, though such a categorisation would not do justice to the incipient ecological tendencies clearly apparent within his work. This given, David Seamon's assessment seems hard to better: 'Wordsworth believed that human feelings work in a kind of virtuous circle whereby contact with nature fosters emotional awareness, which in turn enhances nature's significance' (Seamon 1984a: 765).

'Love of nature', and locality — but also nation: these are his essential values. Wordsworth 'asserts the moral basis of nationalism as in accord with Nature' (Somervell 1965: 63). We will return to this, but it is worth noting here the dangers of combining nationalism with nature worship; the result of such a conjunction, as Bookchin (among others) has pointed out (1986a: 10; 1987a: 4; 1987b: 6; see also Vincent 1993), is the totalitarianism of irrationalist fascism. There are dangers in romanticism, then, and they are not insignificant.

The American nature romantics, the so-called 'Transcendentalists', came together as a group of high-minded young men in the 1830s, and for thirty years they exercised considerable influence upon the intellectual, religious and cultural life of New England.

Ralph Waldo Emerson emerged as their leader. Emerson expounds the notion that a higher 'Reason' of intuitive insight, through which it is possible to attain oneness with God and God's natural creation, is the path to true understanding, rather than the deduction from history and science urged by the philosophers of the Enlightenment. Among the philosophically inclined romantics there is a notion that nature is an aspect of God — to study nature in her concrete manifestations is, therefore, to know God and His creation — and Emerson provides the clearest expression of this. 'Within these plantations of

God', he wrote, 'the currents of the Universal Being circulate through me; I am part and parcel of God' (nd: 311). His romanticism thus proclaims a supreme and sovereign individual — one that is almost god-like — and the function of nature is to serve as the medium for the individual's attainment of a state of high exaltation of the spirit, so much so that 'the whole of nature is a metaphor of the human mind' (nd: 317). Here, then, is an advanced individualism and, as such, it stands little comparison with current, ecologically informed environmentalism.

Ultimately more important is Henry David Thoreau. At the age of twenty-eight Thoreau embarked on his famous experiment — living for two years at Walden Pond, in the woods outside his home town of Concord. This experiment is described in his greatest work, *Walden*, which graphically portrays the zest for life experienced by Thoreau in his time at Walden Pond, and which continues to strike challenging resonances today.

Walden is often mistakenly dismissed as an exhortation to return to a life in the caves and the forests. But Thoreau wrote to contrast the drear complacency of a materialist society with the simple enjoyment he took from his time in the woods. He wished the lessons from his time there to be used to refashion society, rather than for society to vanish in a mass exodus back to the wilds. The key lesson to be learned was the need for simplification: 'simplicity, simplicity, simplicity! I say, let your affairs be as two or three, and not a hundred or a thousand' (1965: 70).

Worster spends much time discussing Thoreau's ecological sensibilities and he does this painstakingly and well. Thoreau was an outstanding natural historian and pioneer ecologist; as Worster points out (1985: 70–75), he did excellent pioneering work on the subject of forest succession. Thoreau, writes Worster, was an 'arcadian': it was Thoreau who was 'most responsible for the development of the arcadian ethic into a modern ecological philosophy' (1985: 76). By 'arcadian', Worster means the outlook that holds that we must learn to accommodate the natural, wild or primitive order rather than seek to overwhelm or transform it. In fact, for Worster, Thoreau is a fully fledged ecocentrist, who sees nature as a vast community of equals (on ecocentrism, see chapter 2).

Again, I am slightly out of step with Worster. There is *almost* this to Thoreau. Certainly, he had no reverence for the institutions of civilisation, which would, like toadstools, disappear with the night. But in Thoreau, too, the romantic's extreme individualism is apparent. His desire to immerse in wilderness was less that of an ecological sensibility than its opposite: a mechanism of individual salvation. He saw it, observes Nash, as 'the source of vigor, inspiration and strength', that is to say, of *human* vigour, inspiration and strength: for Thoreau 'human greatness of any kind depended on tapping the primordial vitality' of wild nature. Nash, thus, seems closer to the mark than Worster when he notes:

> the crucial environment was within. Wilderness was ultimately significant to Thoreau for its beneficial effect on *thought*. Much of Thoreau's writing was only superficially about the natural world. Wilderness symbolized the unexplored qualities and untapped capacities of every individual (1982: 89).

If Thoreau's *primary* concern is to urge us towards a perfection of our inner selves, he is, nevertheless, the closest of the nineteenth century thinkers to present-day environmentalism — particularly in his assertion of the value of wildness, and in his advanced practical understanding of ecological processes. It is no criticism of Thoreau to point out that all this is qualified by his failure to transcend the high individualism that so suffuses the romantic tradition. It is rather a criticism of those who, like Worster, in arguing that 'Romanticism is fundamentally biocentric' (1985: 85), would claim too much for that tradition.

Is the environment movement of today simply recycled nineteenth century romanticism? There is a certain superficial attraction to this idea. The environment movement of today constitutes history's second great articulation of the ecological impulse and, as with romanticism, that expression has sought intellectual clothing appropriate to its sensibilities. It is true, too, that much green literature identifies the Enlightenment value heritage as ultimately responsible for the current ecological malaise. It would not be too difficult to construct a case arguing that the defining feature of modern environmentalism has been its opposition to the 'eternal human progress' agenda of the Enlightenment; the romantic movement, as we have seen, is also best understood as a reaction against Enlightenment values and their practical consequences.

But, in articulating its reaction, romanticism became more complex. It came to stand *for* certain ideas and values as well. This complexity is enhanced by the fact that different romantic ideas and values were emphasised by different writers.

Some of these ideas and values emerge from the outline above of the thought of Wordsworth, Emerson and Thoreau. Prominent among them was a return to nature, but a return to nature specifically as a source of *heightened imaginative sensibility*. It was, thus, individualist rather than collective, and expressive of intuitive or mystical modes of knowing rather than rational. It emphasised the simple life, which tended to be more easily realised in the country than in the city. There was, thus, a tendency to celebrate primitivism, to oppose industrial society, and towards nostalgia for the 'truer' life of the past. And this belief in the past as the repository of accumulated wisdom frequently (but not always) found expression in a change-resistant conservatism and a fervent nationalism, the nation being seen as the embodiment of that common historical wisdom.

How do these characteristics 'fit' today's environment movement? I have argued elsewhere that the 'fit' is not particularly close (P.R. Hay 1988).

As we will see in chapter 2, from the late 1970s until recently most environment movement theorists have been overtly *ecocentric*, placing humankind within an ecological network of inter-species relationships. The point of such a perspective has emphatically *not* been the cultivation of a heightened, *less* wordly, individual imaginative sensibility. It has rather been to reassert the *corporeality* of life; to celebrate its earthliness rather than its ethereality.

There is rather more congruence on the matter of romanticism's anti-rationalism and its faith in unmediated intuitive understanding. There is a marked anti-intellectualism within some present-day forms of green activism (P.R. Hay 1992): a hostility to 'learnedness' that resonates with the romantic notion that all wisdom lies latent within each of us, requiring only the appropriate — which is to say, 'natural' — conditions to become manifest. It is a small step from here to advocacy of the superiority of intuitive and mystical modes of knowing over rational ones. In much green literature this position is strongly articulated (for example, Adler 1989; Eisler 1987: 157–59; Macy 1993; Roszak 1972; 1981). On the other hand, the environment movement also has a base in science. Science has, of course, a place for intuition, but the demands of the scientific method place rationality firmly in the driving seat. Environmentalists who derive their commitment primarily from science are often much puzzled by the priorities and preoccupations of other sections of the movement.

On this criterion, then, there *is* rather more common ground between romanticism and the environment movement of today.

Again, there is within the environment movement an identifiable trend to favour the simple over the complex; the rural over the urban. But here, too, there are complications. Nostalgia for a golden past is only intermittently discernible in this new case for greater simplicity. There is, rather, a tendency to see history as a catalogue of disastrous, options-closing, environmental mistakes. Unlike romanticism, most current environmentalism is oriented firmly to the future, with only some of its thinkers and even fewer of its activists evincing much interest in the past. A consequence of this is that there is more ambivalence on the question of technological advance than was the case with the arcadian romantics. Deep scepticism about technological change *is* widespread today; but at least as prominent is a looking forward to the development of a new life-friendly, *post*-industrial technology that will take us beyond the calamitous age in which we presently live into a more ecologically (and socially) benign time.

Finally, critics on the old political left, alarmed, perhaps, at the potency of the green challenge to the socialist tradition's flagship status within dissident politics, have tagged environmentalism a spurious radicalism; the claim that the green movement is nothing more than a resurgent nineteenth century romanticism — and, so, deeply conservative — figures prominently within this charge. (The label 'romantic' is a damning one in an even more general sense; as Theodore Roszak has written: 'I cannot recall that I have heard the word "Romantic" applied approvingly to a contemporary work of art or thought. The adjective, well exercised in abusive criticism, drips contempt or condescension'; 1972: 277.) But the fervent nationalism and change-resistant conservatism characteristic of the most prominent brand of romanticism strikes few chords within present-day environmentalism, which rejects as ecologically irrational both the unfettered market of the new conservatism and the exalted nation-state of old conservatism. In any case, as we have seen, there is no *necessary* connection between romantic opposition to industrialisation and a change-denying politics. Such a connection is explicit in the case of Wordsworth; in the case of William Morris and John Ruskin, on the other hand, a romantic sensibility is manifested in a commitment *to* change — and to socialism. Even Marx and Engels were susceptible to the romantic pull. How else are we to assess the envisioners of a society that 'makes it possible to hunt in the morning, fish in the afternoon, rear cattle in the evening, criticize after dinner, just as I have a mind ...' (Marx and Engels 1965: 44–45)?

The evidence is mixed, then. There are elements of romanticism within environmentalism, and these are more to be found in some sections of the movement than others. At base, though, the environment movement proceeds from essentially non-romantic values: a holistic ecologism rather than a transcendent individualism; an orientation to the future and lack of interest in history rather than a backward-looking idealisation of a past golden time; a base in a portentous ecosystemic science rather than a 'nature study' aesthetics; and an orientation to social and political change and impatience with the ecological irrationalities of a global system based upon outmoded nationalisms, rather than a change-resistant, nation-focused conservatism.

But romanticism remains important as a first expression of the ecological impulse within the industrial age, and individual romantic thinkers, particularly Thoreau, continue to speak movingly to the environmentalist of today.

MAINTAINING SPACE FOR OTHER LIFE: 'WILDERNESS' PRESERVATION

Romanticism may have been the *first* articulated expression of the ecological impulse, but its status in relation to subsequent articulations is that of undeveloped prototype.

In more recent times the *purest* political expression of the impulse that is the key trigger to a green commitment is to be found in the battles to preserve remaining 'wilderness' areas within such 'new world' countries as the United States, Canada, and Australia; a battle that has more recently been extended to the rapidly disappearing rainforests in countries of the tropics. In these immigrant societies the preservation of wilderness is demonstrably the point of inspiration for modern environmentalism, and the impulse to defend the existence rights of the species contained therein is the main psychological urge behind an environmental commitment. To defend wilderness, in these countries, is the environment movement's first-order goal.

Historical attitudes to wilderness have not always been unambiguously positive. Today wilderness is valued. People want to preserve what is left of it. Until recently, though, the word connoted 'fell, blasted places', and this view of wilderness has a long pedigree. For pre-industrial peoples the prime question faced daily was that of survival. Wilderness was thought to be hostile, threatening, pervasive, and it had to be fought and conquered. Thus, the history of civilisation can be seen as a history of escaping from wilderness; of establishing mastery over it through fire, clearing, cropping, domestication of animals, and so on.

Certainly, tribal peoples value most in wilderness that which provides sustenance. Whilst almost all early cultures had an Eden, it was invariably a wilderness of the useful, with animals living in harmony with humanity. Fear and want disappear. 'Heaven', in most cases, is a 'luxurious garden'. But such a 'paradise' is not really wilderness. It is leached of its 'wild-ness'; it is nature tamed, not nature wild. Similarly, indigenous religions characteristically include celebratory cults of nature. But it is pastoral rather than wild nature that is sanctified. Wilderness, on the other hand, is peopled with the sinister and the supernatural. In folklore and nursery tales, for example, 'the wood' is a fearful place. There lurk the trolls and the dark magicians to waylay the chivalrous knight and ensnare the fair maiden, and there is to be found Red Riding Hood's wolf, and the blood-drinkers of *Beowulf*. In such tales, most of which trace back at least as far as the Middle Ages, wilderness is portrayed in the worst possible light — dank, cold, gloomy, unvisited, immensely threatening.

There is a dissenting tradition, however, which has primitive *homo sapiens* living in harmony with wilderness, at home in it, respecting it as the bountiful provider. This view reached its apogee in the eighteenth century with Rousseau's dissident notion ('dissident' within the larger context of Enlightenment thought) of the 'noble savage'. According to this perception, humanity was now in steady decline from the state of nobility and grace that had existed prior to 'civilisation', and our response to our alienation from the wild is to yearn for a quality missing from our lives, a quality that we can no longer clearly identify.

This ambivalence — a dominant tradition that places humanity at war with other life, and a minority tradition that lodges humanity harmoniously within it — seems to run through western culture. Such ambivalence is obvious, for example, within the Judeo-Christian tradition. Wilderness is an important symbol in that tradition. It was where one was beset with trials and terrors. Wilderness is cursed — even evil. It is Eden after the fall, when God's civilising hand is withdrawn and His creation 'runs wild'. The Old Testament is replete with references, as Nash notes (1982: 14), to a 'great and terrible wilderness'. An Old Testament patriarch who wished to induce a wayward people to mend their ways would threaten, like as not, that God would reduce their habitat to 'wilderness': 'I will make waste mountains and hills, and dry up all their herbs ...' (Isaiah 42: 15).

Another view can, nevertheless, be discerned. Moses went to the wilderness to seek inspiration and guidance from God; here we have a secondary, a 'sanctuary' concept of wilderness. This is even more prominent in the New Testament. John the Baptist went to the wilderness to purify his faith and prepare for the coming of the Messiah (Nash 1982: 17). The two traditions — wilderness as sanctuary, a place where one is close to God; and wilderness as the abode of evil, where one is close to Satan — are here found side by side, for Jesus was subsequently led into the wilderness for forty days to undergo Satan's tempting.

Two events make our current ambivalence towards wilderness explicit: large-scale emigration to the 'new world' in the eighteenth century, proceeding apace in the nineteenth century, and the industrial revolution.

Taking the impact of ex-European colonisation first, the most profound ambivalence over wilderness can be seen to have occurred in North America. On the westward-moving frontier the agenda of taming and mastering nature was refined to its greatest practical expression in the history of the planet. The landscape was made over with extraordinary speed, and the dominant values in the United States (and other societies founded upon European emigration) today affirm the rightness and greatness of that project at a most fundamental level. Roderick Nash (1982: 23–42) writes of the potency of the sense that the wilderness was a loathsome obstacle to be overcome: 'constant exposure to wilderness gave rise to fear and hatred on the part of those who had to fight it for survival and success. American frontiersmen rarely judged wilderness with criteria other than the utilitarian or spoke of their relation to it in other than a military metaphor' (1982: 43). Yet, much American art in the nineteenth century — the landscapes of Thomas Cole, for example — portrayed the vanishing wilderness in idyllic terms, and the nineteenth century writers sang its praises. Emerson, Thoreau, Washington Irving, and James Fenimore Cooper all portrayed wilderness as a source of beauty and moral inspiration (Nash 1982: 44–82).

As the American wilderness vanished at an ever-more alarming rate in the late nineteenth century, so did impetus to preserve parts of it gather steam, and in 1872 the world's first national park — Yellowstone — was proclaimed. The 'high priest' of the preservation movement, founder of the Sierra Club, and the world's first great publicist for wilderness, was John Muir.

Muir was a Scot from a rigidly Calvinist family, and his upbringing found expression in a fire-and-brimstone language in which the virtues of wilderness and religion were potently conjoined. In wilderness God is made manifest on earth. It is 'full of God's thoughts' (Muir 1970: 78; on Muir's wilderness theology, see Oelschlaeger 1991: 182–92), speaking to people spiritually through their intuitive capacity to apprehend the very soul of the universe. Wild country has 'a mystical ability to inspire and refresh' (Nash 1982: 128), and there is in wilderness 'an ancient mother-love' (Muir 1971: 98) that is central to the bodily, intellectual, and, above all, spiritual health of the individual — as against civilisation, which has distorted our sense of 'relationship to other living things' (Nash 1982: 182). Nature romanticism reached its apex in Muir; but, long before 'ecology' — the word or the science — was devised, he also articulated a philosophy 'consistent with the idea of nature-as-an-organism and an ecological worldview' (Oelschlaeger 1991: 194). Like Thoreau, he anticipated the ideas of ecocentric philosophy.

Muir devoted his life to convincing Americans to preserve wilderness and to worship at its 'temple' (a favourite signifier). He also threw himself into political activity, and he was a man for the times, for his activism inspired the first great preservationist movement. His advocacy took him as far as the American president himself, and Theodore Roosevelt became a personal friend.

This was the first great period of national-park creation. But the creation of parks offered little protection from exploitation, because Muir's view of wilderness had a rival. Gifford Pinchot — a professional forester — saw a different purpose for reserves: not preservation, but conservation. Originally close friends, Muir and Pinchot fell out whilst jointly preparing a Forestry Commission survey of woodland that merited protection. For Muir this meant preservation, in perpetuity, from commercial exploitation. For Pinchot it meant 'wise management'. Pinchot pioneered the development of such concepts as 'sustainable yield': a philosophy of 'use in perpetuity, for human consumption'. Pinchot's view prevailed upon Roosevelt, and this was instrumental in the adoption, in 1911, of a plan to turn Hetch Hetchy, part of the Yosemite National Park, into a water-supply reservoir for San Francisco.

The protracted struggle over the fate of Hetch Hetchy was the world's first great wilderness battle. Though lost to the preservationist cause, it gave crucial impetus to the first major wilderness organisations, the Sierra Club and the Audubon Society, which remain important foci of wilderness preservation activity in the United States.

The first of the great professional ecologists-cum-public figures was the American forester Aldo Leopold. Leopold's lovingly observed chronicles of ecological relationships, collected as *A Sand County Almanac, and Sketches Here and There* (published first in 1949), not only constitute the benchmark for modern nature writing; they also take their place among the twentieth century's greatest English language writings – period. And Leopold was alive to the wider biospheric and political significances of his particularised observations. No writer of the present-day, when these matters are much closer to the forefront of public attention, has expressed the great tragedy of imminent and massive species extermination more evocatively than Leopold did in the 1940s.

As a professional forester Leopold was an early disciple of Pinchot, though he was soon to be found fighting within the United States Forest Service for preservationist values — and with considerable success. By the 1930s he was America's leading wilderness campaigner.

What I have described as an 'ecological impulse' Leopold termed an 'ecological conscience' — a respect for life in all its manifold forms. To the sentimentality and spirituality of the Muirist wilderness movement Leopold added a hard-edged science, with ethical principles flowing therefrom. The science of ecology emerged in full flower during his lifetime. For Leopold the perception of 'land and the life it sustains as constituting a large and complex entity functioning through the interaction of its components' was 'the outstanding discovery of the twentieth century' (Nash 1982: 195). Humankind's technological capacity had caused it to lose sight of this discovery, with the result that entire species had been, and were still being extirpated (Leopold called this 'the impertinence of civilisation'; Nash 1982: 196). This would continue, and ecological disharmony would gain momentum, unless a change in attitude could be effected. He therefore devised what he called his 'land ethic': 'a thing is right when it tends to preserve the integrity, stability and beauty of the biotic community. It is wrong when it tends otherwise' (Leopold 1968: 224–25).

Leopold was, thus, the first to argue for a widening of the sphere of ethics to include the natural world, so that it would be deemed unethical — wrong — to regard the environment as our slave, just as in the last 200 years we have come to regard it as wrong to treat other human beings as slaves. 'The land ethic', he wrote, 'simply enlarges the boundaries of the community to include soils, waters, plants and animals or, collectively: the land'. *Homo sapiens* ceases to be the conqueror of 'the land community' and becomes a 'plain member and citizen of it' (1968: 204), with the obligations of respect for the community and its individual members that membership of a community entails.

'WILDERNESS' AND THE PRESENT-DAY ENVIRONMENT MOVEMENT

The question is, then: how did this 'ecological impulse' — this instinctive dismay at the observed impact of technological advance upon the earth's non-human denizens (and perhaps, too, upon the last peoples with lives still embedded in wilderness rhythms) — shape the concerns and values of the environment movement that came to political prominence in the second half of the twentieth century?

Aldo Leopold died in 1948. His books stayed in print through the 1950s and 1960s, and the nature societies kept the Leopold spirit alive through these years. It was not until well into the 1970s, though, when ecologically inspired ethical and social philosophy began to diverge at a bewildering pace, that Leopold's thought, particularly his 'land ethic', returned to prominence.

In a sense, then, he was forgotten — along with Muir and the romantics — because the modern environment movement marks a historical discontinuity, and the land ethic and other Leopoldian ideas had to be subsequently rediscovered by an action-focused movement belatedly seeking a theory. Little of the membership of the burgeoning environment movement of the late 1960s and early 1970s came out of established nature-observing groups (where residual romanticism might still have lingered). The new membership that flooded existing groups and led to the flourishing of new organisations in those years had virtually no familiarity with Muir and Thoreau, and hence cannot be said to be the lineal heirs to the traditions of these men. The politicising spur in the 1960s was, rather, the threat suggested by scientific assessments of pollution and population problems. Those who point to similarities between modern environmentalism and the thought of people such as Muir and Thoreau are right to do so. But such observations obscure the extent to which the post-1960s environment movement was discontinuous with its predecessors. The philosophers and theorists of today's environment movement were preceded by the doom-preaching scientists and their issue-focused followers. Insofar as thinkers from the past have been noted as having relevance to today's concerns — Muir, Thoreau, Morris, Peter Kropotkin, even a figure as recent as Leopold — they have been discovered *post facto* by people seeking a theory for a scientifically inspired movement born largely in a social theory vacuum.

Instead of a seamless development through time, then, the modern environment movement can be quite precisely dated: to the publication in 1962 of Rachel Carson's *Silent Spring*. Though the central concern of that work was the destruction of wildlife, Carson's aim was to demonstrate the culpability of pesticides and other chemicals in that destruction, rather than the impact of vanishing habitat on wildlife. A concern for pollution, population growth and doomsaying thus initially set the tone for the modern environment movement, and the issue of wilderness preservation submerged, to re-emerge in the mid-1970s.

It was in North America and Australasia that a concern for wilderness returned to centre stage, for in Europe (with the exception of the northern reaches of Scandinavia) tracts of land no longer existed upon which the transforming hand of humankind was little in evidence. In Europe the environment movement swung around the twin pivots of the peace and anti-nuclear power movements; these were concerns that could, with little difficulty, be accommodated within the socialist, the liberal, and even the conservative political traditions (Hay and Haward 1988).

But a wilderness-inspired environmentalism is qualitatively different from an otherwise-sourced environmentalism, for the former cannot be derived from the great established traditions of western political thought. Thus, the wilderness movement, at the outset, faced a formidable difficulty. The political debate it engendered was not one of differing conceptions of how resources could best be used in the public interest. Its position was, instead, that a certain course of action — or rather *non*-action — should be taken even if that course of (non-) action could be demonstrated as being *against* the public interest (where 'public interest' is defined as 'activity contributing to the total stock of *human* capital'). To defend wilderness as having its own justification for being, in and of itself — without reference to its use-value to humans — required the asking of some fundamental questions, for the assumption that the natural world is to be reacted to primarily as human 'resource' is a key, and hitherto unquestioned axiom of western history and the economic and technological systems woven into that history.

The impulse to defend the existence rights of wilderness in precedence over human-use rights has thus led to a spirited challenge to what is possibly *the* most fundamental tenet of western civilisation: the belief that moral standing is strictly a human quality, and that no countervailing principle exists to bar humanity from behaving in any way it deems fit towards the non-human world. The challenge posed to that sleepy old branch of western philosophy — ethics — by philosophical developments inspired by the mobilisation to defend wilderness against exploitation for human use has thus opened up the most genuinely new field of ethical disputation that has been seen in many years. It constitutes nothing less than a challenge to the very basis of western value systems. It is a challenge to the assumption — an assumption which is common to all major traditions of western thought — that issues of 'right behaviour' are exclusively questions of human relations, all other entities having no moral standing.

For many years this revolution in moral philosophy was largely a North American and Australasian phenomenon; not surprisingly, given that it is the need to find an in-principle defence of wilderness that has engendered this development. There are important exceptions, but most of the pioneer 'ecophilosophers' of the 1970s and early 1980s developed their positions in non-European intellectual environments — they include animal rights philosophers, of whom the best known pioneer is the Australian philosopher, Peter Singer, 'intrinsic value' theorists such as the American Baird Callicott and the Australian Richard Sylvan, and the more phenomenologically oriented 'deep ecologists' such as Americans Bill Devall and George Sessions and Australians John Seed and Warwick Fox.

The common touchstone to most of these positions is *ecocentrism*, the belief that the earth and its bounty are not the sole preserve of a single species, *homo sapiens*, and that the key ecological insight of the interconnectedness of life should inform conceptions of what is 'good behaviour'. The current state of green thought could hardly be more dynamic, however. In the recent literature it is possible to discern something of a backlash against ecocentrism, the tendency being to find within anthropocentrism all that is necessary to establish the background conditions and inter-species relationships under which non-human life should be able to flourish.

It is, thus, possible to discern a shift away from questions of humankind–other nature relationships; certainly, away from questions of the status of 'wilderness', and probably away from the focus, evident through the 1980s, upon 'pure' questions of ethics. Jostling more and more for space on the agenda of green theoretical concerns there is a number of fresh issues, to do with: social structure; state, corporate and global political power; social justice; economic processes and arrangements; and, perhaps most noticeably, emergent democratic forms of civic process, and the relevance to them of a green perspective.

It is likely that theoretical preoccupation will remain fluid, its orientation somewhat unpredictable. What we can predict with confidence is that activists will, by contrast, remain outcome-focused rather than process-focused, and that the act of instinctive ecological compassion that has here been called the 'ecological impulse' will remain the most potent source of green recruitment.

SOME PROBLEMS, INCLUDING THE QUESTION OF THE 'CONSTRUCTION' OF NATURE

What arguments are customarily raised against the case that has been developed in this chapter?

We need only concern ourselves briefly with romanticism. We have already seen that there are some eloquent dissenters to the qualified assessment of the ecological credentials of romanticism here presented. Worster (1985), in the case of the romantics generally, and Bate (1991), in the case of Wordsworth specifically, both dispute the view that their subjects did not possess a fully developed ecological sensibility. In addition, Nash, earlier adduced in support of the position advanced here, in a subsequent work endorses the Worster position (Nash 1990: 36). But there is much support for the position I have adopted: that, for the romantics, nature was the medium through which the inspirited individual could attain a this-world personal transcendence, rather than the medium through which a corporeal living-in-world wholeness was to be achieved. Thus, John Carroll writes: 'romanticism shared with the Enlightenment a radical individualism ... It kept freedom as its pivotal ideal, an unbounded free will' (1993: 125). And Paul Thompson: 'the romantic movement had been a reassertion of human individuality and emotion, a protest against oppression, whether by despotic government, rigid rules of taste, hypocritical social and religious convention, or industrial squalor and distress' (1977: 167). And Andrew Dobson, also testing romanticism for its green fit:

> as far as romanticism is concerned, Green politics has little time for individualism or for geniuses, and one suspects (though this will be disputed by members of the movement) that the nonconformity so beloved of romantics would be a pretty scarce commodity in Green communities (1990: 10).

Let us leave romanticism, then, and return to the question with which we started — that of meanings endowed to key terms. It can be argued that a movement basing itself upon regard for 'other nature' builds upon flawed foundations. Terms such as 'nature' and 'wilderness' are not unproblematic; they are vigorously contested.

It is often maintained, for example, that such a term as 'wilderness' is saturated with Eurocentric arrogance — that the notion of pristine space 'untouched by human hand' denies the very humanness of the tribal peoples who invariably dwelt therein, and often still do. Furthermore, in this view, when the existence of such peoples *is* acknowledged, it is often via an insulting and sentimentalised interpretation of an idealised harmony in which tribal peoples purportedly live (or have lived) in relation to their environment.

The first of these difficulties — the arrogance held to spring from a denial of human presence — is more apparent than real, being a classic instance of the setting up of a straw man. It is very difficult to find, *within* the environment movement, wilderness definitions of the 'untouched by human hand' type. More typical is the Devall and Sessions definition: 'wilderness is a landscape or ecosystem that has been minimally disrupted by the intervention of humans, especially the destructive technology of modern societies' (1985: 110); or that supplied by Snyder: 'a large area of wild land, with original vegetation and wildlife' (1990: 11). Such definitions may be flawed in other respects — what, for example, can an evolutionary perspective make of 'original vegetation and wildlife'? — but they clearly do not preclude humans in wilderness, neither in the present nor historically. Rather than setting up an objective standard for the hard determination of 'wilderness', a more subjective and flexible notion of 'wild *quality*' is envisaged. In any case, the political potency of the term 'wilderness' (within political systems where a wilderness movement exists) is such that it is unlikely to be soon abandoned.

The second problem — that of the idealisation of the purported harmony of aboriginal peoples' relationships with the land — is the weightier problem, though it is hardly one of definition. Much green thought probably *is* 'guilty' of idealising conceptions of non-European human–environment harmony, though it must be noted that indigenous peoples usually claim that harmony for themselves, especially in contrast with the pathological relationship that exists between western industrialism and the sustaining earth. We will return to this question.

Difficulties with the term 'nature' are rather more formidable and, given the wider applicability and ideological centrality of the term, probably of greater import. '"Nature"', writes Giddens (1994: 204), 'is as important to ecological thought as "tradition" is to conservatism'. Yet it is often argued that 'nature' has no objective existence — that it is merely a 'social construct'.

The claim that nature is a 'social construct' emanates from several distinct — indeed, mutually hostile — perspectives. Currently the most prominent of these is 'postmodernism', but Marxism has the longest pedigree. Marxists and ex-Marxists within the green movement tend to be the most forthright critics of a preoccupation with nature/wilderness.

The Marxist position on nature as a 'mere' social construct is succinctly put by Neil Smith (1990; see also Pepper 1993a: 107–09; 1993b). As Smith sees it, 'nature' is 'socially produced' by human labour via the medium of technology. He is aware that, to some, this is a 'red-rag-to-a-bull position':

> there will be those who see this analysis, indeed the very idea of the production of nature, as sacrilegious effrontery, and a crude violation of the inherent beauty, sanctity and mystery of nature ... But it is real ... Through human labour and the production of nature at the global scale, human society has placed itself squarely at the centre of nature. To wish otherwise is nostalgic (1990: 64–65).

It is important not to read 'dominance over' for 'production of' nature:

> the production of nature should not be confused with control over nature. Although some control generally accompanies the production process, this is by no means assured. The production of nature is not somehow the completion of mastery over it, but something qualitatively quite different ... Thus the industrial production of carbon dioxide and of sulphur dioxide into the atmosphere have had very uncontrolled climatic effects (1990: 62–63).

A variation on the theme 'all that exists is a product or by-product of human culture' is the notion that this present state of affairs was not always so, but that industrial society has brought about 'the death of nature'. Thus, Smith, from his base within Marxism, notes Marx's distinction between 'first' and 'second' nature. 'First' nature was 'prior' nature, unmediated nature, and it was this upon which human labour worked to produce 'second' nature, which can thus be defined as the product of social interaction with first nature. For Marx, 'this unceasing sensuous labour and creation, this production [is] the basis of the whole sensuous world as it now exists' (Marx and Engels 1965: 58). Marx was of the opinion that — even back in the nineteenth century — 'first nature' no longer exists: 'nature, the nature that preceded human history ... today no longer exists anywhere (except perhaps on a few Australian coral islands of recent origin)' (Marx and Engels 1965: 58).

Other observers, perhaps unwilling to assign so early an end to 'first nature', see the end of nature occurring in the age in which we presently live. Giddens states: 'the paradox is that nature has been embraced only at the point of its disappearance. We live today in a remoulded nature ...' (1994: 206). And for Ulrich Beck, nature has become 'a concept, norm, *memory*, utopia, counter-image. Today more than ever, now that it no longer exists, nature is being rediscovered, pampered' (1995a: 38).

A second perspective from whence a 'real-world', tangible realm of nature is denied is that of postmodern deconstructionism. Postmodernism's 'foundational concept', George Sessions argues, is 'cultural relativism', the notion that there is 'no standpoint beyond human cultures'; that reality is nothing more than the separate perceptions of it reached through the prisms of different cultural lenses (1996: 33). Furthermore, as cultures are many and various, fragmenting into sub-cultures and sub-sets of sub-cultures, no single view of reality should be privileged. There can never be an 'objective truth': all theories, 'wisdoms', and schemes of knowledge reflect the interests of the elites that have propagated them.

Given these premises, says Sessions, nature itself is reduced to being a 'social construct', and as a 'mere' social construct it has no greater meaning, no overriding claim upon our care, reverence, delight, awe, respect. Humans can, in fact, 'reinvent nature' at their mere whim. Sessions finds this utterly dismaying. For him the existential interests of nature are beyond human/cultural categories. In the sense that there is a planetary interest in the maintenance of the integrity of biological systems that transcends the vagary of cultural perception, here is a 'grand narrative' — a totalising claim, the validity of which postmodernism denies. We will return to questions arising from postmodernism's intersection with environmental thought in chapter 10.

A third perspective from whence the victory of culture over nature is proclaimed comes, with reluctance, from the mainstream of environmental thought itself. The most prominent expression of this is advanced by Bill McKibben, whose gloomy book, *The End of Nature*, made quite an impact when published in 1989. McKibben argues that the reach of human impact is now so all-encompassing that it has taken over the inner workings of nature — in particular, that the by-products of industrial processes have so changed the composition of the earth's atmosphere that it has become an artefact. And, as that reconstructed atmosphere is omnipresent, it follows that no place remains wherein the realm of culture has not colonised the realm of nature.

Perhaps excepting the McKibben position, the credence accorded the foregoing arguments seems dependent upon the ideological and geographical locus of the observer. Thus, commentators from the Marxist tradition will, almost invariably, portray the notion of the 'social construction of nature' as unproblematic. Observers writing from within landscapes that are undeniably artefactual tend to argue similarly, whether Marxist or not. In parts of the planet where large tracts of wild country are not obviously the products of human enterprise there is more ambivalence. A case can be made along the lines that — whilst Beck and Giddens may well be right in twenty, thirty or fifty years hence — as yet most of Antarctica, the utter depths of the oceans, and much of the world's rainforested areas are patently not artefacts. This is arguably the case even within the wild lands not far from my own back-door in wet and temperate Tasmania. The local Aborigines practised fire-stick farming, and the button-grass plains of the south and west and the dry eucalypt forests of the east are their 'social creations'. But the still-extensive Gondwanaland rainforests are not. A similar argument can be made in relation to the tropical forests. It is true that the world's rainforests have long been home to indigenous peoples, and that those peoples have modified the landscape in their own interests. The scale of modification, though, is the crucial thing. In most cases — there are certainly exceptions — rainforest ecosystems would have differed only a little if humans had never set foot inside them.

I therefore disagree with those who deny the real-world, corporeality of nature. Whilst the time may come when nature is reduced to a sub-set of culture, we have not reached that time yet. And there is another line of argument, according to which it is most unlikely that we ever will. Why should it be assumed that the smallest incursion of culture into nature constitutes the end of nature? It is just as logical to argue the opposite — that, because trees grow in London's parks, and geraniums in its window boxes, London has ceased to be part of the realm of culture, and has become nature. The fact is that there are *natural processes* and there are *cultural processes*, and in any place the mix is likely to be uneven. To focus on the notion of 'process' as the key site-designator for the terms 'culture' and 'nature' thus rescues 'nature' in an important way. Humans might well determine the locations and boundaries of natural systems — settlement here, fauna-rich plain there, rainforest beyond — but they do not, in the two latter instances, construct the intricate biophysical processes and patterns of species interactions that such terms signify. These are natural processes — and likely to remain so for a good many years to come. The only plausible objection to this is the case from McKibben, for his argument has cultural processes invading the most basic of 'biophysical processes'. Again, though, the mix of determining elements within such processes is the crucial factor. And, to date, those human modifications, potentially far-reaching though they be, cannot be said to be the defining constituents of the earth's biophysical systems.

Within environmentalism there is little support for the proposition that there is no real-world referent for 'nature'. There is, however, little agreement over the status that should be accorded some key concepts — particularly 'nature'. For some, the term carries too much unfortunate historical baggage: 'nature has continually been the preferred sign', write Chaloupka and Cawley, 'for the justification of authority' (1993: 5); whilst Phelan notes that 'nature has served as a barrier to social and political change', because 'the claim that a given feature is natural is a way of refusing to consider the possibility of different ways of doing, thinking or being' (1993: 45). Some argue that the term is also abused and debased within present-day discourse. Anyone who has wished to defend the environment interest will have come across the clever-dick riposte, 'if people are part of nature, then everything they do is natural, and so human-induced species extinction is "natural", and so are chemical weapons, and so was Chernobyl, and so was Bhopal, and so was the Exxon Valdez oil spill ...'. Snyder thinks the 'bush logic' of this position should be acknowledged, and he therefore wishes the currency of the term 'nature' restricted. Because its meaning, 'the physical world and all its properties' (1990: 9), ultimately leads to the observation that '*everything* is natural' and, this being so, 'there is nothing unnatural about New York City, or toxic wastes, or atomic energy, and nothing — by definition — that we do or experience in life is "unnatural"' (1990: 8), 'nature' thereby loses definitional precision and personal and social potency. He prefers 'wild'.

Others wish to do away with the concept 'nature' for precisely the opposite reason. Just as Snyder dislikes its all-encompassing shapelessness, others wish to abandon it because, far from being all-encompassing, it is sharply dichotomised with the concept 'culture' in an unnecessary and destructive dualism. Neil Evernden, tracing the evolution of the term 'nature' *as a social construct*, argues that 'one feature has dominated: an ever-deepening sense of separation between the human subject and the surrounding field of natural objects' (1992: 102). Evernden maintains that this dualism has been put to unfortunate political effect, for the human–nature tension so introduced has been made a fulcrum of conflict: 'we must make it *ours*, by one means or another. In the most literal sense, we make nature ours in the domestication of plants and animals' (1992: 120). Thus, '"Nature" is a perilous device, all too easily employed in the quest to dominate others. To consign something to Nature — including ourselves — is to submit it to domination and control' (1992: 132).

It will be seen that we have moved right away from interrogating the 'fact' of nature's real-world existence. Neither Snyder nor Evernden are in any doubt that something we are wont to call 'nature' (what Marx called 'first nature') does, objectively, exist. Within environmental thought, few people have difficulty with the real world/social construct divide. Barry's position has wide support: 'whereas epistemologically humans can never discover nature-in-itself ... the ontological existence of nature-in-itself is an indisputable fact' (1994: 391). Paul Bryant states the distinction thus:

> there is a physical 'world' of which all of us are parts, which exists whether or not I or any other human is aware of it or thinks about it or 'constructs' it. What we perceive and how we perceive it is influenced/limited/shaped by the acuity of our senses, by our level of attention, by our cultural background and knowledge and attitudes, and by many other factors. We 'construct' our own perception of the world, including whatever we mean by 'nature', in our individual consciousness. But that construction does not of itself alter the physical world. It only alters what each of us thinks it is. Our activities affect the natural world, of course. So do those of gypsy moths and any other biological organism (10 February 1995).

There are, though, people who are loth to concede the social construction of nature even in Bryant's semantic sense, on the ground that to make such a concession is to risk the reduction of the natural world to the status of a cultural sub-set; precisely the subordinate position to which Marxists would consign it. It is of a piece with the relegation of the corporeal to the status of a sub-set of mind; or reality to a secondary aspect of abstraction. Thus, Karl Kroeber insists upon a dialectical relationship between the world 'out there' and the mind, itself a product of real-world processes, that appertains the world through the medium of 'social constructs'. To privilege the abstracting mind in this dialectic is to partake of 'the current critical banality that nature is a social construct' (Kroeber 1994: 17).

Others taking this position voice precisely the opposite fear to that articulated by Evernden. Whereas Evernden saw the concept 'nature' as participating in the destruction of its real-world referent, others believe that to acknowledge the social construction of 'nature' is to connive in its real-world destruction. Bryant writes:

> if we turn our regard for nature more and more into clever philosophical word games, if we begin to think that we are intellectually creating nature

> rather than physically participating in it, we are in danger of losing sight of the real wolves being shot by real bullets from real aeroplanes, of real trees being clearcut, of real streams being polluted by real factories (10 February 1995).

The cornerstone of the environment movement may well be the impulse to defend, under conditions of unprecedented threat, the existential interests of other lifeforms. There are, however, substantial difficulties associated with the concept central to that project — the concept of 'nature'. It is apt, then, that much of the intellectual energy within environmental thought should have focused, since the mid-1970s, on the search for a theory to underpin defence-of-nature activism. It is to the philosophical developments engendered by the modern environment movement that we now turn.

2 ECOPHILOSOPHY

TAKING SHAPE IN THE 1970s

The modern environment movement, at the time of its creation in the 1960s, lacked a sense of lodgement within an ongoing tradition — in thought as well as action. It set out on a path that it thought to be wholly novel, the subsequent discovery of precursors being an unlooked-for surprise. Not until the early 1970s did environment movement literature begin to take note of Baruch Spinoza and Alfred North Whitehead; of Aldo Leopold, Peter Kropotkin, and others. The impetus for this new interest in precursors was the need to provide an in-principle underpinning for an activist politics. Consequently, it is only to be expected that, of the fields of intellectual endeavour in which the environment movement made a prompt and substantial impact, it was in moral philosophy — ethics — that the felt effect was most immediate (certain branches of science aside, and acknowledging that the impact within non-mainstream economics was also profound at this time). And, given the challenge posed by new-minted theories of ecocentrism to centuries-old assumptions concerning the scope of the moral commonwealth, this effect had (and still has) the potential to be profoundly revolutionary.

In 1967–68 two of the classic environmentalist texts were published. These were Lynn White Jr's seminal indictment of the Christian religion as *primarily* responsible for the environmental crisis, 'The Historical Roots of our Ecologic Crisis' (1967), and Garrett Hardin's 'The Tragedy of the Commons' (1968), in which he argued that property-in-common and democratic liberty were the chief villains. Hardin's thesis chimed in with the doom-laden prognostications of environment movement science to create a distinctly illiberal politics. But, because it raised the question of human pre-eminence among the species, within *philosophical* circles White's paper had the greater impact.

At the University of Oslo Arne Naess, influenced by Gandhi and Spinoza, had already begun to focus on the human–nature dialectic. Naess resigned his philosophy chair in 1969 and embarked upon a period of intellectual production and environmental activism (W. Fox 1990: 89–91). In 1973 he published a paper entitled 'The Shallow and the Deep, Long-Range Ecology Movement: A Summary'. This paper was to contribute greatly to the burgeoning of environmental ethics, the concepts of 'shallow' and 'deep' giving rise to the variant of environmental thought known as 'deep ecology', which predominated within ecophilosophy through the 1980s and remains a major strand of ecophilosophical thought. Other important elucidations of deep ecology were provided by George Sessions, whose six-issue newsletter, *Ecophilosophy*, first published in 1976, provided a crucial focus for the early shaping of deep ecology in North America, and Bill Devall (Sessions 1974; Devall 1977; 1980). We will revisit deep ecology shortly.

Also in 1973 (published in 1976), Richard Routley (from 1985, Richard Sylvan) presented his essay, 'Is There a Need for a New, an Environmental Ethic?', and the search was on in earnest for just such an ethic, with John Passmore (1974), Holmes Rolston III (1975), Kenneth Goodpaster (1978; 1979), William Godfrey-Smith (1979), William Frankena (1979), and J. Baird Callicott (1979) making significant early contributions. Callicott played no small part in redirecting attention towards Leopold's land ethic and its contemporary potential. In 1979 Eugene Hargrove commenced production of the journal *Environmental Ethics*, which served as both catalyst and focus for scholarship in the field defined by the journal's title, and helped sustain the ethics orientation of extra-scientific environment movement scholarship through the 1970s.

So, too, through the 1970s the ecological application of anarchist principles developed by Murray Bookchin under the rubric of 'social ecology' proceeded apace, whilst, from the 1975 publication of Rosemary Radford Ruether's *New Woman, New Earth*, ecofeminist thought also began to exert strong impact. Bookchin, indeed, could claim to have been in the field from the very first, his *Our Synthetic Environment* having been published in 1962. In the 1970s the divisions that were to open between deep ecology, social ecology and ecofeminism had not yet become manifest. More about this later; but we should note here that, at least as far as the two rival critiques of deep ecology are concerned, much of the problem stemmed precisely from deep ecology's preoccupation with ethical refurbishment. Any scheme of environmental thought abstracted from a social context and lacking a theory of political power, it was argued, is inadequate and misguided. Both social ecology and ecofeminism engage with ethics, but in each case this engagement is subordinate to a theory of power. On this ground neither is taken up as subject matter for this chapter.

Yet another strand was added in 1975 when Peter Singer published his pathfinding work on the ethical status of animals, *Animal Liberation*. Singer did not develop his ideas within the still-young context of environmental value theory, but took impetus from the observed situation of certain feedstock animals, and applied these conditions against the pleasure–pain principle of the English utilitarian philosopher, Jeremy Bentham (1748–1832): that an act is moral if it results in an overall increase in pleasure, and immoral if its consequence is an overall increase in pain. Such a principle establishes that all entities have a capacity to experience pleasure and pain as moral 'subjects'. Although subsequent Benthamites assumed this principle had an application solely to humans, the higher animals have a demonstrable capacity for the experience of pain, as Singer points out, and their interests must therefore be incorporated within an application of the pleasure–pain principle. Indeed, Bentham himself had intended such an application of his principle: 'the question is not, can they *reason*?', he wrote of animals in 1789, 'nor can they *talk*? but can they *suffer*' (1970: 283). Singer's book had considerable impact, and a movement was born.

By the late 1970s, then, the intellectual terrain of environmental thought contained the following major features:

- theories of power to which an ethics was merely an adjunct (social ecology, ecoMarxism, ecofeminism);
- ecocentric axiologies (an 'axiological' system of thought posits an objective and universal ground, or grounds, for value);
- a particularly influential ecocentrism that rejected value-based thinking in order to find a base, in Livingston's apt phrase (1984: 61), 'not in moral guidance but experiential knowing' (deep ecology);
- an ethical system that certainly widened the scope of moral subjectivity, but retained a focus upon the individual as the morally relevant unit, and was thus only in a rudimentary sense 'ecological' (animal liberation).

We have seen that the generation of ecocentric schemes of thought may be attributable both to an 'ecological impulse' and to a desire to find, for the defence of tracts of wild land and wild beings, in-principle arguments on non-contingent grounds. It can also be argued that developments in the science of ecology rendered the emergence of philosophical ecocentrism an obvious consequence: 'the ecological world view' suggests that, 'as the world is based upon systemic processes and relationships, our values and actions should be consistent with systemic reality' (Sterling 1990: 81). And it can also be argued that, politically, such a project is reactionary rather than radical. These contentions will be considered later. But we have also noted that *philosophically* this project was extraordinarily radical. 'For virtually the entire duration of Western intellectual history', Warwick Fox has written, 'ethical discussion has not had any direct concern with moral obligations that humans might be thought of as having toward the nonhuman world' (1996: 1–2).

Earlier we noted that one possible engagement with this project — the non-axiological engagement — is that taken by Livingston (1981), who could find no rational argument that unambiguously met his need. He concluded that the compelling case for wildlife preservation is to be found in the sub-rational sense, lodged within the very core-of-being of unalienated humans, of a deep complicity 'in the beauty that is life process' (1981: 117).

An opposing axiological position was adopted by William Godfrey-Smith, who, in a much-read paper published in 1979, identified four grounds upon which arguments in defence of wildness are customarily made. He labelled these: the 'cathedral' argument, the 'laboratory' argument, the 'silo' argument, and the 'gymnasium' argument. The 'cathedral' argument was that 'wilderness areas provide a vital opportunity for spiritual renewal, moral regeneration, and aesthetic delight' (1979: 310). According to the 'laboratory' argument, 'wilderness areas provide vital subject matter for scientific inquiry which affords us an understanding of the intricate interdependencies of biological systems', important because 'if we are to understand our own biological dependencies, we require natural systems as a norm, to inform us of the biological laws which we transgress at our peril' (1979: 311). The 'silo' argument, which values wild places as 'a stockpile of genetic diversity', is important 'as a back up in case something should suddenly go wrong with the simplified biological systems which, in general, constitute agriculture', and which will almost certainly prove to be repositories of important medicinal and food sources as yet unidentified (1979: 311–12). Finally, there is the 'gymnasium' argument, which 'regards the preservation of wilderness as important for athletic or recreational activities' (1979: 312). Godfrey-Smith found all these justifications inadequate and partial, arguing the need to acknowledge that biological systems themselves 'possess intrinsic value ... that they are "ends in themselves"' (1979: 318). Of course, one need not adopt an axiological position to accept Godfrey-Smith's categorisation of arguments typically used in defence of wild nature. The most philosophically rigorous of the deep ecologists, Warwick Fox, refines Godfrey-Smith's four arguments into no fewer than nine in-the-interests-of-humankind reasons typically used to support the preservation of wild nature (1990: 155–61).

Our survey of the swirling clouds of ecophilosophy's shape-taking days in the 1970s must not end without acknowledging several extraordinary contributions from a California-based political philosopher, John Rodman. It is a pity that Rodman is little read today, because the brilliant series of papers that he produced in the 1970s and early 1980s remains as vigorous and challenging as it was at the time of its production. Rodman developed a case for an environmentalism based within liberalism's non-market stream, with its stress upon such values as tolerance, diversity, citizenship and virtue. In early papers he located himself within the tradition of John Stuart Mill (1977a; 1978) and T.H. Green (1973), and against that of Robert Nozick (1976a); a persuasively mounted case that has had, nevertheless, little lasting influence. He argued a case for ecologically grounding political science (1980) — seed that also fell upon stony ground. He produced possibly the first detailed critique of the animal rights position from an ecocentric perspective within environmental thought, and this 1977 paper, 'The Liberation of Nature?', remains one of his two most enduring contributions. He produced one of the first theories of ecological resistance, a theory grounded within his increasingly marginalised liberal tradition. Rodman here emphasised that 'diversity is natural, good and threatened by monoculture', and that 'the struggle between diversity and monoculture is perceived to occur in different spheres of experience':

> a good historical example is provided by John Stuart Mill, who defended biological diversity against the threat of a wholly humanized planet, West Indian blacks against European racists, women against the tyranny of a patriarchal tradition, and the many-sided personality against the totalitarian claims of economic/technological rationality (1978: 53).

Work in progress by the Canadian political theorist, Douglas Torgerson (for example 1998), may go some way towards redirecting attention to Rodman's remarkable achievement.

Of course, not all the above papers are of a purely philosophical nature. Though his writings are imbued with an overt and non-axiological ecocentrism, one of the claims I would make for Rodman's importance is that he was an early and skilful drawer of cross-disciplinary insights. This holds for the two papers that are, in my view, his most outstanding accomplishments: academic papers both, obeying all the protocols of academic writing and published in professional journals, but taking the form of stories written in the genre of science fantasy. These are 'The Dolphin Papers' (1974) and 'On the Human Question: Being the Report of the Erewhonian High Commission to Evaluate Technological Society' (1975). In 'The Dolphin Papers' Rodman provides a flamboyant critique of the basis of modern science as it is found within classical thought and in the philosophies of Francis Bacon and René Descartes. This is a critique which has since entered the mainstream of environmental thought, and to which we will have recourse in chapter 5. In 'On the Human Question' he develops a long, brilliantly written critique of the technological imperative within modern European thought.

But the second of Rodman's papers which, along with 'The Liberation of Nature?', continues in public prominence is 'Four Forms of Ecological Consciousness Reconsidered' (1983, updating a project commenced with 1976b). The historical importance of this paper is that it was the first taxonomy of ecophilosophical thought. It was not, however, the first creation of environmental typologies; as Warwick Fox (1990: 22–35) points out, several of these were generated in the 1970s, usually presented as a single pair of opposites. But we will start with Rodman.

TAXONOMIES OF ENVIRONMENTAL THOUGHT

In his 1983 paper Rodman organised environmental thought into four categories. To some extent these categories represented phases of historical unfolding, with a progression towards a satisfactory maturity in the final, still-emergent phase, 'Ecological Sensibility' (1983: 82).

His first category — the least normatively satisfactory — is 'Resource Conservation', the position associated with Gifford Pinchot, in which 'the reckless exploitation of forests, wildlife, soils, etc.' was replaced by 'ethical and legal requirements that "natural resources" be used "wisely"', in the interests of humanity at large rather than in the interests of a mere few, and 'considered over "the long run"' rather than the short term (1983: 82). Rodman's second category, 'Wilderness Preservation', is that associated with John Muir and the view that 'certain natural areas were sacred places where human beings could encounter the holy' (1983: 84). His third category is 'Moral Extensionism', the view that humans have duties 'directly to (some) nonhuman natural entities, and that these rights are grounded in the possession by the natural entities of an intrinsically valuable quality such as intelligence, sentience or consciousness' (1983: 86). Rodman cites Singer's 'zoocentric sentientism' (1983: 87) as exemplar of this position, and it is a position of which he is trenchantly critical: 'we are asked to assume that the sole value of rainforest plant communities consists in being a natural resource for birds, possums, veneer manufacturers, and other sentient beings' (1983: 87).

All these modes of environmentalist thought will be rendered obsolete by the coming of 'Ecological Sensibility' — 'a complex pattern of perceptions, attitudes and judgments which ... would constitute a disposition to appropriate conduct that would make talk of rights and duties unnecessary'. Nevertheless, the first of the three components of 'Ecological Sensibility' is 'a theory of value that recognizes intrinsic value in nature without engaging in mere extensionism'. The other components are 'a metaphysics that takes account of the reality and importance of relationships and systems as well as of individuals', and an ethics that 'includes such duties as noninterference with natural processes, resistance to human acts and policies that violate the noninterference principle ... and a style of coinhabitation that involves the knowledgeable, respectful, and restrained use of nature', for 'one ought not to treat with disrespect or use as a mere means anything that has a *telos* or end of its own — anything that is autonomous in the basic sense of having a capacity for internal self-direction and self-regulation' (1983: 88). From such a standpoint the Resource Conservation position is revealed to be 'an ideology of human chauvinism and human imperialism' (1983: 89).

Fox developed Rodman's typology into a more comprehensive taxonomy. Again, the types are organised on a continuum moving from least satisfactory to most satisfactory.

The least satisfactory categories fall under the organising subheads of 'Anthropocentric Approaches' (W. Fox 1992: 2–7) and 'Instrumental Value Theory' (1990: 150–52; 1996: 3–18). These are the frameworks within which 'the vast majority of environmental discussion is couched' (1992: 2), and they are in evidence 'when we see the nonhuman world as there to be dammed, farmed, pulped, slaughtered, and so on', or 'whenever we argue that the nonhuman world should be *conserved* or *preserved* because of its use value to humans' (1992: 2).

Fox's first category is 'Unrestrained Exploitation and Expansionism'. This is the position that has dominated modern western history and has only recently come under serious challenge: that there is only value in the non-human realm when it is physically transformed by human agency into economic resource for human consumption; that 'resources' are infinite (Fox terms this 'the myth of superabundance'); that no problem is beyond technological resolution; and that, therefore, there is no need to consider the lot of future human generations, let alone the future of non-human forms of life (1990: 152–53; 1992: 2–3).

It can be objected that this is not a form of environmentalism at all — quite the opposite, in fact — and that it, therefore, has no place in a typology of environmental thought. Fox concedes that this is so, adding:

> it represents a way of thinking about the environment in terms of its use value to humans — or what philosophers refer to as its instrumental value — that has been of the first importance for the situation in which we find ourselves today. Indeed the influence of this approach has been and is still so pervasive that people will start to claim that they are thinking 'environmentally' as soon as they begin to acknowledge any kind of restraint upon this free-for-all, frontier ethic approach (1992: 4).

Environmental thought proper thus begins with Fox's second category, 'Resource Conservation and Development', which remains firmly within the anthropocentric tradition that only human interests count, and that value only enters the natural world at the point of its transformation into product for human consumption. But it concedes that nature is not inexhaustible: accordingly, that there are limits to material growth, and that husbandry must therefore be practised because it becomes necessary to consider the interests of human generations as yet unborn when determining courses of action. This position is fast supplanting the 'Unrestrained Exploitation and Expansionism' category as the orthodoxy of late-capitalist modernity.

Fox's third 'anthropocentric' or 'instrumental' category of environmental thought ('instrumental' in that 'the nonhuman world is considered to be valuable only insofar as it is *instrumental* to human ends'; 1990: 150) is 'Resource Preservation'. This is a stance which does seek grounds for preservation of the non-human world (as distinct from its husbanded use), but which stays within the framework of assumption that is adopted in the two previous positions: right and appropriate action is deemed to be right and appropriate from the standpoint of human interest. It is the category within which the four historical justifications for wilderness preservation identified by Godfrey-Smith are located. Fox differentiates 'Resource Preservation' from his previous category in this way:

> the resource *preservation* approach tends to stress the instrumental values that can be enjoyed by humans if they allow presently existing members or aspects of the nonhuman world to follow their own characteristic patterns of existence. The emphasis in the resource conservation and development position, on the other hand, is on the values that can be realized by dramatically altering (physically transforming) the characteristic patterns of existence of presently existing members of the nonhuman world ... albeit in a way that is presumed to be reasonably sustainable (1992: 5).

The three anthropocentric positions 'represent progressively more sensitive approaches to environmental conservation and protection' (1992: 7), but even the Resource Preservation category remains, for Fox, an unsatisfactory way-station on the road to ecological responsibility. He thus moves from anthropocentric to ecocentric forms of environmental thought; these he finds much more satisfactory, though they again continue within his hierarchical arrangement of least to most satisfactory. Fox has two overarching classes of ecocentric thought, and these roughly correspond with the distinction made above between axiologically and experientially grounded systems of thought.

He identifies three grounds upon which an axiological (intrinsic value) claim is usually made (1992: 8–11; 1996: 8–18): *sentience* (a capacity to feel sensations); having cell-based *life*; and *autopoiesis* (a capacity for self-renewal). Elsewhere, he conflates the biological and autopoiesis approaches and adds two more — ecosphere/Gaian ethics and 'cosmic purposes'/theological ethics (1990: 162) — but, as we consider these positions in later chapters, we will stay for the present with his 1992 three-approach classification.

Sentience, the first of these, is explained by Fox thus: 'if an entity is sentient then it may be said to have interests — in particular it seeks pleasurable states of being and seeks to avoid unpleasurable states of being', and beings that have interests 'ought to have their interests taken into account in the context of actions regarding them irrespective of the species to which they belong' (1992: 8).

The second grounding of intrinsic value — having cell-based life — Fox characterises in this way: 'interests are interests, these advocates argue, irrespective of whether the bearer of these interests may be said to be aware of them'. This being so, all biological organisms 'are actively *concerned* with seeking some states of affairs rather than others regardless of whether or not they are sentient' (1992: 9).

The third basis for according intrinsic value is autopoiesis (or the 'holistic integrity approach'; 1996: 14), which Fox traces to the work of Chilean biologists, Humberto Maturana and Francisco Varela. It involves the provision of a formal definition of life, one that avoids problems at the margins. 'The concept of autopoiesis', writes Fox, 'refers to the fact that the essential feature of living systems is that they continuously strive to produce and sustain their own organizational activity and structure' (1992: 10). This has significant advantages over the earlier groundings of intrinsic value, says Fox. Firstly, a capacity for regeneration constitutes an entity as an end in itself rather than as an instrumental constituent of extrinsic ends, and 'this amounts to the classical formulation of intrinsic value: by definition ... any entity or process that is an end in itself has intrinsic value' (1992: 11; see also 1990: 172). Secondly, the autopoietic grounding of intrinsic value shifts the focus from individual components of biological life to '*all* process-structures that continuously strive to regenerate their own organizational activity and structure', and this includes 'not only individual plants and animals but also species (or at least gene pools), ecosystems, and the ecosphere considered as entities in their own right'. Thus, 'the criterion of autopoiesis is clearly the most ecocentric criterion of the three criteria of instrinsic value' (1992: 11). Fox instances Leopold's 'land ethic' as 'the outstanding "early" statement of this approach' (1996: 14).

He then moves 'beyond' axiological or intrinsic value schools of ecophilosophy to an approach that I have termed 'experiential', and Fox calls 'psychological' (1992: 11). Here he has only one sub-category: 'deep ecology' (Fox prefers 'transpersonal ecology', though we will stay with the older name for the present). This involves widening out one's biological or bodily sense of self into a larger, perhaps even cosmological field of identification. Deep ecologists denote this extended sense of self as a capital-S 'Self'. More about this shortly:

here, we need merely note that this is an alternative ethical approach to an axiological one. The difference is described by Fox in this way: 'arguments for intrinsic value ... imply certain codes of conduct. In contrast, transpersonal ecologists explicitly reject approaches that issue in moral "oughts"'. They are seeking, instead, a 'sense of commonality with all that exists', with the result that 'one seeks, within obvious kinds of practical limits [here Fox is acknowledging that life must feed upon other life to survive], to allow all entities (including humans) the freedom to unfold in their own ways' (1992: 12–13). Thus, this approach posits an integrated ecological self rather than an individual with a consciously appended code of values.

There are other taxonomies, but Fox's seems to have been the most influential. The taxonomy used by Sylvan and Bennett (1994: 61–159), for example, is similar to Fox's, even though they are critics of deep ecology. The main difference is that, when Sylvan and Bennett progress to 'deep' environmental positions — those that 'run deeper than instrumental considerations or mere extension' (1994: 92) — deep ecology is *not* the apex of their hierarchy of ethical validity. That position is reserved for another 'deep' position, their own axiological one, which they refer to as 'deep-green theory' (1994: 137–59). Sylvan and Bennett capitalise 'Deep Ecology' but not 'deep-green theory', to give point to their claim that deep-green theory is pluralist, de-centred and inclusivist (1994: 153–59) in ways that 'Deep Ecology', despite its intent in this regard, is often not.

The salient point here is that these roughly comparable taxonomies (and there are others that could have been adduced as well) differ mainly in that, for each, the particular taxonomist's preferred system of thought is positioned at the head of the hierarchy. Neither Fox's nor Sylvan and Bennett's is a disinterested formulation; on the contrary, both taxonomies are designed to place the favoured schema in the best possible contextual light, and on this score both are vulnerable to criticism.

However much we might find one or other of these hierarchically arranged schemes of thought discursively persuasive, perhaps the most satisfactory taxonomy for our purposes would be a simple, non-hierarchical list of the forms of environmental thought generated to date. Such a list, arranged alphabetically to avoid any suggestion of hierarchy, might look as follows.

- animal rights (for example, Regan, Singer)
- anthropocentric ethics (for example, Passmore, O'Neill)
- axiological ('intrinsic value') theory I: 'deep-green theory' (for example, Bennett, Sylvan)
- axiological ('intrinsic value') theory II: the Gaia hypothesis (for example, Goldsmith, Lovelock)

- axiological ('intrinsic value') theory III: 'holistic integrity'/'land ethic' approaches (for example, Callicott, Leopold)
- axiological ('intrinsic value') theory IV: life-based ethics (for example, Goodpaster, Taylor)
- Christian ecology (for example, Attfield, Cobb, Thomas Berry, Matthew Fox)
- deep ecology (for example, Devall, Warwick Fox, Naess, Sessions)
- ethics derived from power theory I: bioregionalism (for example, Sale)
- ethics derived from power theory II: 'doomsday' ethics (for example, Hardin)
- ethics derived from power theory III: ecofeminism (for example, Plumwood, Salleh, Shiva, Karen Warren)
- ethics derived from power theory IV: ecoMarxism (for example, Benton, Enzensberger, O'Connor, Pepper)
- ethics derived from power theory V: social ecology (for example, Bookchin)
- 'new science'-based ethics (for example, Capra, Shepard)
- place-based ethics (for example, Lopez, Norton, Relph, Snyder)
- postmodern ethics (for example, Cheney, Haraway)
- spiritualist ethics (for example, La Chapelle, Spretnak, Starhawk)
- sustainability ethics (for example, Brundtland Report, Rio Declaration)

Of these, only positions derived *primarily* from philosophical speculation are considered in this chapter.

ANIMAL LIBERATION/ANIMAL RIGHTS

Peter Singer has written that he became interested in the ethical status of animals as a graduate student in 1970. When he looked to the philosophical literature, he found that human equality was customarily defended on the ground that *all* humans possess interests, but that scant consideration was ever given to the question of whether interests cease to exist at the boundary of the species. In fact, he was struck by how patently this was not so. He considers the claim, 'all humans are equal because they all have interests'. 'This means', he writes, that 'their lives can go well or badly, they can suffer or be happy, they have interests we can harm or help them to fulfil', and this, he thinks, is the strongest justification for equality of consideration. But:

> how could these writers not notice that this applied also to nonhuman animals, at the very least to vertebrate animals, who also have lives that can go well or badly, can suffer, and hence have interests that we can affect? If it is wrong to discount interests on the basis of sex, can it be right to discount them on the basis of species (1993a: 63)?

Singer's two key ideas are contained within this passage.

Adopting Bentham's pleasure–pain principle, Singer argued for *sentience*, and in particular the capacity to suffer, as the criterion for moral considerability. Put simply, animals can feel pain, which makes them moral subjects. Animals that can suffer have an interest in avoiding pain, and thus pain in a non-human is no different in moral significance to pain in a human. Lawrence Johnson provides an elegant summary of the Singer position that runs thus: 'pain is bad. That pain happens to an animal is irrelevant to its badness. We ought to minimize the occurrence of badness in the world. Therefore we ought to avoid causing unnecessary pain to animals' (1991: 48).

Other possible criteria for according moral considerability are discussed by Singer, and all are found wanting: self-consciousness, the possession of linguistic skills, the capacity for rational thought. Using the 'argument from marginal cases', Singer shows how some humans fail each of these tests (and some animals pass at least some of them); so, each test fails as a criterion for drawing the line of moral considerability at the *homo sapiens* species boundary. The power-to-reason criterion is the defence of human privilege most often used, and Singer pays this particular attention. Some people (he points out) have greater reasoning capacities than others. Does this mean that moral worth increases as you move up the IQ scale? And what of the brain-damaged human infant? No rational capacities here. Yet, few people would be willing to deny moral considerability to the very young, or the senile, or the mentally enfeebled. No criterion for human specialness works, then, and to insist on maintaining a clearly inadequate ground of difference amounts to mere 'speciesism', 'a prejudice or attitude of bias towards the interest of members of one's own species and against those of members of other species' (1975: 26). Only the pleasure–pain principle stands the test for moral subjectivity, and so 'the capacity for suffering and enjoyment is *a prerequisite for having interests at all*', with the consequence that this is the 'only defensible boundary of concern for the interests of others' (1975: 8–9, see also 1979a: 90). It is not an easy boundary to draw, as Singer acknowledges — in *Animal Liberation* he tentatively located it 'somewhere between a shrimp and a mollusc' (1975: 178) — and he expressed even less confidence in the second edition (1990: 174).

Singer's second foundational principle can be called 'the principle of equal consideration of like interests' (Sylvan and Bennett 1994: 85). Because all entities with a capacity to suffer have an interest in avoiding suffering — of equal moral standing in each case — each such entity has a claim to equality. But this does not mean equal or identical treatment. Because interests vary and are not identical across living beings — they differ, indeed, across humans — it is more accurate to say that animals are entitled to equal *consideration*, and 'equal consideration for different beings may lead to different treatment and different rights' (P. Singer 1975: 22).

A second early contribution of significance was made by Tom Regan. There is a case for distinguishing between *animal liberation* and *animal rights* (Fox 1990: 165), in order to account for the more restrictive moral domain defended by Regan (hence 'animal rights') when compared to Singer ('animal liberation'). Although Regan also deems sentience to be of significance, he does not source his case to utilitarianism. In an early paper Regan (1976) argued the need for a more rights-based focus than could be found within Singer's animal liberation, on the ground that it may not be possible to argue convincingly the case for animals unless they are held to possess a right to life (see Regan 1980, for his full case against Singer). Regan accords some animals moral standing because they are, as he puts it, 'subjects-of-a-life'. This criterion is only met by entities that have 'beliefs and desires, perception, memory and a sense of the future, including their own future' (1983: 243) — a more stringent test than Singer applies. Beings that meet the criterion, however, are ends in themselves and possess 'inherent worth', and on this ground they can be said to possess *rights*.

Contributions to this field of thought have burgeoned over the years. One such, that of Stephen Clark, is contemporaneous with Singer's *Animal Liberation*, though it did not have the popular impact of the latter. Whereas Regan is more restrictive than Singer, Clark seems more inclusive. Although he, too, locates within the utilitarian position of minimising suffering to others, Clark (1975; see also 1997) places more stress on the notion that animals have a right to strive for the fulfilment of their particular potentialities. Mary Midgley is another who has written challengingly of animal ethics over two decades. Midgley finds in the demonstrable sociality and evolved consciousness of the higher animals a case that 'animals matter'; that they have interests that merit human respect; and that early human communities, though drawing boundaries against other humans, were species-mixed, and involved in intricate relationships of trust and reciprocity (for example, 1983a; 1983b; 1992a; 1992b). More recently, a specifically ecofeminist concern for the status and wellbeing of animals has emerged (Adams 1992; 1994; Adams and Donovan 1995), with a focus upon perceived links between meat-eating, hunting and butchering and the institutionalisation of patriarchy ('patriarchy' is discussed in chapter 3). There are even re-evaluations of the resource-for-humans status of animals within Christianity (Linzey 1994; Sobosan 1991; Wessels 1989) and within Marxism (Benton 1993b; Benton and Redfearn 1996).

We have here a prominent perspective within environmental thought; it is, however, regarded as a partial, less developed, or *less radical* perspective by those arguing from a 'deep' position. This is because of the method of arriving at the ethic: by 'mere' extension of the scope of the ethically familiar and acceptable. Singer is particularly anxious to lessen the distance between 'what is' and his expanded moral realm. He does this by stressing that modern history has seen many ethical extensions, each resisted in its day, but since attaining broad acceptance. Thus, the once-accepted grounds for denial of moral value to people on the basis of race and gender have, within comparatively recent history, been judged invalid. Singer views the exploitation of animals as:

> a vast social practice in which the most powerful group exploits the less powerful and builds ideological justifications for what it does. From this perspective there were familiar analogous situations, foremost among them the enslavement of Africans by Europeans. Here too, if one looked at the historical literature, it was easy to find intelligent and otherwise civilised writers who accepted the most evidently spurious reasoning to justify a practice that was convenient for them and the society in which they were living (1993a: 63).

Non-human animals are, on this view, another aggrieved group being subjected to unjustifiable discrimination by a privileged group (humans) with the power to indulge their urge to discriminate.

Thus, the extension of ethical protection to this aggrieved group is merely the next historical stage in the widening of the ethical scope; there is even a hint of the inexorable about it. But to portray the extension of accepted ethical criteria to cover some other animals as a step of only limited radicalism is to fudge the enormous practical consequences of such a development — and elsewhere Singer does *not* fudge this issue. He has, himself, a long record of activism, and he does not make light of the practical consequences of his proposed ethical extension:

> animal liberation will require greater altruism on the part of human beings than any other liberation movement. The animals themselves are incapable of demanding their own liberation, or of protesting against their condition with votes, demonstrations or boycotts. Human beings have the power to continue to oppress other species forever, or until we make this planet unsuitable for living beings (1990: 247–48).

The practical consequences include an end to research involving animals and a massive shift away from meat-eating, with the implications that such a move would have for global agricultural practices and the food-processing and supply industries, as well as perhaps even more fundamental change to scientific and industrial processes. It would be a revolution as profound as any since — and perhaps not even excluding — the industrial revolution.

Perhaps, if the extension envisaged by Singer does take place, it might be conceded in stages, with research on animals the first of the stages to fall. The case for such research is particularly tenuous, given the logical bind it involves; for, as Sylvan and Bennett perceptively note, 'much medical and psychological experimentation is based on the assumption that animals are in given respects analogous or homologous to humans in their capacities' (1994: 84).

A blanket ban on the killing of animals, to eat, say, is another matter. If Singer's ethical extension becomes generally accepted, it is difficult to see how the killing of a higher animal could be separated out, in moral terms, from the killing of a human, just as it becomes difficult to see how cannibalism should attract any greater condemnation than the eating of a non-human animal. Yet, Singer does provide the grounds for such a human-privileging distinction:

> the killing of nonhuman animals is itself not as significant as the killing of normal human beings. Some of the reasons [are] ... the probable greater grief of the family and friends of the human, and the human's greater potential. To this can be added the fact that other animals will not be made to fear for their own lives, as humans would, by the knowledge that others of their species have been killed. There is also the fact that normal humans are beings with foresight and plans for the future, and to cut these plans off in midstream seems a greater wrong than that which is done in killing a being without the capacity for reflection on the future (1985: 484).

Regan, by contrast, may draw the boundaries of moral considerability more restrictively, but within those boundaries he seems less inclined than Singer to human-privileging distinctions. For Regan it is only permissible to hurt animals in situations in which it would also be permissible to harm humans. Thus, 'people do not have to refrain from shooting a lion that is attacking them or from killing deer who are eating crops upon which the people depend for survival' (Wenz 1988: 145).

In an earlier case for moral differentials in the killing and eating of humans and other animals, Singer also makes vegetarianism non-mandatory:

> given that an animal belongs to a species incapable of self-consciousness, it follows that it is not wrong to rear and kill it for food, provided that it lives a pleasant life and, after being killed, will be replaced by another animal which will lead a similarly pleasant life and would not have existed if the first animal had not been killed. This means that vegetarianism is not obligatory for those who can obtain meat from animals they know to have been reared in this manner (1979b: 153).

This is not a position with huge support within the animal liberation movement and, given the tenor of his subsequent writings, it may not be a position to which Singer would still subscribe himself. Those who reject such a view can, moreover, point to an increasing array of evidence to suggest that the difference between humans and some other animals may be far less than we have confidently supposed. Much of the evidence is biological. 'The genetic difference (1.6%) separating us from pygmy or common chimps is barely double that separating pygmy from common chimps (0.7%)', writes Jared Diamond. It is, moreover, 'less than that between two species of gibbons' (1992a: 19). Potentially even more compelling is the emerging evidence that many animals have a developed sense of their own wellbeing and a capacity to make choices that work to their good rather than to their detriment, and that some have a capacity for thought, although we may need to adopt a less tunnelled view of what constitutes thought before we can recognise it as such (for example — Birch 1995; Coren 1994; M.S. Dawkins 1993; DeGrazia 1996; de Waal 1996; M.A. Fox 1997; D.R. Griffin 1992; Masson and McCarthy 1995; Noske 1989; Rodd 1992; L.J. Rogers 1997; Sapontzis 1987). 'If an animal is thought to be a sort of organic wind-up toy, people are unlikely to go far out of their way for it', writes DeGrazia, 'but if an animal is believed to be self-aware or rational, or to have a rich emotional life, different responses are likely' (1996: 1). If this evidence holds — even, perhaps, obtaining still further support — it will become increasingly difficult to contest the observation of Michael Fox that 'one of the greatest tasks humans face is to work out a new set of relationships, of ways to coexist, with other animal species on this planet' (1997: 46).

DEEP ECOLOGY

The term 'deep ecology' entered the discourse in 1973, when Arne Naess published the paper which was to become one of environmental philosophy's foundational texts, 'The Shallow and the Deep, Long-Range Ecology Movement: A Summary'. But deep ecology only began to attain real prominence in the late 1970s. This was largely as a result of the collaborative effort of two California-based academics, the philosopher George Sessions and the sociologist Bill Devall.

In Naess's conceptualisation 'deep, long-range ecology' had three components. One of these can be briefly treated. This is deep ecology as descriptive of a *method of inquiry*, a method based upon the asking of 'deep' questions, of interrogating one's assumptions concerning ecological relationships in ever 'deeper' sequence until the point of personal philosophical bedrock is reached (Naess 1988; 1995a). This sense of deep ecology has had comparatively little influence on either the popular ecological imagination or the philosophical mind. Only the durable pioneers of deep ecology continue to feature it with any prominence (for example, Devall's description of the basic ideas of deep ecology, 1988: 11 — but an exception is Anker 1996). Warwick Fox, on the other hand, wishes to dispense with this sense of deep ecology, arguing that it is logically unsatisfactory. If deep ecology is to be defined in such a methodologically formal way, he writes, then 'even an environmentally destructive view must be characterized as a deep ecological philosophy if it is derived from fundamentals', that is to say, 'from a coherent philosophy asking deep questions' (1990: 95).

We are left with two interlocked but distinct and not always mutually comfortable senses of the term 'deep ecology'. The first of these is deep ecology as *movement*, the second is deep ecology as *philosophy*. In the former sense 'deep ecology' is an umbrella term, its meaning somewhat smeared. Fox notes that at this level it can become almost synonymous with 'ecocentrism', designating a broad-based ecological politics loosely coalescing around opposition to 'anthropocentrism'. Although its most prominent spokespersons are philosophers, deep ecology has tended as a movement to avoid philosophical honing. Despite much being made of its genealogy, and despite the presence of philosophers, most notably Spinoza, among deep ecology's key influences, it is difficult to find among the writings of the deep ecologists any detailed exposition of these philosophies. Discussion tends to remain at the level of the digestible short-paragraph 'grab'.

Sessions is a notable exception here. His early papers are conspicuously erudite contributions to an understanding of the contemporary relevance of such foundational influences upon deep ecology as Spinoza (for example, 1977; 1985a). Baruch Spinoza (1632–77), a seminal influence upon both Naess and Sessions, was a metaphysical pantheist: all particularity is an expression or manifestation of an immanent God. There is, thus, no hierarchy in nature, but rather an ecological democracy. On the other hand, Sessions's influential collaborator, Bill Devall, has always evinced much more interest in deep ecology as 'movement', and the jointly authored book that brought their collaboration to a close, *Deep Ecology: Living as if Nature Mattered* (1985), reflects Devall's priorities more than it does those of Sessions.

This work was the definitive deep ecology text for the 1980s (which is not to denigrate the substantial contributions made at the time and since by, for example, Abram 1996; Drengson 1980; 1989; Drengson and Inoue 1995; McLaughlin 1993; 1995; Reed and Rothenberg 1987; Rothenberg 1995, and others). It appeared at a crucial time in the development of ecophilosophical ideas — a time, moreover, when deep ecology commanded the field — and it largely established the priority of the 'movement' sense of 'deep ecology' over its more focused philosophical dimension. In *Deep Ecology* formative influences are given the 'survey' treatment typical of writings that treat deep ecology as 'movement'. One is struck by the eclectic way in which influences were taken aboard. Thus, among western philosophers, Spinoza is important but the late-Renaissance savant Michel de Montaigne (1533–92) is not. Yet Montaigne advocated living the simple life (1991: Bk III, chap. 13); he maintained that humankind is embedded within a commonwealth of species, all of which, including forms of botanical life, merit respectful treatment by humans (1991: Bk II, chap. 11; Bk II, chap. 12); he also argued for the unity of the body, the emotions and the reasoning mind (1991: Bk II, chap. 12). Similarly, from the rich North American literary tradition of nature-inspired writing, Gary Snyder is identified as an important influence, along with the poet Robinson Jeffers, active in the period 1910s–1940s. It is easy to see why the strong nature pantheism of Jeffers would have appealed to the early deep ecologists. But he also wrote poems that can only be read as hymning the redemptive and cleansing character of acts of violence, and has been widely seen as profoundly misogynist. Jeffers also professed a 'doctrine of Inhumanism'. He meant, though, by this doctrine 'the rejection of human solipsism and a recognition of the transhuman magnificence' (1977: xxi) rather than a simple misanthropy, the unfortunate title of his 'doctrine' has not helped deep ecology ward off the persistent attacks that have been made upon it on this score.

Other key influences sketched by Devall and Sessions are native spiritualities and eastern religion, particularly Buddhism and the Tao; 'western process metaphysics', in which tradition Spinoza stands and which also prominently includes Alfred North Whitehead; phenomenology, and particularly the dwelling-in-place thought of Martin Heidegger; alternative Christian traditions; the American romantic tradition of Thoreau, Emerson and Muir; the 'new physics' and ecological science, particularly the 'super' ecology of the Gaia hypothesis; the ethics-from-ecology of Leopold; the 'real world' of wilderness; and the writings of some ecologically inclined feminists (*Deep Ecology* appeared just before the rift between deep ecology and ecofeminism). To this list can be added Gandhi, a major influence upon Naess, and Santayana, an important influence upon Sessions (1985b).

As with key influences, so with key principles. In keeping with its 'movement' orientation, much deep ecology writing borrows the form of the manifesto, its principles enumerated, outlined, and put in the form of guides to action.

In his 1973 paper Naess listed seven basic principles: 'rejection of the man-in-the-environment image in favour of the relational, total-field image'; 'biospherical egalitarianism - in principle'; 'principles of diversity and of symbiosis'; 'anti-class posture'; 'fight against pollution and resource depletion'; 'complexity, not complication'; and 'local autonomy and decentralization' (1973: 95–98). In 1984 this list was superseded by an eight-point 'platform', jointly devised by Naess and Sessions, and quite different from the 1973 formulation. Widely reprinted (for example — Birkeland, Dodds and Hamilton 1997: 131; Brennan 1996: 4; Devall 1991: 249–50; Drengson 1995: 143–44; McLaughlin 1993: 173–74; 1995: 86–89; Naess 1989: 29; Sylvan and Bennett 1994: 95–96), the platform has come to be seen as the definitive principles of deep ecology; a status not dissimilar to that of the

Ten Commandments within Christianity. It reads:

> 1. The well-being and flourishing of human and non-human Life on Earth have value in themselves (synonyms: intrinsic value, inherent value). These values are independent of the usefulness of the non-human world for human purposes.
>
> 2. Richness and diversity of life forms contribute to the realization of these values and are also values in themselves.
>
> 3. Humans have no right to reduce this richness and diversity except to satisfy *vital* needs.
>
> 4. The flourishing of human life and cultures is compatible with a substantial decrease of the human population. The flourishing of non-human life requires such a decrease.
>
> 5. Present human interference with the non-human world is excessive, and the situation is rapidly worsening.
>
> 6. Policies must therefore be changed. These policies affect basic economic, technological, and ideological structures. The resulting state of affairs will be deeply different from the present.
>
> 7. The ideological change is mainly that of appreciating life quality (dwelling in situations of inherent value) rather than adhering to an increasingly higher standard of living. There will be a profound awareness of the difference between big and great.
>
> 8. Those who subscribe to the foregoing points have an obligation directly or indirectly to try to implement the necessary changes (Naess and Sessions 1984).

One of the first ecophilosophers to break ranks, Richard Sylvan, described deep ecology as 'a normative and policy- and lifestyle-oriented theory' (1985: 43), and these dimensions are clearly apparent in the platform. That its normative aspects were treated, through the 1980s, in much the same way as key influences — as a series of snapshots — must be attributed in large part to Naess himself. From the start Naess was anxious to promote deep ecology as a broad-church movement; even his papers that explicated 'normative principles' tended to be short and popularly accessible. One thing supporters of deep ecology share, he has written, 'is that the articulation of our views is, and must be, fragmentary' (1995b: 215). His own 'deep ecology' he labels 'Ecosophy T' — 'ecosophy' literally meaning 'household wisdom' (but better translated as 'ecological wisdom'), whilst the 'T' is a modest disclaimer of philosophical priority within deep ecology, and an acknowledgement that there is space for many lodgers within the halls of deep ecology:

> to avoid unfruitful polemics I call my philosophy 'Ecosophy T', using the character T just to emphasize that other people in the movement would, if motivated to formulate their world view and general value priorities, arrive at different ecosophies: Ecosophy 'A', 'B',..., 'T',... 'Z'. By an

'ecosophy' I here mean a philosophy inspired by the deep ecological movement (1985: 258).

It is difficult to dispute Naess's contention that 'a philosophy, as articulated wisdom, has to be a synthesis of theory and practice' (1985: 258), for the global social and environmental conditions within which deep ecology has developed surely render practical action a more urgent priority than rarefied philosophical abstraction. But the credibility of deep ecology as a movement has been made vulnerable to challenge, in part because its normative underpinnings through the 1980s remained insufficiently developed. A good case in point is the principle of 'biospherical egalitarianism', present in Naess's 1973 list ('biospherical egalitarianism - in principle'), but absent from the 1984 'platform'. The notion of 'biospherical egalitarianism' caused problems from the start. Are we really to believe, it was pointedly asked, that the interests of humans are to be equated with those of the smallpox virus? 'This way of thinking is hard to apply convincingly', wrote Mary Midgley, 'to locusts, hookworms and spirochaetes' (1983a: 26).

Naess's response was to acknowledge that the term 'biospherical egalitarianism - in principle' has perhaps caused 'more harm than good' (1984: 202), and to qualify it in two ways. One involved a retreat from full-blown egalitarianism by introducing the notion of *vital needs*: 'a high level of identification does not eliminate conflicts of interest', for 'our vital interests, if we are not plants, imply killing at least some other living beings ... Identification with individual species, ecosystems and landscapes results in difficult problems of priority' (Naess 1985: 262).

It is this important qualification upon 'biospheric egalitarianism' — this acknowledgement of an entitlement to pursue the satisfaction of 'vital needs' — that is recognised in the 1984 platform: 'humans have no right to reduce ... richness and diversity except to satisfy *vital* needs'. On one level this is difficult to contest. Without it, any act taken by any individual organism to maintain itself in existence violates a first-order principle. But there is a sting in the tail here. If Naess is to claim that 'a rich variety of acceptable motives can be formulated for being more reluctant to injure or kill a living being of kind A than a being of kind B' (1984: 202), then, argue Sylvan and Bennett, 'egalitarianism does not mean what it seems. For it now must be understood that value is not equally spread throughout the ecosphere' (1994: 100–01), and thus 'egalitarianism is seriously eroded in principle as well as in practice' (1994: 101). This is of some import, for the principle of biospheric egalitarianism is, Sylvan and Bennett point out, crucial to the *depth* claim within deep ecology, and 'without it Deep Ecology may slide into yet another intermediate position' (1994: 102).

Precisely this sort of uncertainty led Warwick Fox to undertake the task of elaborating philosophically the key normative principles of deep ecology. The project culminated in his 1990 book, *Toward a Transpersonal Ecology: Developing New Foundations for Environmentalism*. Although such a task by prior claim belonged to Naess, with his modest philosophical pluralism and his commitment to the 'movement' conception of deep ecology he was reluctant to assume this role, and it fell to Fox to tackle the problems inhering within the vexed term 'biospherical egalitarianism'.

He argues that, for Naess, the word 'life', as implied by the prefix 'bio', is not to be taken in an orthodox, individual-organism sense, but in a broad enough way to gather in such life collectivities, processes and contexts as species, ecosystems, and habitats (W. Fox 1990: 117; the prefix 'eco' is thus to be preferred to 'bio', and from the mid-1980s the one rapidly displaces the other from the discourse). Here we come to the second qualification introduced by Naess upon the concept of 'biospheric egalitarianism', this one upon the 'biospheric' component of the formulation. Part of the confusion stemmed from Naess's deployment of terms such as 'biospherical egalitarianism' and 'intrinsic value': these seemed to imply a moral 'ought', though deep ecology claimed to stand in a quite contrary philosophical tradition, one which eschews the axiological project of according rights. Drawing on Naess's unpublished papers and comparatively obscure observations made elsewhere in his writings, Fox shows that a term such as 'intrinsic value' is used in 'an expressive, metaphorical, non-technical, everyday sense' (1990: 222), whilst the phrase 'biospherical egalitarianism' is similarly not used in a formal, prescribing-rights-to-individual-organisms way, but is intended, in an accessible and generalised sense, to stand as 'a statement of non-anthropocentrism' (Naess, cited in W. Fox 1990: 224). Thus, the term refers more accurately 'to the ecosphere as a whole' (Naess and Sessions 1984: 5). Once this is appreciated, Fox argues, confusion vanishes and, when Naess's other writings (for example, 1979) are taken into consideration, it becomes apparent that Naess does reject 'formal intrinsic value theory approaches' (W. Fox 1990: 224).

It is important for Fox to establish this, because one of his central claims is that 'transpersonal' ecology is the philosophical development of principles inherent in the 'deep ecology as movement' writings of Naess and other deep ecology pioneers. 'The shortcomings', Fox writes of deep ecology as ecophilosophy, 'consist primarily in its lack of elaboration. Its original proponents have not developed this emphasis into a coherent theory' (1995a: 166). He is at pains to stress that, in pursuing his project of elaborating the principles of transpersonal ecology, he is not at odds with Naess but is embarked upon a complementary project. There has been much laudatory material written about Naess (for example — Devall 1992; Drengson 1995; Inoue 1992; Langlais 1992; Rothenberg 1989; 1992; Sessions 1992a; 1995a; Witoszek and Brennan 1999) and Fox's voice is prominent among these tributes (for example, 1990: 79–118), though he does argue that

Naess's asking-deeper-questions sense of deep ecology is 'untenable' (1990: 118). What, then, are the central insights of Naess's thought that Fox seeks to elaborate into a more philosophically compelling system of thought?

In 1985 Naess listed the 'five themes' of his 'Ecosophy T' as:

> the narrow self (ego) and the comprehensive Self (written with capital S). Self-realization as the realization of the comprehensive Self, not the cultivation of the ego. The process of identification as the basic tool of widening the self and as a natural consequence of increased maturity. Strong identification with the whole of nature in its diversity and interdependence of parts as a source of active participation in the deep ecological movement. Identification as a source of belief in intrinsic values (1985: 259).

Elsewhere he has stated that 'Ecosophy T has only one ultimate norm: "Self-realization!"' (1993: 209). And it is here, Fox argues, in the normative principles of 'Self-realization!' and identification, that deep ecology's philosophical distinctiveness is to be sought. Much that is associated with deep ecology as a movement, he maintains, is more or less common across the range of environmentalist positions. This is true, for example, of the practical consequence of the 'vital needs' hedge upon biospheric egalitarianism, which is to draw upon the earth's living and finite systems with prudent circumspection (Naess 1990: 91). This theme is all-but universal within environmental value-sets. This from Henryk Skolimowski, a critic of deep ecology: 'frugality is an act of reverence. You cannot be truly reverential toward life unless you are frugal in this present world of ours in which the balances are so delicate and so easy to strain ... To understand the right of others to live is to limit our unnecessary wants' (1990: 101).

By contrast, the concept of 'Self-realization!' — the construction of as wide a sense of self as possible through a process of identifying out and including an enlarged scope of life and living process within one's sense of (S)self — Fox holds to be *the* central and distinctive idea of deep ecology, that which 'most distinguishes the work of deep ecological writers from that of other ecophilosophical ones' (1993: 75). This focus 'sets aside the nondistinctive and untenable aspects of deep ecology, zooms in on the identification approach, and elaborates that approach substantially beyond that previously found in the deep ecological literature' (1995a: 167). Thus, transpersonal ecology 'points to the realization of a sense of self' that 'extends beyond any narrowly biographical or egoic sense of self' (1993: 75).

'Self-realization!' is attained by a process of 'identification', and the latter concept shares equal billing with 'Self-realization!' within Fox's schema. There is nothing mystical about this notion; the commonsense definition is the one that prevails. One extends the perception of self by identifying the interests of entities with which one experiences a sense of commonality as one's own interests; thus, 'our sense of self can be as expansive as our identifications' (1993: 233). The notion of commonality is given prominence: identification is 'the experience not simply of a sense of *similarity* with an entity but a sense of *commonality*' (1990: 231). To experience commonality is not to lose one's self in an ego-dissolving unitariness: 'the realization that we and all other entities are aspects of a single unfolding reality — that "life is fundamentally one" — does not mean that all multiplicity and diversity is reduced to homogeneous mush' (1990: 232). Individual entities are 'relatively autonomous' but not 'absolutely autonomous' — a relational complexity for which Fox, the most scientifically inclined of the deep ecologists, finds independent corroboration within physics, evolutionary biology, ecological science and new systems theory. There are also resonances here with the eastern philosophies upon which deep ecology draws: 'all things in the biosphere have an equal right to live and blossom and to reach their own individual forms of unfolding and self-realization within the larger Self-realization!', write Devall and Sessions (1985: 67), an injunction that Fox intuits to apply to 'all identifiable *entities* or *forms* in the ecosphere' (W. Fox 1989: 6) — species, ecosystems, perhaps even geophysical processes.

At the same time, via this creation of what Dobson has called a 'state of being' ethics (1995: 56), traditional 'code of conduct' morality is rendered redundant. This is because the creation of the expansive ecological self will cause one to act in appropriate ways without need of guidance from an externally imposed set of 'thou shalts' and 'thou shalt nots': 'if one has a wide, expansive, or field-like sense of self then (assuming that one is not self-destructive) one will naturally protect the spontaneous unfolding of the expansive self (the ecosphere, the cosmos) in all its aspects' (W. Fox 1990: 217). Such a spontaneous earth-friendliness is to be preferred to a moral code because it does not require the intervention of a third party — the intellect — bringing to bear 'oughts' and 'ought nots' which, as they do not lodge reflexively within the core of one's being, one is ever tempted to transgress. 'The history of cruelty inflicted in the name of morals has convinced me', Naess has written, 'that the increase of identification might achieve what moralizing cannot: Beautiful actions' (1985: 264).

SOME ENVIRONMENTAL AXIOLOGIES

We have noted the case deep/transpersonal ecology has made against axiologically based environmental ethics. Not that it is only deep ecologists who take this position. Peter Marshall, a critic of deep ecology, is also a critic of rights-based ecophilosophy. 'The language of rights is unfortunately vague and confusing', he writes. 'All too easily it can degenerate into complacent rhetoric' (1992: 437). Rights are also 'individualistic and legalistic', requiring 'the state to enforce them', as well as being 'patronising', as it is 'usually a question of those holding power extending their privileges to others' (Marshall 1992: 434). How can a radical advance possibly emerge from such a paradigm? It is a familiar discourse, one couched 'in principles and concepts that already have common currency', writes Andrew McLaughlin (1993: 170), and would thereby seem unsuitable to a radical, nonhuman-regarding ethics:

> as long as we assume that we are fundamentally different from the rest of nature, then we would seem to be left only with the possibility of raising a claim that there is some sort of intrinsic value in nonhuman nature as a restraint on humanity's actions. However, if we accept that we are part of nature, that we are submerged within the unfolding of life, then the need for the theoretical construct of 'intrinsic value' diminishes (1993: 170).

It may be that some of the intrinsic value theorists whose work is outlined below would agree, in part, with this critique. Callicott, Rolston and Sylvan, for instance, would agree that rights-based language has hitherto assumed that moral considerability stops at the boundary of species *homo sapiens.* Where they will disagree is in the insistence that it must ever be so; that the rights discourse is incapable of adaptation to more radical ethical ends. Sylvan and Bennett, for example, see deep ecology's failure to develop 'a proper ethical theory' as one of its most significant weaknesses, because the 'development of a satisfactory axiology as a prelude to a fuller ethics comprehending the range of ethical actions assumed is an important outstanding requirement of adequacy' (1994: 167). An axiology is thus necessary if radical environmental ethics are to be competitive with 'shallow environmental ethics'. It is a question of being taken seriously. In the absence of a value-based ethic, conventional ethics will retain a 'competitive edge': 'rights claims provide a powerful way of stating principles, a way that is important and influential, especially in North America, where rights matter and count' (1994: 166). There is no point in pooh-poohing the legalistic dimension of rights-thinking, then. Ethics must be made legally relevant or they will be ineffectual, and rights are needed to provide a basis within environmental law.

Further compounding the pluralistic nature of ecophilosophical discourse is the diversity of environmental axiologies, with disagreement between them expressed as fiercely as it is between other disputing schools of environmental thought. Regan's is an axiological position after all, and his conflict with Baird Callicott (see below) was almost as bitter as the debilitating Barry Commoner–Paul Ehrlich debate in the early 1970s over whether the base cause of environmental degradation was human population-increase or technological imperatives (on the Commoner–Ehrlich debate, see Passmore 1974: 128), or the attacks by various ecoMarxists, social ecologists and ecofeminists upon deep ecology. Only some of the more prominent environmental axiologies are sampled below.

The school of thought that sits closest to animal rights/liberation is 'biocentric individualism': 'the perspective that all individual living beings have intrinsic moral worth and that species and ecosystems count morally only as collections of individuals' (Sterba 1995: 191).

Most assessments of this school of thought see the great Christian pastor, Albert Schweitzer (1875–1965), as standing at its head. Schweitzer proposed a principle of 'reverence for life', the significance of which has been articulated by Skolimowski: 'Schweitzer's ethic of reverence for life is a remarkable anticipation of the ecological ethic; we should say, perhaps: the ecological ethic-in-the-making' (1981: 65). Schweitzer himself described his ethic thus:

> just as in my own will-to-life there is a yearning for life, and for that mysterious exaltation of the will which is called pleasure, and terror in the face of annihilation and that injury to the will-to-life which is called pain; so the same obtains in all the will-to-life around me, equally whether it can express itself to my comprehension or whether it remains unvoiced. Ethics thus consists in this, that I experience the necessity of practicing the same reverence for life toward all will-to-life as toward my own. Therein I have already the needed fundamental principle of morality. It is good to maintain and cherish life; it is evil to destroy and to check life (1923: 254).

Here is an injunction to move through the world in a way that 'tears no leaf from its tree, breaks off no flower, and is careful not to crush any insect' (1923: 255). Living things are 'teleological'; they have interests and ends — particularly in maintaining the integrity of their own life's unfolding. In this view the criterion that determines moral considerability is not sentience but *conation* — the directionality, *strivingness* ('will-to-life') that all life exhibits (on conation, see Callicott 1986a: 154).

But these philosophical endpoints remain largely undeveloped by Schweitzer. In particular, Schweitzer provides no criterion to take us through the 'what to consume' problem of practical ethics, since we must live, and so consume at least some life. In the no-nonsense words of Edward Johnson: 'in the end it just doesn't come to much. Its basis is obscure and its implications are useless, at best, and incoherent at worst. Though Schweitzer seems to say that we are not to rank lives, that is, practically, impossible' (1984: 346). Moreover, as Singer (1979a: 92) points out, had Schweitzer faced the full implications of his ethic he could not have continued to privilege the saving of human life, which was his indefatigable mission as a doctor in Africa, and Skolimowski makes a similar point concerning Schweitzer's commitment to Protestant Christianity:

> he considered the value of the Christian ethic to be universal ... The point of fundamental importance is this: he did not see that it was the other way around, that the principle of the Reverence of Life and its conse-

> quences spell out a much more general ethic, of which the Christian ethic is a particular manifestation. Starting from Reverence for Life, the Christian ethic follows as a consequence, but the converse is not the case; the pronouncement of all life sacred does not follow from the Christian ethic (1981: 66).

Responding to the narrowly drawn biological individualism of Singer and Regan, the first notable attempt to develop a more coherent life-based ethic was provided in the late 1970s by Kenneth Goodpaster (1978; 1979). Goodpaster argued, against Singer, that sentience is not a first-order value. It is, rather, derivative of what *is* primary — and that is the possession of a life. It is not necessary to be sentient in order to have interests that can be furthered or impaired, and this is so even if that impairment does not result in felt pain.

This sounds very much like Schweitzer. The significant difference is that Goodpaster does not portray life in the mystical 'life is sacred' terms of Schweitzer. He utilises instead the scientifically respectable language of systems theory. A living organism is a 'homeostatic feedback mechanism', its feedback processes geared towards its striving for self-maintenance. 'The core of moral concern', he writes, 'lies in respect for self-sustaining organization and integration in the face of pressures toward high entropy' (1978: 323).

Whilst Goodpaster makes the philosophical content of biocentric individualism more coherent, it is not clear that the practical dilemma of choosing which forms of life to destroy in the interests of self-sustenance has been much advanced. 'The world is full of feedback mechanisms that "strive" to do their thing, yet there seems to be nothing wrong with interrupting the "striving" of the machine', writes Edward Johnson. 'The machine doesn't care if its "purposes" are thwarted, so why should we?' (1984: 350)

A stronger case for biocentric individualism has been made by Paul Taylor. For Taylor, 'respect for nature' is what he calls an 'ultimate moral attitude' (1987: 96); not derivative, not to be sourced to something more fundamental. To the terms 'intrinsic value' and 'inherent value', both of which are attributions bestowed upon entities by humans, Taylor adds another — 'inherent worth'. Acording to Taylor, 'inherent worth' is not *granted* but inheres within entities that can be said to have a good — and a bad — of their own, without reference to any other entity (1987: 71–80). The 'ultimate moral attitude' of 'respect for nature' is only given point by 'the biocentric outlook', the 'core' beliefs of which are:

> (a) The belief that humans are members of the Earth's Community of Life in the same sense and on the same basic terms in which other living things are members of that Community.
>
> (b) The belief that the human species, along with all other species, are integral elements in a system of interdependence such that the survival of

each living thing, as well as its chances of faring well or poorly, is determined not only by the physical conditions of its environment but also by its relationships to other living things.

(c) The belief that all organisms are teleological centers of life in the sense that each is a unique individual pursuing its own good in its own way.

(d) The belief that humans are not inherently superior to other living things (1987: 99–100).

Taylor acknowledges the significance of the interconnectedness of all life — the central lesson of ecological science — whilst retaining the moral individualism of the western philosophical tradition. He is no methodological radical, and in this respect his system shares marked similarities with those of Singer and Regan, though Taylor's is a 'search for an environmental ethic, rather than a merely humane one' (Norton 1987: 263). Taylor's biocentrism rests, says Christopher Stone, 'on the rationality, which he supports at some length, of recognising the non-instrumental worth of all living organisms' (1987: 94). Here, too, is the view of organisms as 'teleological centers of life' that is at the heart of Schweitzer's and Goodpaster's thought. Each organism strives for self-maintenance, maturity and reproduction — the realisation of its individual *telos*. It does not have to be conscious of that interest; an innate, programmed capacity for goal-seeking is all that is required. And it certainly does not rely upon the existence of a valuer; though a human can '*take an animal's standpoint* and, without a trace of anthropomorphism, make a factually informed and objective judgment regarding what is desirable or undesirable *from that standpoint*' (P.W. Taylor 1987: 67).

Humans with 'the attitude of respect for nature' will 'subscribe to a set of normative principles and hold themselves accountable for adhering to them. The principles comprise both standards of good character and rules of right conduct' (1987: 169). There are four of these, the most far-reaching being the first, 'The Rule on Nonmaleficence', explained as 'the duty not to do harm to any entity in the natural environment that has a good of its own. It includes the duty not to kill any organism and not to destroy a species-population or biotic community' (1987: 172; in the case of the latter collectivities this is not because they are themselves moral subjects, but because such destruction transgresses massively upon the rights of the individual moral subjects within them).

But what of the vexed question of how to maintain one's own life, given that this can only be done at the expense of other teleologically striving moral subjects? Taylor sets up '*priority rules*' to adjudicate conflicts between human interests and the interests of other organisms (1987: 263). These he lists as:

- 'the principle of self defence',

- the principles of 'proportionality' and 'minimum wrong' ('basic interests must prevail over nonbasic interests'; 1987: 270),
- the 'principle of distributive justice' ('provides the criteria for a just distribution of interest-fulfillment among all parties to a conflict when the interests of all are basic'; 1987: 292),
- the 'principle of restitutive justice' ('some form of reparation or compensation is called for if our actions are to be fully consistent with the attitude of respect for nature'; 1987: 304).

As Bryan Norton notes, 'Taylor's practical problems are substantially more severe than are those of animal liberationists, who can opt for vegetarianism. Eating plants is for Taylor, however, equally in need of justification with eating animal flesh' (1987: 265). One form of biocentric individualism for which this is not a perceived difficulty is provided by Gary Varner, who qualifies his primary position with an ethical non-egalitarianism: 'a certain class of human interests', he argues, 'is more important than the interests of ... almost every nonhuman organism' (1998: 8). But does Taylor himself resolve these problems satisfactorily? We will come back to this.

According to Baird Callicott, the limited biocentrism of the animal liberation/animal rights schools and the more inclusive biocentrism of Taylor and Goodpaster represent no challenge to 'the standard paradigm of traditional moral philosophy', being mere extensions of accepted human moral attributes (1993: 9; see also 1980). A 'theoretical paradigm' is required that is 'different from that institutionalized in contemporary moral philosophy' (1987: 207). Callicott proposes such a paradigm, one based on 'hints and suggestions offered by Aldo Leopold' (1993: 9). Callicott is Leopold's current great interpreter and champion. Only the land ethic, with its base in evolutionary science and ecological science, can provide the substance of a qualitative ethical breakthrough (Callicott 1980; 1987: 194). However, Leopold's 'brief but suggestive foray into ethics is not theoretically well formed or fully argued', and to 'construct the theoretical superstructure that occasionally shows through the informal texture of Leopold's prose' (1993: 10) is the task Callicott sets himself.

The focus of Leopold's land ethic was upon ecosystem integrity. As Callicott (1980: 314) and Lawrence Johnson (1991: 236) point out, Leopold was a hunter and a carnivore; the focus upon larger interdependencies rather than individual life units thus results in behavioural prescriptions that differ from those advanced by Singer and Regan. For Leopold, the passing of a complex ecosystem or an entire species is more deplorable than the passing of an individual organism, and ethical injunction should be so shaped. Callicott sharpens this focus. The insights of scientific ecology supply a capacity to view the natural environment as a community, and thus 'an ecocentric environmental ethic may be clearly envisioned' (1986b: 407). Much emphasis is placed upon the *communal* implications of the 'land community': 'all contemporary forms of life are represented to be *kin, relatives, members of one extended family*. And all are equally members in good standing of one *society* or *community*, the biotic community or global ecosystem' (1986b: 407). Moreover, the perceived boundaries of a society are also the perceived boundaries of its moral community' (1986b: 405–06).

In developing his ethical holism Callicott sets two additional influences — Charles Darwin and David Hume — beside Leopold. Here is Callicott, describing his own position (but using the third person!):

> according to Callicott, Charles Darwin suggested that such feelings [benevolence, sympathy, loyalty] were naturally selected in many species, including our prehuman ancestors, because without them individuals could not bond together into mutually beneficial communities. Rudimentary ethics emerged as human beings evolved to the point that they could articulate codes of conduct conforming to their social sentiments (1993: 10).

Sounding a little like one of those extensionists whose methodology he has explicitly eschewed, Callicott, following Leopold (and as we have seen, Leopold was also an ethical extensionist), portrays the land ethic as 'the next stage of human moral evolution', one in which 'a universal ecological literacy would trigger sympathy and fellow-feeling for *fellow-members* of the biotic community' (1993: 10). For this he claims the authority of Darwinian science, this 'fellow-feeling' having been engrained within humans through long ages of evolutionary adaptation. Certainly, the evolution of feelings occurred much earlier than reason and, Callicott argues, as a consequence they are more trenchantly embedded (1986b: 404–05).

It is, thus, slightly ironic that Callicott claims the authority of science for an ethical system based, not in cognitively derived principles, but upon 'feelings'. It is here that his third formative influence comes in. In not sourcing morality to reason, Callicott is following the Scottish philosopher David Hume (1711–66), for whom 'morality is grounded in feelings, not reason; although reason has its role to play in ethics, it is part of the supporting cast'; and for whom 'altruism is as primitive as egoism; it is not reducible either to enlightened self-interest or to duty' (Callicott 1992: 253). Read against ecological science in general and Leopold's thought in particular, the view that emerges of the Earth's linked communities of life 'can actuate the moral sentiments of affection, respect, love and sympathy ... noticed by both Hume and Darwin, that we have for the *group as a whole* to which we belong' (1986b: 407). Callicott thus concludes, against Regan and Singer, that collectivities (such as ecosystems) are appropriate objects of value.

They may be *objects of* value, but are they also *subjects with* value? Callicott's theory is, unlike deep ecology, a *value* theory, but one which seems nevertheless to reject the notion of *intrinsic* value, for value is sourced instead to (human) feeling. We have the apparent paradox that Callicott claims to have constructed a revolutionary new ethical paradigm, one of 'holism with a vengeance' (1987: 196), only to retreat to an apparently subjective and individualistic anthropocentrism as his ground of value. But Callicott sees value as 'anthropogenic', not 'anthropocentric'. He rejects the notion that there can be value independent of a human valuer, and hence there can be no 'intrinsic' value. But he distinguishes 'intrinsic' value from 'inherent' value, the latter obtaining 'when something is valued for itself, independently of any instrumental value that it may, or may not, have for a valuing consciousness', and so, McLaughlin writes of Callicott, 'though there would be no value without human consciousness that does not mean that all values are homocentric' (1993: 161). It is audaciously argued, and McLaughlin is not alone in finding the distinction not entirely convincing (see also W. Fox 1990: 264–65).

Holmes Rolston III is one of the pioneers of environmental ethics, having made his first significant foray into the field in 1975. His thought is not easy to reduce to a few paragraphs, demanding extraction from the two books and numerous specifically focused papers that he has published over twenty years (for example, 1975; 1982; 1985; 1986; 1987; 1988; 1990; 1994). Unlike Callicott, he does find a ground for value in nature ('autonomous intrinsic value' rather than 'anthropogenic intrinsic value'; 1988: 114), arguing that 'some values are objectively there — discovered, not generated, by the valuer' (1988: 116). This objective value is revealed within nature by the insights of ecological science, and so 'ecology is an ethical science' (1986: 55).

Despite his emphasis upon the ethically imbued insights of ecology, Rolston constructs a hierarchy of value. Sentient life is more intrinsically valuable than non-sentient life, and animals capable of self-reflection, ethical construction and rational choice are more valuable still. Where there is conflict then, human interests take precedence (1988: 62–71, 335–37). Not that there *is* any fundamental conflict between human and non-human interests. Human subjectivity ought to be 'spirit incarnate in place, where the passage of consciousness through nature in time takes narrative form' (1988: 341). In any case, humankind's reflective capacity not only endows humans with superior *rights* (in the unnecessary event of an inter-species conflict of interest) but it also, perhaps more significantly, imposes responsibilities: 'humans are in the world *ethically*, as no other animal is. Humans are in the world *cognitively* at linguistic, deliberative, self-conscious levels equalled by no other animal. Humans are in the world *critically*, as nothing else is' (1988: 71). Herein lies their responsibility.

And Rolston finds a place for collectivities in this hierarchy. Species and ecosystems also possess intrinsic value. 'Subjects count', Rolston argues, 'but they do not count so much that they can degrade or shut down the system, though they count enough to have the right to flourish within the system' (1987: 272). In *Philosophy Gone Wild* Rolston outlines several maxims which extend intrinsic value beyond individual biological units to ecological interdependencies. The 'reversibility maxim' decrees that we ought not to cause irreversible change to natural systems, the 'china shop maxim' maintains that 'the more fragile an environment, the lighter it ought to be treated', and the 'scarcity maxim' holds that we ought to take particular steps to preserve rare collectivities (1986: 155–56, 158, 159–60). Rolston is perhaps best viewed as the great synthesiser within ecophilosophy. His is primarily an attempt to resolve the dilemmas of the place of interdependent wholes within the biocentrism of Taylor (and Singer and Regan) and of the place of the individual within the holism of Callicott.

Like Rolston's, Richard Sylvan's contribution to the early emergence of environmentally informed philosophy was of considerable importance. His 'deep-green theory', originally forged in collaboration with Val Plumwood (as Val Routley) and later elaborated with David Bennett, can be traced through his writings back to 1973.

Much influenced by Taoism (Sylvan and Bennett 1992), 'deep-green theory' is nevertheless 'emphatically a philosophical approach', one which 'combines elements of traditional ethical consideration with a radically new ecologically-oriented consideration for the environment ... a comprehensive alternative environmental philosophy, a detailed (if not total) philosophical paradigm' (1994: 137). Sylvan (with Bennett) proceeds largely by comparing deep-green theory with deep ecology, the better to highlight the strengths of the former when the comparison fails. Thus, though Naess's eight-point platform is largely accepted, Sylvan and Bennett withhold endorsement from Naess's 'extreme holism, biospherical egalitarianism or the ultimate norm ... of self-realization' (1994: 137–39).

The stated objective of 'deep-green theory' is to rework philosophy 'in a way free of chauvinism', which is to say 'without any specially privileged place for humans', nor with any place for intra-human (such as class or gender) chauvinism. Moreover, 'no single feature, such as sentience or life, serves as a reference benchmark for determining moral relevance and other ethical dimensions' (1994: 140). Sylvan proposes an ethic of *eco-impartiality* — which is posited as superior to deep ecology's problem-plagued principle of biospherical impartiality (1994: 142–43) — though 'natural features' such as sentience and diversity still 'afford criteria for value' (1994: 143). Eco-impartiality then, but value is still assessed against *criteria* (such as 'goodness';

1994: 143). These criteria are not sourced to a valuer, however: values 'do not depend upon the presence of valuers ... they are not secondary or tertiary or response-dependent properties' (1994: 143).

How, then, are values to be identified? Here Sylvan and Bennett's argument becomes somewhat complex, but the kernel of it is that one initially intuits value via 'emotional presentation': 'a valuer feels raw value and disvalue ... apprehension of value is seated in emotional, and especially visceral, presentation' though 'what is apprehended is not to be confused with its apprehension' (1994: 145). This is simply the *basis* of critical engagement, and there is a clear need for 'checks on emotional presentation, such as constancy over time and reflection' (1994: 146). It is here that those 'traditional' elements of ethical consideration become relevant. This is not a schema that rejects the rationality and the interest-based approach of mainstream ethics. Foremost among these front-of-brain 'checks on emotional presentation' is the behavioural and norm-adopting 'check' against the key 'constraint' of eco-impartiality.

ANTHROPOCENTRISM MAKES A COMEBACK

When John Passmore mounted his attack upon the attribution of moral significance to nature in 1974, only the initial cases for a broadening of the scope of moral subjectivity yet existed. Passmore attempted to ensure that these initiatives were stillborn. For him there could be no basis for rights in nature, not for individual items of non-human life, and certainly not for collectivities such as species and ecosystems. Moral obligation, he maintained, only arises within the interaction of human communities:

> ecologically, no doubt, men form a community with plants, animals, soil, in the sense that a particular life-cycle will involve all four of them. But if it is essential to a community that the members of it have common interests and recognise mutual obligations, then men, plants, animals, and soil do not form a community. Bacteria and men do not recognise mutual obligations nor do they have common interests. In the only sense in which belonging to a community generates ethical obligation, they do not belong to the same community (1974: 116).

Moreover, there was no need to re-found the basis of morality, because western ethics comes ready furnished with the conceptual wherewithal to provide nature with the requisite protection.

For the next decade-and-a-half, as ecophilosophical system-building burgeoned, the main intra-movement opposition to a broadening of the ground for western ethics came from ecoMarxism, social ecology and, more ambiguously, ecofeminism — all branches of ecological thought to be discussed in later chapters. Purely *philosophical* defences of a human-based grounding for ethics within environment movement writings are somewhat uncommon during this period, but an anthropocentric environmental ethics in the Passmore tradition bubbled unobtrusively along. W.H. Murdy, for example, mounted the plausible argument that 'species exist as ends in themselves. They do not exist for the exclusive benefit of any other species ... To be anthropocentric is to affirm that mankind is to be valued more highly than other things in nature — by man. By the same logic, spiders are to be valued more highly than other things in nature — by spiders' (1983: 13). This given, 'our current ecological problems do not stem from an anthropocentric attitude *per se*, but from one too narrowly conceived. Anthropocentrism is consistent with a philosophy that affirms the essential interrelatedness of things and that values all items in nature' (1983: 20).

Warwick Fox identifies two prominent classes of argument within anthropocentric environmentalism's rebuttal of ecocentrism during the 1980s. The first of these revolves around the claim that ecocentrism is misanthropic, or conduces to misanthropy. Fox refers to this as '*the fallacy of misplaced misanthropy*'. It is fallacious because 'being opposed to human-*centeredness* is logically distinct from being opposed to humans *per se*' (1990: 19). This, nevertheless, is the charge most persistently made against ecocentrism in general and deep ecology in particular, it being pivotal to the hostile critiques mounted within ecoMarxism and social ecology (as we will see, there are other more specific grounds upon which a claim of misanthropy can be made against deep ecology).

The second broad argument is that it is impossible to be anything other than anthropocentric 'since all our views are, necessarily, human views' (1990: 20), a charge more likely to be encountered within ecophilosophical thought itself (for example, Norton 1991: 251; Ferre 1994), and given credence by the fact that it is conceded by some prominent environmental axiologists (Taylor and Callicott, for example). Fox, however, will have none of it. This is his '*perspectival fallacy*', so called because 'it conflates obvious and inescapable facts about a speaker's perspective with the substance of the speaker's view' (1990: 21–22). To make this argument, Fox distinguishes between a 'weak' and a 'strong' sense of 'anthropocentrism':

> the tautological fact that everything I think and do will be thought and done by a human (the weak, trivial, tautological sense of anthropocentrism) does not mean that my thoughts and actions need be anthropocentric in the strong, informative, substantive sense, that is, in the sense of exhibiting unwarranted treatment of other beings on the basis of the fact that they are not human — which, again, is the sense that really matters ... [The perspectival fallacy] confuses the inescapable fact of human identity, the trivial sense of anthropocentrism, with the entirely avoidable possibility of human chauvinism or human imperialism, the significant sense of anthropocentrism (1990: 20–21).

Others concur. Mary Midgley, though far from Fox's position, and finding the word 'anthropocentrism' unhelpful in other ways, nevertheless echoes Fox's sense of its meaning within current ecophilosophical discourse: 'what is commonly *meant* by the word "anthropocentric" today ... is simple *human chauvinism, narrowness of sympathy*, comparable to national or race or gender chauvinism' (Midgley 1994: 111).

Those who continue to argue a case for anthropocentrism are at pains to differentiate anthropocentrism from 'human chauvinism'. They do this by also distinguishing between 'weak' and 'strong' anthropocentrism, but employing these terms differently to Fox. In this typology, 'strong' anthropocentrism is equated with 'human chauvinism'; the 'weak' sense is that which can serve as an adequate foundation for a nature-regarding ethic. Eugene Hargrove, for example, argues that a 'weak anthropocentric value theory' is not only plausible but is actually all we yet have, as 'most nonanthropocentric value theories are in various ways really anthropocentric' (1992: 184). Thus, the professed non-anthropocentrism of Taylor, Rolston and Callicott is 'infected with anthropocentrism' (1992: 202). Hargrove's own anthropocentrism is admittedly conditional. He holds that 'weak anthropocentrism ... can serve environmental ethics well until such time, if ever, that a convincing nonanthropocentric theory appears' (1992: 191). But he is not holding his breath.

It is Bryan Norton, though, who is most associated with the defence of 'weak anthropocentrism'. Norton developed a theory of 'convergence' (1984; 1986; 1991; 1992a; 1992b), in which it is argued that the apparently unbridgeable chasm between anthropocentrism and deep ecology (which Norton uses as paradigmatic for ecocentrisms generally) is rendered irrelevant when it comes to devising regimes for policy-making and practical environmental management. In examining 'the philosophical hypothesis that some natural objects have nonanthropocentric, intrinsic value', Norton insists upon 'keeping in mind the more fundamental practical problem — that of proposing an adequate theory to explain and support the activities of environmental, as opposed to mere resource, managers', and, from such a perspective, 'at least in the important area of moral obligations regarding fauna and flora, the policy implications of a broadly formulated and farsighted anthropocentrism and nonanthropocentrism will be identical' (1992a: 209; see also 1995). Moreover, as the base claims of deep ecology seem alien and radical to people not accustomed to ecophilosophical discourse, and as these base claims are in any case not necessary to the achievement of the sorts of policy outcomes that flow from deep ecological principles, it makes much more sense to adopt a 'weak anthropocentric' framework and proceed therefrom. In moving from the world of principle to the world within which policy is framed, deep ecologists must, in any case, make intuitive assumptions about the relative priority claims of unit of life against unit of life. In reality, what matters is that the integrity of ecological wholes be maintained, even if this requires the death or suppression of some individual components within those wholes. And, given 'the central insight of ecology — that all things in nature are interrelated' (albeit not equally so), it follows that 'no long-term human values can be protected without protecting the [ecologically robust] context in which they evolved' (1991: 240). And so, the management and policy imperatives arising from a commitment to deep ecological values are identical to those arising from an enlightened 'weak' anthropocentrism (1991: 226–27; see also Wells 1993: 525; for a riposte to Norton, see Stevenson 1995; for a case arguing for a flexible and contextual deployment of 'anthropocentrism' and 'non-anthropocentrism', see Katz 1999).

Despite the efforts of such proponents of 'weak' anthropocentrism as Hargrove and Norton, the tide continued to run for ecocentrism until, in the 1990s, it became evident that anthropocentric environmentalism was winning substantially more support than it had enjoyed through the 1980s. Whilst ecopolitical theory was much stronger in Europe than in North America through the 1970s and 1980s, the reverse was clearly the case with ecophilosophy (Arne Naess being the outstanding exception), but this began to change in the late 1980s. And, as a general rule, European ecophilosophers have shown a comparative reluctance to jettison traditions of thought stemming from respected intellectual forebears.

Perhaps the best known of the European anthropocentrists is an Aristotelian, John O'Neill. Here is an anthropocentrism that can coexist with the attribution of intrinsic value. O'Neill considers the sense of 'intrinsic value' hitherto assumed in grounding an environmental ethic — 'that non-human beings are not simply of value as a means to human ends' (1993: 10) — and finds it inadequate. The corollary, for example — that non-human beings have non-instrumental value for humans — will not do. Some non-humans clearly have a negative value for humans, and it is impossible that it could be otherwise: 'it is not the case that goods of viruses should count, even just a very small amount. There is no reason why these goods should count at all as ends in themselves' (1993: 23).

The path to an environmental ethic is rather to be sought in 'an Aristotelian conception of well-being' (1993: 3): 'human beings like other entities have goods constitutive of their flourishing, and correspondingly other goods instrumental to that flourishing. The flourishing of many other living things ought to be promoted because they are constitutive of our own flourishing' (1993: 23–24). In response to the obvious objection that this would seem to return us to 'a narrowly anthropocentric ethic' (1993: 24), O'Neill argues that 'it is compatible with an Aristotelian ethic that we value items in the natural world for their own sake, not simply as an external means to our own satisfaction':

> for a large number, although not all, of individual living things and biological collectives, we should recognize and promote their flourishing as an end itself. Such care for the natural world is constitutive of a flourishing human life. The best human life is one that includes an awareness of and practical concern with the goods of entities in the non-human world (1993: 24).

O'Neill is thus able to argue that 'there is no incompatibility between a concern for human well-being and the recognition of and care for the intrinsic value of the non-human world' (1993: 3), and this is a position for which there would seem to be growing support.

SOME PROBLEMS: MEANWHILE, IN THE REAL WORLD ...

Here we must be modest. Each philosophical position contains a critique of other positions, sometimes tacit, often overt. In addition, philosophical enterprise at the general level has been called into question by some strands of environmental thought. To complicate matters further, there are keenly contested differences even within ecophilosophical schools, and some of these are complex and finely drawn. Only the most persistent and fundamental criticisms are noted here: many of these have already emerged in part, and want only for more explicit recognition.

There is, too, the question of the use of ecological science to underpin systems of ecophilosophy. Carolyn Merchant notes: 'for Rolston, science is objective truth whose sphere continually expands with greater human knowledge' (1990: 79). Callicott, too, 'accepts the evolution of science as providing ever greater access to truth' (Merchant 1992: 80), and there are others who claim the authority of science for their systems of thought. One philosopher who has criticised this tendency is Andrew Brennan. Brennan believes that the perceptions of ecology upon which much ecophilosophy has been based are essentially ideological: 'metaphysical ecology' rather than 'scientific ecology' (1988: 31). He identifies Naess (and deep ecology generally) and Callicott as being particularly at fault. Brennan is also more generally critical of attempts to develop a philosophical holism from ecology, arguing that a true reading of scientific ecology does not render such a project tenable.

The first notable within-the-movement riposte to animal liberation came in 1977 with the publication of Rodman's 'The Liberation of Nature?'. Singer's insistence that one of animal liberation's strengths is that it requires no new ethical basis but a mere logical extension of existing ethical precepts was labelled 'ethical extensionism' by Rodman. He criticised it as involving an unacceptable anthropomorphising of animals; a denial of their biological especialness by transposing upon them essentially human characteristics:

> the process of 'extending' rights to nonhumans conveys a double message. On the one hand, nonhumans are elevated to the human level by virtue of their sentience and/or consciousness ... on the other hand nonhumans are by the same process degraded to the status of inferior human beings, species anomalies: imbeciles, the senile, 'human vegetables' (1977: 93–94).

Rodman also attacked the ecologically tenuous credentials of an ethics which had, after all, emerged from a concern for cruelty to domestic animals, rather than from an engagement with ecological insights. This is apparent in 'the moral atomism that focuses on individual animals', a stance that 'does not seem well adapted to coping with ecological systems' (1977: 89). 'I need only to stand in the midst of a clear-cut forest, a strip-mined hillside, a defoliated jungle, and a dammed canyon', he observes, to feel the inadequacy of the argument that human actions cannot 'make any difference to the welfare of anything but sentient animals' (1977: 89).

The ecological shortcomings of animal liberation have remained a recurrent theme in critical environment movement assessments. It was the key point argued by Baird Callicott in his important 1980 paper. For Callicott animal liberation and emergent environmental thought were as remote from each other as each is from mere 'humanism' ('a triangular affair'), a position echoed in 1984 by Mark Sagoff ('bad marriage, quick divorce'). Callicott argued that animal liberation, as it is incapable of taking cognisance of ecological relationships, holds no promise for environmental thought (unlike the holism of Leopold's land ethic). The individualistic basis of animal liberation has led, in Callicott's view, to the adoption of positions that are, from an ecological perspective, downright silly. He instances animal liberation's failure to morally privilege life in the wild over domesticated animals, whilst he has this to say about animal liberation's exhortation to vegetarianism: 'from the ecological point of view, for human beings to become vegetarians is tantamount to a shift of trophic niche from omnivore with carnivorous preferences to herbivore', and as this would, 'by bypassing animal intermediaries', result in an increase in available food for humans, 'the human population would probably ... expand in accordance with the potential thus offered', resulting in even fewer nonhuman animals. Meat-eating may, thus, be 'more ecologically responsible than a wholly vegetarian diet', as 'a vegetarian human population is *probably* ecologically catastrophic' (1980: 335).

More recently, W. Fox has most lucidly summed the case against animal liberation/rights. He outlines several objections, most of which refine and develop the 'failure from ecology' case outlined above. Animal liberation fails to acknowledge any difference between non-sentient life and non-life: 'from the perspective of this approach, a world that is luxuriantly rich in plant life but that possesses no sentient life-forms whatsoever is considered to be just as intrinsically valuable (or, rather, value*less*) as a barren asteroid or a tin can' (1996: 9). Animal liberation is also 'blind to the existence' of an ecosystem, and even of 'a species *per se*', and so 'the *last* one hundred members of a species' are of no greater account than one hundred members of an abundant species (1996: 9–10). In similar vein, animal liberation has no capacity to distinguish between monoculture and diversity: 'there is just as much intrinsic value in a world of, say, a trillion sentient creatures of the *same* kind as there is in a world of a trillion sentient creatures made up of a million *different* kinds of creatures' (1996: 9). Animal liberation cannot distinguish between introduced and native or between wild and tame animals (1996: 10). Finally, the logic of its position warrants such ecological absurdities as the fencing off of the lion from the wildebeest:

> if it is bad for humans to cause suffering by killing other animals then it is also bad for nonhuman animals to cause suffering by killing other animals. From the point of view of the animal being killed it doesn't matter whether it is a human or some other animal that is causing the suffering (1996: 10).

One final point here. There is clearly a vast gulf between the individualistic ethics of animal liberation/rights and the various ethical holisms. But it may be that the extent of this gulf will only really be manifest at the level of philosophical abstraction. Mary Ann Warren (1992) and Dale Jamieson (1998) argue that the differences between animal liberation and ethical holism are somewhat exaggerated, and even Callicott (1992) now sees scope for animal liberation and ecological thought to come 'back together again'. At the level of policy prescription ecological holists, though unlikely to place the same priority on reforming feedlot practices, are likely nevertheless to approve animal liberation-generated reform processes. Similarly, animal liberationists are likely to be supportive of attempts by ethical holists to defend the integrity of ecosystems, even if they will do so for reasons that holists find unsatisfactory. In my experience, even holists who argue against vegetarianism on ecological grounds tend not to be consumers of animal products on any great scale.

Several points of criticism have been directed at the life-based ethics of Schweitzer, Goodpaster and Taylor. Much of the criticism that has been levelled at animal rights also applies to biocentric individualism. The ethical extensionism involved in each case is blind (runs the criticism) to the imperatives of ecology. Again, Warwick Fox is the most trenchant critic, listing the consequences thus:

> the life approach is (i) incapable of distinguishing between monoculture and diversity, since one can have just as much life and, hence, intrinsic value in both cases; (ii) blind to the existence of species per se, since

> species per se are not alive on this understanding; (iii) blind to the existence of ecosystems per se, since ecosystems per se are not alive on this understanding; (iv) incapable of distinguishing between introduced and indigenous life-forms, since the former are just as alive and, hence, intrinsically valuable as the latter; (v) incapable of distinguishing between domesticated and wild life-forms, since, again, the former are just as alive and, hence, intrinsically valuable as the latter; and (vi) capable of being pressed into service as an argument against killing any forms of life — even for food (1996: 13).

The last of these criticisms is the most persistently made. If all organisms from the smallest and biologically simplest to elephants and whales are of equal intrinsic worth, the task of putting such an ethic into practice becomes 'diabolically difficult' (Callicott 1993: 8): 'if rigorously practised, it would seem to require a restraint so severe that it would lead if not to suicide by starvation, at least to a life intolerably fettered' (Callicott 1986a: 154).

We closed our earlier discussion of Paul Taylor's thought having left open the question of whether Taylor had satisfactorily resolved this problem of the necessity for life to feed upon life. Taylor, it will be remembered, put forward several principles and 'priority rules', which he believed overcame this difficulty. But biocentric individualism's critics remain unconvinced. Peter Wenz, for instance, sees Taylor's 'principle of minimal wrong' as a flat contradiction of his ethical position rather than a means of rendering it operational (1988: 286–87). Taylor is saying, in effect, that the principles of biocentric individualism cannot be put into practice, so we must limit damage to the best achievable level. 'Biocentric individualism is so confining', writes Wenz (1988: 287), 'that even Taylor, its foremost proponent, fails to apply it consistently'. Others have found Taylor's 'rule of restitutive justice' similarly deficient. How does one offer restitution to an individual moral subject that is dead? As Norton perceptively notes (1987: 266), any restitution offered is to the kind, not the deceased individual: the principle is therefore plausible only if it is assumed that there are interests and moral claims at the level of the collective — a clear contradiction of Taylor's own carefully constructed individualism.

Finally, Warwick Fox makes some more general criticisms of the life-based approach. Biocentric individualism argues that to have intrinsic value is to have an interest in self-maintenance; to *strive* to that end. But this, says Fox, implies *intentionality* — a 'goal in mind'. Terms such as 'striving', 'aiming' and 'desiring' are normally reserved for human psychological attribution, not for processes in the physical or biological realms. They are 'explanations for things in terms of goals and reasons' (1996: 12), and not physical 'forces' or 'causes'. Biocentric individualism, thus, wrongly bestows 'interest' — perhaps even *telos* — to all living entities, by erroneously characterising biological processes as 'striving'.

Biocentric individualism also employs a non-tenable understanding of 'life'. The notion of life inhering within 'individual plants and animals' only works at the level of 'everyday, conventional understanding' (W. Fox 1996: 14); in a more formal sense 'living systems' are those which are 'primarily and continuously involved in making and remaking themselves on an ongoing basis' (1996: 13). On such a definition an individual organism is certainly a living system — but so too are 'ecological wholes such as ecosystems and the ecosphere' (1996: 14). Biological *individualism* is, thus, untenable. 'Taylor writes', says Fox, 'in the absence of any particular knowledge of, or references to, the recent developments in theoretical biology' (1990: 174). At least one Taylorist, James Sterba (1995), bowing to the force of this argument, has sought to rescue Taylor by retreating from full-blown individualism to a position that he calls 'biocentric pluralism'. Thus, hybridised biocentrisms are now emerging.

We have seen that Callicott was a prominent early critic of the individualist 'ethical extensionism' of Singer, Regan, and the life ethics of Goodpaster and Taylor. It should come as no surprise to learn that the most trenchant criticisms Callicott's ethical holism has received have come from precisely these quarters, and in particular from Regan.

The advanced holism of the Callicott position leaves it subject to the charge of misanthropy: the view that 'bacteria and plankton carry more ethical weight than human beings because of their importance to the wellbeing of the ecosphere' (Marshall 1992: 442). It is certainly reasonable to assume that Callicott's 'holism with a vengeance' implies that the interests of the individual — the human individual, say — are of no ultimate moral account, but must be submerged within the interests of the collective whole.

Regan takes just such an implication and, in the strongest of words to pass between rival ecophilosophical schools, has argued that holistic natural ethics 'might be fairly dubbed "environmental fascism"':

> the rights view ... denies the propriety of deciding what should be done to individuals who have rights by appeal to aggregative considerations, including, therefore, computations about what will or will not maximally 'contribute to the integrity, stability, and beauty of the biotic community'. Individual rights are not to be outweighed by such considerations ... Environmental fascism and the rights view are like oil and water: they don't mix (1983: 362).

Lawrence Johnson concedes a pinch of validity to Regan's case, but argues that 'we should not overreact and go to another extreme, finding moral importance *only* in the concerns of individuals' (1991: 176). Whilst the tendency to lose sight of individual worth within a focus upon collective needs may well be a danger to guard against, then, at the extreme of Regan's formulation the case is difficult to sustain.

Setting aside the quibble that 'fascism' has a much more precise meaning than simply 'the subordination of the interests of the individual to the interests of the whole', Callicott can in any case argue plausibly that his holism does not sacrifice the interests or compromise the moral standing of individuals. The criticism would be more telling if Leopold and Callicott used the metaphor of organism to characterise ecosystem interdependencies. But they do not. Callicott's collectivity is explicitly a *community*, and the notion of community accords a measure of autonomy to its constituent parts. There are, in fact, 'nested overlapping community entanglements' — some of which are, of course, exclusively human — and thus 'we are still subject to all the other more particular and individually oriented duties to the members of our various more circumscribed and intimate communities' (1993: 247).

Of course, the linked charges of misanthropy and ecological fascism can be applied to any form of ecocentrism, not just Callicott's variant, and it has been deep ecology that has borne the brunt of this charge — as it has borne the brunt of most charges that can be levelled against ecocentrism generally.

In large part, this is the consequence of Naess's ambition to have 'deep ecology' serve as a generic rallying cry for all political activism that takes its inspiration from environmental holism. Thus, much criticism of deep ecology has been targeted at the avowedly misanthropic stance of Earth First! under the leadership of Dave Foreman (Foreman himself employing 'deep ecology' to describe his position). For example, its newsletter of the same name, *Earth First!* published, under the pseudonym of 'Miss Ann Thropy', what came to be an infamous and much-cited letter welcoming the large-scale incidence of AIDS in Africa as a solution to perceived human overpopulation. 'If the AIDS epidemic didn't exist', wrote 'Miss Ann', 'radical environmentalists would have to invent one' (1987: 32). The association of deep ecology with Earth First!'s overt misanthropy has caused considerable problems for, and some disagreement within deep ecology. George Sessions has defended Earth First!'s right to identify with deep ecology (1992b: 66; see also Manes 1990: 155), whilst Michael Zimmerman has questioned the 'fit' between Earth First!'s views and the principles of deep ecology (1991: 124). Murray Bookchin, deep ecology's most unyielding critic, has been particularly scathing on the matter of its alleged misanthropy: '"humanity" is essentially seen as an ugly "anthropocentric" thing — presumably, a malignant product of natural evolution — that is "overpopulating" the planet, "devouring" its resources, destroying its wildlife and the biosphere' (1987a: 3).

A variation on the alleged anti-humanness of deep ecology is that it is, if not broadly anti-human, certainly set against the interests of the poor nations of the world, perhaps even guilty of a tacit racism on this account. The most oft-cited case made along these lines is that of

Ramachandra Guha, who argues that deep ecology's 'anthropocentrism/biocentrism distinction is of little use in understanding the dynamics of environmental degradation'. He argues that 'the implementation of the wilderness agenda is causing serious deprivation in the Third World' (1989: 71; for a response, see Naess 1995c), and that what is needed, and what is missing from deep ecology, is 'a greater emphasis on equity and social justice (both within individual countries and on a global scale) on the grounds that in the absence of social regeneration environmental regeneration has little chance of succeeding' (1989: 81–82).

On the general question at the heart of these debates, I am inclined to side with those who argue for a less all-encompassing definition of 'deep ecology', one that would place the anti-humanness of the early Earth First! outside its rubric, and one that would be less vulnerable to the charges Guha makes against it. And I would draw attention to Naess's insistence (1985, for example) that humans are no less entitled than any other species to seek satisfaction of their 'vital needs'.

This is, though, just one of the heads of criticism stemming from deep ecology's definitional imprecision. Postmodernists criticise it as yet another 'grand narrative', constraining individual action and denying individual autonomy, whilst critics of deep ecology who retain an allegiance to progressive modernism (Bookchin, and the ecoMarxists, for example) are wont to type deep ecology as the very essence of sloppy and reactionary New Ageism. Yet, many writings in deep ecology carry overt *rejections* of New Ageism, on the ground that 'to New Age thinkers humans occupy a special place in the world because we possess consciousness, reason, morality, and any number of privileged traits that make us fit to be stewards over the natural processes of this planet' (Manes 1990: 145; see also Devall and Sessions 1985; Sessions 1995b). Some variations on this New Ageist theme (repugnant to deep ecologists) portray humans as the medium of planetary consciousness — the privileged mechanism whereby the planet reflects upon itself.

A further consequence of deep ecology's definitional imprecision is the inevitable criticism of it as intellectually half-baked. Thus, Richard Sylvan has written of deep ecology as a 'conceptual bog' which 'has been presented as a metaphysics, as a consciousness movement (and as primarily psychological), and even as a sort of (pantheistic) religion' (1985: 1–2; note also W. Fox's 1986 response to Sylvan), whilst Peter Van Wyck finds it 'both curious and troubling that deep ecology has been so influential, while at the same time remaining so theoretically underdeveloped' (1997: 1–2).

Devall notes that 'deep ecology has increasingly become associated with radical environmental activism, or the use of tactics such as ecotage, sit-ins, guerrilla theater, demonstrations, and other forms of direct action' (1991: 251; see also Devall and Sessions 1985: 17–39; Naess 1989: 130–62). Christopher Manes constructs a fully fledged political program around radical confrontation in the name of deep ecology, claiming of ecotage that it is 'the activity that more than anything else defines radical environmentalism' (1990: 21). Despite such contributions, an enduring criticism of deep ecology has been that it is difficult to get a credible *politics* from it. Naess would not agree. He argues that 'the ecological movement cannot avoid politics', and that 'in principle, it is desirable that everyone in the ecological movement engage in political activity' (1989: 130–31). But, of course, this is Naess using 'deep ecology' as synonymous with 'ecology *movement*'. On the other hand, even Warwick Fox has argued that the deep ecology platform 'is limited in terms of political scope', noting that 'of the widely known "four pillars" of Green politics — ecology, social justice, grassroots democracy, and nonviolence ... the deep ecology platform essentially picks up on just one: ecology' (1995a: 165). Critics have been even harsher, saying, with Sylvan and Bennett:

> there is no new political vision forthcoming from Deep Ecology or ecosophy. Similarly for any accompanying economic vision. Present arrangements are highly incompatible with Deep Ecology, yet no alternatives are really offered. The deep ecopolitics promised is but empty rhetoric, attractive noise signifying nothing (1994: 123).

These deficiencies exist because (it is usually argued), by identifying an amorphous cultural disposition to 'anthropocentrism' as the root cause of environmental decay, deep ecology insufficiently acknowledges the reality of exploitative power relations *between* people, from which environmental degradation follows as a sure *consequence*. Put baldly, the deep ecology position is that culturally embedded anthropocentrism drives humans to impose dominance upon the non-human world. Its critics, on the other hand, reverse the causal direction: for them, the desire of some humans to dominate others leads to the dominance and degradation of non-human nature as the latter is harnessed to serve the project of human-over-human domination.

In the case of ecoMarxism, the intra-species axis of domination that has environmental degradation as one of its inevitable consequences is that of social class by social class (for example, Pepper 1993a); whilst for eco-anarchist/social ecologist Murray Bookchin the missing intra-species dynamics are the patterns of dominance that stem from the production of social *hierarchies*. Bookchin's falling out with deep ecology was spectacular and, for the movement as a whole, extremely debilitating. In their 1985 book, Devall and Sessions identified (and praised) Bookchin's thought as an important feeder component *within* deep ecology, and Bookchin had himself published under the deep ecology banner (he has a paper in Michael Tobias's book, *Deep Ecology*). The break occurred without warning at a conference in 1987, when, to the astonishment of bioregionalist/deep ecologist Kirkpatrick Sale, who was also on the platform, Bookchin launched an assault upon deep ecology:

> until that moment, I sincerely and naively thought that Bookchin and I were on the same wavelength (indeed, friends), that there was really only one big ecology movement and that we shared an essentially similar position on the environmental destruction of the earth. But I suddenly realized that ... the social ecologists ... were actually out to destroy the influence of those thinkers and activists they found distasteful: the deep ecologists, in particular (Sale 1988: 670).

Bookchin followed with an attack in print, in which deep ecology was decribed as 'flaky spiritualism' (1987a: 22), 'a "black hole" of half-digested and ill-formed ideas' (1987a: 4), 'an ideological toxic dump' (1987a: 5), and a 'self-contradictory and invertebrate thing' that 'has parachuted into our midst from the Sunbelt's bizarre mix of Hollywood and Disneyland' (1987a: 3). Although Bookchin subsequently claimed the role of the injured victim, making reference to 'a veritable hurricane' of 'vituperative attacks upon me as a person and as a social ecologist!' (1988: 5), an uninvolved observer of the 'debate' could only conclude that the extravaganza of abuse has been largely fuelled from Bookchin's side. Invective aside, though, he does make several substantial claims against deep ecology. He rejects the 'wilderness worship' and the tendency to mystical and intuitive knowing he discerns in deep ecology (1987a: 22). Most significantly for our present purposes, he attacks deep ecology for evincing 'very little concern for the manipulation of human beings by each other' (1987a: 22). Of all deep ecology's shortcomings, its greatest is its failure to make 'a resolute attempt to fully anchor ecological dislocations in social dislocations', and particularly its failure 'to analyze, explore, and attack hierarchy as a *reality*' (1987a: 22).

For ecofeminists, the largely unacknowledged fulcrum of social domination (again having, as a sure consequence, the domination of nature) is the multifaceted dominance relationship that stems from culturally embedded attitudes of masculine gender bias. The dominance of men over women (and the male principle over the female principle) and the dominance of 'man' over nature are entwined processes, the inferiorising of the female taking reinforcement from the view that women partake more fully of nature than men, and the degradative manipulation of nature taking legitimacy from its characterisation as 'she'. Against deep ecology's identification of an *anthropocentric species bias* as the underlying cause of the exploitation of nature, ecofeminism sets *androcentrism*: the male gender domination and elevation of masculinist values that underlies social and environmental exploitation. Janet Biehl, for example, holds that 'for ecofeminists the concept of anthropocentrism is "deeply" problematical. It assumes that humanity is an undifferentiated whole, and it does not take into account the historical and political differences between male and female, black and white, rich and poor' (1987: 24). Marti Kheel, too, has written that ecofeminists and deep ecologists differ significantly 'in their understanding of the root cause of our environmental malaise. For deep ecologists, it is the anthropocentric worldview that is foremost to blame ... For ecofeminists, it is not just "humans" but men and the masculinist worldview that must be dismantled from their privileged place' (1990: 128; see also Salleh 1984; 1992; 1993a).

For all the heat, a rapprochement may be possible. Warwick Fox (1989), Zimmerman (1987; 1990a) and Capra (1997) have argued for this, with Capra including social ecology in his ecumenical wishfulness, insisting that 'each of the three schools addresses important aspects of the ecological paradigm and, rather than competing with each other, their proponents should try to integrate their approaches into a coherent vision (1997: 8). Most ecofeminists, though, remain unimpressed. Deborah Slicer describes Fox's case as one of deep ecological co-option of ecofeminism rather than a genuine search for common ground (1995: 153). For her, it is 'only marginally true' that any scope exists for agreement (1995: 164). But Ariel Salleh, whose criticism of deep ecology has been as trenchant as any, has offered a qualified 'endorsement of the deep ecological project' (1992: 199; see also 1984: 339), and there are some ecofeminists who are pleased to identify with both deep ecology *and* ecofeminism. Freya Mathews is one. In a series of 1994 papers she explicates the 'complementary ethical perspectives' she sees between the two (1994a: 162–69; 1994b: 39–40; see also 1994c); a mutual compatibility in which 'each captures an important aspect of our metaphysical and ethical relationship with Nature' (1994a: 162). Each can, too, compensate for the 'blind spot' of the other. In the case of deep ecology this blind spot is an unwarranted balefulness towards the human species; in the case of ecofeminism it is an unwillingness to move beyond the notion of wholes as agglomerations of individuals to a notion of wholes as moral entities in themselves. A similarly non-oppositional stance is taken by pioneer ecofeminist Rosemary Radford Ruether, who writes: 'what is ecofeminism? Ecofeminism represents the union of the radical ecology movement, or what has been called "deep ecology", and feminism' (1993: 13).

One final observation before we move on. We have already noted that intellectual energy within environmental thought seems, since the early 1990s, to have moved away from ecophilosophy to what we might crudely call more 'real world' matters. There has always been the odd lonely voice within radical environmentalism decrying the preoccupation with ethical abstraction. Livingston, for example, wrote in 1981: 'in speaking of ethics in the non-human context, we are jabbering into a void. Nature does not need ethics; there is no one to hear' (1981: 54). But a more widespread impatience with the 'luxury' of pure, unapplied thought in a time of massive biophysical stress now seems evident. Philosophy conference attendees report a decline in the number of 'straight' ecophilosophy papers, and an increase in papers that address questions of *applied* environmental philosophy. The need, Roger Paden writes, is to 'turn away from grand theories to "theories of the middle range" while adopting a more "empirical" approach to moral philosophy' (1994: 61).

Nor am I alone in discerning a trend along these lines. Andrew Light and Eric Katz have noted the 'decreasing importance of theoretical debate and the placing of practical issues of policy consensus in the foreground of concern' (1996: 5). Affirming Bryan Norton's position, Light and Katz argue that, despite the significant achievements at the level of theory, 'it is difficult to see what practical effect the field of environmental ethics has had on the formation of environmental policy' (1996: 1). The very relationship between environmental ethics and policy is thus 'problematic' (Light and Katz 1996: 2; see also Light 1996).

An ambivalence concerning the status of pure theory has always been apparent in ecofeminism. There we will go.

3

ECOFEMINISM

PATRIARCHAL DUALISMS

Like the varieties of thought considered in chapter 2, ecofeminism is a radical ecophilosophy; but it merits separate treatment, for it is both that and more. Its ethics are subordinate to a theory of power, and the perceived lack of an adequate account of social structure and political power in radical ecophilosophies has attracted sustained ecofeminist criticism. Some ecofeminists, Ariel Salleh for example, also oppose the 'philosophication' of ecofeminism, on the ground that women do not need ethical abstraction as a basis for an authentic grounding within nature (Salleh 1984; see also Davion 1994: 18) — on the ground, indeed, that a preoccupation with abstraction directly partakes of dominance-imbued, masculine modes of thought. In this vein Irene Diamond and Lisa Kuppler proclaim that 'the strength of ecofeminism is in the streets', where 'close attention to the practices of ecofeminism recovers what much academic discourse loses' (1990: 160, 176).

There is another reason, too, for according ecofeminism separate consideration. It was earlier suggested that radical ecophilosophy may no longer be at the cutting edge of green thought. This observation certainly does not apply to ecofeminism, which, as a theory of power relations and political action first and foremost, remains particularly robust. Although a culture of embattlement prevents most ecofeminists from acknowledging it, a survey of the current state of green intellectual and activist development can only lead to the conclusion that ecofeminism is now the predominant position within ecological thought.

The simplest definition of ecofeminism holds it to be an ecologically informed feminism, though some would find this breathtakingly inadequate. For instance, Ariel Salleh argues that 'the embodied materialism of ecofeminism is a "womanist" rather than a feminist politics. It theorises an intuitive historical choice of re/sisters around the world to put life before freedom' (1997: ix–x). There are, of course, several varieties of feminism; in its stress upon an ecological dimension, ecofeminism is, I think, rightly considered distinct from other forms of feminism. But I will come at this obliquely.

As we have seen, the main point over which ecofeminism has taken issue with deep ecology is the latter's identification of anthropocentrism as the factor responsible for the destruction of nature. Against this, ecofeminism argues for an explanation from *androcentrism*: that the destruction of nature is consequent upon structures of exploitation embedded *within* human society. And the particular structure to which ecofeminist attention is directed is that based around gender.

There is nothing particularly *eco*feminist about such an analysis. All feminisms identify *patriarchy* as the most significant of the constraints upon the fulfilment of human potential. Patriarchy is a gender-privileging system of power relations that is subtly embedded within dominant social structures, at all social levels, across almost all cultures, and sustained throughout most of history. The explanation for its tenacity is to be found, not in overt discrimination (though such discrimination *is* often overt), but within conceptual frameworks that systematically deny access and justice to women. There are other axes of discrimination, too — age, class, religion, race — but the most enduring axis of discrimination is gender. Karen Warren defines the especially potent patriarchal paradigm thus:

> a *patriarchal conceptual framework* is one which takes traditionally male-identified beliefs, values, attitudes, and assumptions as the only, or the standard, or superior ones; it gives higher status or prestige to what has been traditionally identified as 'male' than to what has been traditionally identified as 'female'. A patriarchal conceptual framework is characterized by *value-hierarchical thinking* (1987: 6).

Two things follow. The first is that the key mechanism of patriarchal domination is not overtly institutional; it is, rather, engrained within our most fundamental perceptions, and all the more difficult to combat by virtue of its non-tangibility.

The second is that the patriarchal conceptual framework is inherently *dualistic*, for the value hierarchies to which Warren makes reference are not complex: they consist of opposed pairs of values and interests, corresponding to the gender division itself, in which the 'male' interest/value is accorded superiority. More than just superior, though — the poles that are associated with 'male' are held to be *transcendent* rather than merely gender-specific. Thus, masculine values are regarded as 'species-defining', whilst the feminine is marginalised and trivialised. This has led to the categorisation of women and the values associated with the feminine as 'other', and it has paved the way for the development of systems of power and oppression that have consistently devalued the role and place of women.

There are several axes of dualistic conceptualisation identified within feminist literature as corresponding with a masculine/feminine dualism: mind/body; spirit/corporeality; abstraction/embodiment; sky/earth; competition/co-operation; asceticism/promiscuity; rationality/intuition; culture/nature (a more comprehensive catalogue of the dualisms in western thought is provided by Plumwood 1993a: 43). It is the masculine side of each of these dualisms, feminists argue, that has been elevated and universalised: herein lies its potency as an instrument of domination. Dualism, Plumwood explains, is a 'logic of hierarchy', in which women are reduced to 'inferior, impoverished or imperfect human beings, lacking or possessing in a reduced form the characteristics of courage, control, rationality and freedom which make humans what they are' (Plumwood 1993b: 36). And this is made to seem the unavoidable order of things, a process, in Rosemary Ruether's words, of 'naturalized domination' (1975: 189; see also 1992: 3; King 1989; Plumwood 1986: 122–24; K.J. Warren 1987: 3).

Patriarchal dualism has been central to ecofeminist analysis since Ruether's pioneering work, *New Woman, New Earth*, was published in 1975 (though the coining of the term 'ecofeminism' is usually attributed to a French feminist, Francoise d'Eaubonne, in 1974). Although patriarchal dualism remains prominent within Ruether's thought (1992: 3), its most comprehensive treatment is provided by Val Plumwood. Unlike much ecofeminist analysis, Plumwood's work presents a theory of multi-faceted domination/subordination relationships stemming from the series of interlinked dualisms noted above. 'I use examples from a number of forms of oppression, especially gender, race and colonisation', she writes', to show what this structure is, and discuss its logical formulation. By means of dualism, the colonised are appropriated, incorporated, into the selfhood and culture of the master, which forms their identity' (1993a: 41). Plumwood is thus able to present patriarchal dualism as a central mechanism of social domination whilst addressing other axes of domination besides gender and human/nature: 'the set of interrelated and mutually reinforcing dualisms which permeate western culture forms a fault-line which runs through its entire conceptual system' (1993a: 42).

Plumwood also deals effectively with one of the conceptual difficulties that have plagued feminist thought: if dualism is so pervasive, if it is so robust that it has persisted through all history and most cultures, is it not inevitable that it will always prevail? Plumwood says not. She says that it is precisely because it has always been accepted as the *natural* order of things that it has attained its aura of invincibility (she instances Heidegger and Beauvoir as examples of influential near-contemporary thinkers who saw dualism as 'an inevitable part of the human condition'; 1993a: 61). Recognition that dualism is a political artefact rather than the natural order of things goes a long way towards creating conditions for effective resistance. The problem, after all, is essentially one of perception; of escaping the tyranny of subordinated identity:

> the logic of colonisation creates complementary and, in advanced cases, complicit subordinated identities in and through colonisation. The reclamation and affirmation of subordinated identity is one of the key problems for the colonised, especially in race, class, and ethnic colonisation. The affirmation of women and the feminine falls within this problematic (Plumwood 1993a: 61).

NATURE AND GENDER

Whilst there is nothing specifically *eco*feminist about theorising dualism, such analysis does loom large within ecofeminism. This is because of the focus within ecofeminism on the dualisms that link the subordination of women to the subordination of non-human nature.

If there is one central contention within ecofeminism it is that the exploitation of nature and the exploitation of women are intimately linked. In Ariel Salleh's words, 'the basic premise in ecofeminism is acknowledgement of the parallel in men's thinking between their "right" to exploit nature, on the one hand, and the use they make of women, on the other' (Salleh 1989: 26). Ecofeminists maintain that in patriarchal cultures not only has nature been strongly identified with the feminine but the reverse has also been true. 'In these cultures', writes Janis Birkeland, 'women have historically been seen as closer to the earth or nature' (1993: 18), and this identification has set in process 'a complex morality based on dominance and exploitation' (1993: 19) in which the two linked foci of exploitation are powerfully and mutually reinforcing. Moreover, these twin foci of exploitation are bolstered by other axes of dualist thought. Thus, the nature–feminine principle is associated with the primitive rather than the civilised, with the realm of necessity rather than freedom and high-mindedness, with carnality rather than discipline, with associative, 'non-rigorous' thought rather than rationality, and so on.

That this nature–feminine association has been a persistent core assumption within western culture, and that it has been used to legitimise the domination of both women and nature, seems incontestable. Events in the seventeenth century throw this observation into sharp relief. In licensing the total conquest of nature, in rendering nature inert and insensible in order to morally justify the dissecting table, Francis Bacon, particularly in *Novum Organum* (1620) and *The New Atlantis* (1626), used vivid sexual imagery to describe the force and violence with which nature's secrets would be extracted from 'her' (Merchant 1980: 164–76; 1992: 46–47; Sheldrake 1990: 29–32).

Bacon was aware of (perhaps even involved in) the witch trials prevalent in England and mainland Europe at the time; in fact, 'much of the imagery he used in delineating his new scientific objectives and methods derived from the courtroom, and, because it treats nature as a female to be tortured through mechanical inventions, strongly suggests the interrogations of the witch trials and the mechanical devices used to torture witches' (Merchant 1980: 168). 'Is it', Patsy Hallen asks, 'an accident that modern science was born during "the burning times" when 8–11 million women were killed on charges of witchcraft?' (1994: 19). Almost certainly not. Brian Easlea's investigation of the intellectual and political currents surrrounding the 300 years of witch-hunting led him to the conclusion that there were two related factors that brought about the period of terror in which the witch persecutions took place. The first of these was economic self-interest. Women, particularly older women — witches — were the repositories of the natural medicinal lore that had been handed down in the oral tradition through the centuries. They were the main obstacle to the rapidly growing class of professional — and male — physicians and doctors. But, Easlea concludes, this does not of itself explain the zeal and ferocity with which the persecution of witches was pursued. Also important were 'non-economic factors such as gender identity and sexual attitudes' (1980: 7; see also Mies 1986: 83–87).

Women and nature, then, are deemed inferior categories of beings, existing as raw material for man's physical needs. But perhaps this can be turned around. Perhaps it can be argued that women and nature *do* exist in a special relationship, and that this relationship, far from being an axis of inferiority, actually renders women uniquely capable of ecological insight, and thus places upon them the prime responsibility for devising strategies for planetary healing.

This line of argument — since labelled 'biological essentialism' — characterised the writings of ecofeminists in the 1970s and early 1980s. Thus, Joan Griscom observed in 1981, 'nature feminists are now invoking women's closeness to nature in order to heighten our value. A powerful theme in their work is the idea that women are closer to nature than men' (1981: 8). This biological 'closeness' was said to stem from women's physiology; from the menstrual cycle, the acts of carrying and giving birth to young life, and from nurturing and nursing — all processes said to make for a privileged oneness with Mother Earth and innate attunement to the rhythms of other natural processes. Close involvement in the planetary rhythms acts to ward off the hubris to which the masculine gender is prone, providing constant reminders of mortality and biophysical limits, necessary correctives from which men are in large part shielded (for examples of this line of thought, see S. Griffin 1978; Salleh 1984).

Other arguments claiming a privileged empathy for women with the non-human world rested less on biology than on socio-historical factors. Ynestra King wrote at the time:

> the liberation of women is to be found neither in severing all connections that root us in nature nor in believing ourselves to be more natural than men. Both of these positions are unwittingly complicit with nature/culture dualism. Women's oppression is neither strictly historical nor strictly biological. It is both (1981: 14–15).

King voices a connection between the special *historical* experience of women and their ability to connect empathetically with the natural world. It is a perspective that urges celebration of the devalued reality of being female, a stance that has been described by King as 'critical otherness' (1981: 14). She was one of the first critics of the notion that women's biology makes them more 'natural' beings than men. 'Since all life is interconnected', the argument ran, 'one group of persons cannot be closer to nature' (Birkeland 1993: 22), though 'women have a critical ledge from which to view the artificial chasm male culture has placed between itself and nature' (King 1981: 14). Noting the difference between this position and the biological essentialism of Susan Griffin and the early Salleh, Deborah Slicer (1994: 34) observes that this insights-from-historical-experience position is similar to what has been called in the wider feminist literature a 'feminist standpoint'.

With King's, Joan Griscom's was one of the first voices raised against essentialism:

> I find it difficult to assert that men are 'further' from nature because they neither menstruate nor bear children. They also eat, breathe, excrete, sleep and die; and all of these, like menstruation, are experiences of bodily limits. Like any organism, they are involved in constant biological exchange with their environment and they have built-in biological clocks complete with cycles. They also play a role in childbearing. In reproduction, men's genes are as important as women's (1981: 8).

Through the 1980s the Griscom-King position attained wide acceptance within ecofeminism. Thus, Ariel Salleh, who had written in 1984 that 'women's monthly fertility cycle, the tiring symbiosis of pregnancy, the wrench of childbirth and the pleasure of suckling an infant, these things already ground women's consciousness in the knowledge of being coterminous with nature' (1984: 340), had, by the 1990s, come to see the notion that 'women are superior to men because of innate qualities' as 'naive' (1993b: 98). Whilst leaving some space for biology (1993b: 99; 1994: 116), Salleh now sees ecofeminism's 'profoundly democratic project' as a politics that 'enlists men to join women in reaffirming their place as part of nature and in formulating new social institutions in line with that position' (1993b: 100; see also 1997: 13) — certainly not the project of a biological essentialist.

If not essentialism, what of the notion of a 'feminist standpoint' — a point of cultural vantage based upon women's long history of being treated as 'other'? The argument from culture seems now to be, and to have been since the late 1980s, the position enjoying broadest support within ecofeminism. By the end of the 1980s only a small number of ecofeminist writers — Andree Collard (Collard, with Contrucci 1988), Mary Daly (1987), Sharon Doubiago (1989) and Deena Metzger (1989) come most readily to mind — still held to a strict biological essentialism. Plumwood puts the argument from culture thus:

> to the extent that women's lives have been lived in ways which are less directly oppositional to nature than those of men, and have involved different and less oppositional practices, qualities of care and kinds of selfhood, an ecological feminist position could and should privilege some of the experiences and practices of women over those of men as a source of change without being committed to any form of naturalism (1993a: 34; see also Salleh 1994: 116).

In such a view the experience of women's reproductive role is still important, because this experience includes aspects other than the strictly biological. The different historical experiences from which men and women derive their social roles, their access to power, and their personal identities 'are real, and although not necessarily inherent in biology are not just conventional either'. Women have historically been *assigned* roles that have permitted the development of insights and empathies denied to men — Plumwood uses the phrase '"custodians" of a different culture' (1993a: 201) — but which are 'in principle' accessible to them.

This may be the dominant position within ecofeminism, but it is not uncontested. Christine Cuomo has questioned the extent to which ecofeminism should make any claims to special empathy with nature on the basis of a gender-specific cultural heritage. Both female and male identities are established in part 'through elevation over the natural world.' Thus:

> ecofeminists should not romanticize the connections between women and nature. Many human females have been conceived, and have conceived of themselves, as dominators within the logic of domination — as above nature, and/or as above other members of the human species. Women, especially members of industrial and technological societies, have contributed to the oppression of the nonhuman world, and must admit to this complicity so that they can create alternatives (1992: 356).

A similar view is expressed by Victoria Davion, who argues that some men feel more connected to nature than some women; many women, indeed, feel no connection at all (1994: 25–26).

Ecofeminism, it is clear, is evolving at such a rate that it is difficult to state with confidence what the defining characteristics of the movement are at any point of time, and this is particularly the case with the vexed question of the women–nature relationship. Among the major casualties of this evolution have been the Mother Earth and Mother Nature metaphors. As Joni Seager observes, 'the conceptualization of the earth as a mother has a long and honorable history; Earth as Mother, as a sacred and honored female life force, is a powerful icon in non-Christian, nonEuroAmerican, mostly agricultural cosmography' (1993: 219). But Primal Mother cosmography cannot be imported into western culture while retaining its original meanings intact. In fact, the Mother Earth/Mother Nature icons are now seen as components of the linguistic structure of patriarchal domination. They perpetuate western images of the mother as eternally generous, unceasingly fecund and bountiful to the point of inexhaustibility, whilst meriting no economic value. They are complicit, then, in the women–nature nexus of exploitation.

It is difficult to contest such an observation. It is remarkable how frequently the 'Mother Nature' image occurs in political debate over controversial environmental issues, and how frequently it is invoked by those who defend violence against nature on the ground of Mother Nature's resilience, great absorbing capacities, and healing power. Against this, Seager argues:

> The earth is not our mother. There is no warm, nurturing, anthropomorphized earth that will take care of us if only we treat her nicely. The complex, emotion-laden, conflict-laden, quasi-sexualized, quasi-dependent mother relationship ... is not an effective metaphor for environmental action (1993: 219; see also T. Berman 1994; Merchant 1996).

Other charges against the use of the Mother Earth image are that it allows those most complicit in environmental destruction to escape responsibility, because 'if the earth is really our mother, then we are children, and cannot be held fully accountable for our actions' (Seager 1993: 219); and that 'anthropomorphizing the earth ... is disrespectful ... for it seeks to understand another not on her or his own terms but as a projection of ourselves' (Gaard 1993: 305).

Because the 'Earth as Great Mother' image is entwined with native peoples' cosmologies, particularly the cosmologies of native North Americans, and because these cosmologies are crucial to primal earth/goddess ecofeminism, the fall from favour of the Mother Earth metaphor exacerbated a rift that emerged within ecofeminism in the early 1990s. Why a rift? To answer this question we must note the historically prominent strand of ecofeminist thought that is characterised by a search, through close identification with nature, for a primal earth-based ecofeminist spirituality. It is a position that does still hold to the belief in women's superior *natural* insightfulness.

There has been much ecofeminist interest in primal peoples' religions, particularly in the earth-based religions of the native peoples of North America, Australia and pre-Christian Europe, and in the shamanic practices characteristic of many of those religions (Starhawk

1989: 174). Carolyn Merchant depicts this strand of ecofeminism as:

> celebrating the relationship between women and nature through the revival of ancient rituals centered on goddess worship, the moon, animals and the female reproductive system. A vision in which nature is held in esteem as mother and goddess is a source of inspiration and empowerment for many ecofeminists (1992: 191).

The revival of earth-based spirituality will be revisited in chapter 4, where it seems more appropriately placed given that not all its practitioners are ecofeminists; indeed, one of its most prominent partisans, Dolores La Chapelle, explicitly identifies with deep ecology and against ecofeminism (La Chapelle 1989). What is germane here is the basis on which primal earth ecofeminists hold their ecofeminism. Merchant observes of them that:

> human nature is grounded in human biology. Humans are biologically sexed and socially gendered ... The perceived connection between women and biological reproduction turned upside down becomes the source of women's empowerment and ecological activism. Women's biology and Nature are celebrated as sources of female power. This form of ecofeminism has largely focused on the sphere of consciousness in relation to nature — spirituality, goddess worship, witchcraft — and the celebration of women's bodies (1992: 191–92).

In this reading primal earth ecofeminism partakes of biological essentialism; indeed, earth-based spirituality is the *only* major strand of ecofeminism to remain supportive of biological essentialism. Small wonder that just as essentialism fell from favour, so did Earth goddess spirituality become, for a time, somewhat marginalised within ecofeminism. Other factors also contributed to this. Cuomo (1998: 6), Garb (1990), Plumwood (1993a: 126–28), and Seager (1993: 250–51) detail criticisms of 'goddess pantheism', whilst Birkeland (1993: 47–49) finds no fewer than nine grounds for rejecting mystical spiritualities. One particularly telling criticism centres around the perceived act of cultural appropriation involved in the 'Europeanisation' of native spiritual traditions. Karen Warren describes this process as 'an expression of ethical colonialism, coopting indigenous cultural practices as part of an otherwise unchanged dominant Western worldview' (1993: 124), whilst Carol Adams writes:

> what Eurocentric outsiders identify as the 'spiritual' dimensions of a culture may actually be a thoroughly eviscerated spirituality that a dualistic world-view cannot even perceive. And regrettably, some aspects of Euro-American ecofeminism have 'borrowed' from these cultures those parts that resonate with a noninstrumental view of human nature and have depoliticized the context for these views. In many instances cultures that are struggling for physical survival against genocide are romanticized, their spirituality misappropriated and misunderstood (1993: 3).

Of course, it is important to distinguish between primal-earth ecofeminism and an ecofeminism which promotes a broadly based concern for spiritual matters whilst nevertheless eschewing immersion in primal earth cosmology. Thus, Greta Gaard, who rejects the 'cultural imperialism' involved in 'plundering' the spiritual traditions of native cultures, nevertheless sees ecofeminism as:

> a way of describing a political theory and practice for what we know intuitively to be true. Knowledge and awareness of our interconnectedness provide the impetus for ecofeminist political acts as well as ecofeminist spirituality ... Ecofeminist spirituality is ecofeminist politics is ecofeminism. Without stealing pieces from other cultures and other traditions, ecofeminists can arrive at this answer on their own (1993: 309).

Such a broadly conceived concern for spirituality has always been prominent within ecofeminism.

FEMINISM AND ECOFEMINISM

I have indicated that ecofeminism warrants a distinct status within feminism on the ground of its unique focus on the women–nature nexus of subordination. But this observation requires justification. The use of the prefix 'eco' in front of 'feminism' has been resisted by some feminists. Anne Cameron has written:

> the term 'ecofeminism' is an insult to the women who put themselves on the line, risked public disapproval, risked even violence and jail ... Feminism has always been actively involved in the peace movement, in the antinuclear movement, and in the environmental protection movement. Feminism is what helped teach us all that the link between political and industrial included the military and was a danger to all life on this planet. To separate ecology from feminism is to try to separate the heart from the head. I am not an ecofeminist. I am a feminist (1989: 64).

The difficulty with Cameron's position is this. Think back to the early part of the chapter and consider again the dualisms listed there, particularly the 'feminine' half of those opposed attributes. Excepting 'culture/nature', there is no reason why the feminine attributes should not, in their practical applications, be thoroughly anti-ecological. In fact, as a German student of mine once pointed out, when these dualisms are applied to National Socialism, we find that a majority of the values fetishised by the Nazis are on the feminine side of the dualisms — and from that basis was launched the horror of the Holocaust and the most all-consuming military conflagration the world has yet seen.

It is perhaps not surprising, then, that the twentieth century's most influential feminist, Simone de Beauvoir, having alerted us to the existence of these dualisms in the first place, wished to overcome them, not by championing the distinctively female, but by invading the male sphere, adopting those values for her own, and softening them from within. She sought, in particular, to rid feminism of its ecological concerns:

> an enhanced status for traditional feminine values, such as woman and her rapport with nature, woman and her maternal instinct, woman and her physical being ... this renewed attempt to pin women down to their traditional role, together with a small effort to meet some of the demands made by women — that's the formula used to try and keep women quiet. Even women who call themselves feminists don't always see through it. Once again, women are being defined in terms of 'the other', once again they are being made into the 'second sex' ... Equating ecology with feminism is something that irritates me. They are not automatically one and the same thing at all (Beauvoir in Schwarzer 1984: 103).

Whereas her analysis of the historical association of women and nature as one of mutual subordination largely parallels ecofeminist analysis, Beauvoir's project is not the liberation of both women *and* nature, but the liberation of women via a radical distancing from nature, that 'vein of gross material in which the soul is imprisoned' (1982: 176). Liberation for women lies in rejecting nature for full, productive participation in the realms of culture that have hitherto been denied them. With modern technological achievement, women no longer need to be chained to the tyranny of their bodies. Now they can become as men, masters and possessors of nature, able to transcend nature's enslaving chains through creative immersion within the golden fields of culture.

In her depiction of nature as the realm of necessity and un-freedom — as the dark nemesis of that great liberatory agent, culture, with which nature struggles in zero-sum conflict — Beauvoir herself makes the case for separating ecofeminism out from the other feminisms. Her position on 'tyrannous nature' was widely adopted. Shulamith Firestone, for example, called for full liberation from 'biological tyranny', for 'humanity can no longer afford to remain in the transitional stage between simple animal existence and full control of nature'. Furthermore, 'we are much closer to a major evolutionary jump, indeed to direction of our own evolution, than we are to a return to the animal kingdom from which we came' (1970: 193).

So widespread was this position that it came to characterise what Plumwood has called 'the first, *masculinising* wave of feminism': 'in the equal admittance strategy, women enter science, but science itself and its orientation to the domination of nature remain unchanged' (1988: 19). Plumwood notes that 'the very idea of a feminine connection with nature seems to many to be regressive and insulting', an idea that conveys a view of women as 'passive, reproductive animals ... immersed in the body and in unreflecting experiencing of life' (1993b: 38; see also King 1990).

Thus, it is just not the case that, as Cameron would have it, all feminists share in a common project for the liberation of nature.

It is on this account that some ecofeminists have spelt out the distinctiveness of *eco*feminism. Karen Warren identifies 'four minimal condition claims of eco-feminism' (1987: 7). These place stress on the connections between the oppression of women and nature; an understanding that these linked foci of oppression are sanctioned by a 'patriarchal conceptual framework'; a critique of patriarchal conceptual frameworks that 'is grounded in familiar ecological principles' (this grounding being 'the basis for the uniquely eco-feminist position that an adequate feminist theory and practice embrace an ecological perspective'; 1987: 7–8); and the insight that an ecological politics must similarly 'embrace a feminist perspective'. The stakes are high, Warren says, for what is in the balance is 'the theoretical adequacy of feminism itself' (1987: 9).

She then provides an ecofeminist critique of the most prominent feminisms — liberal feminism, traditional Marxist feminism, radical feminism, and socialist feminism — each of which is tested against her four 'minimal claims of ecofeminism' and found wanting in greater or lesser degree. Liberal feminism, which 'endorses a highly individualistic conception of human nature' (1987: 8) insufficiently challenges the patriarchal conceptual framework, whilst:

> the extreme individualism of a liberal feminist ecological perspective conflicts with the ecofeminist emphasis on the independent value of the integrity, diversity, and stability of ecosystems, and on the ecological themes of interconnectedness, unity in diversity, and equal value to all parts of the human-nature system. It also conflicts with 'ecological ethics' per se. Ecological ethics are holistic, not individualistic; they take the value and well-being of a species, community, or ecosystem, and not merely of particular individuals, let alone human individuals, as basic (1987: 10).

Traditional Marxist feminism fails to adequately conceptualise the connections between the oppression of women and the oppression of nature (1987: 11–13). Radical feminism has the virtue that it does locate the oppression of women unambiguously within patriarchy, which defines women 'as beings whose primary functions are either to bear and raise children ... or to satisfy male sexual desires' (1987: 14). But it incorporates biological essentialism, and Warren finds it inadequate for reasons similar to those already rehearsed. Finally, socialist feminism does recognise 'the interconnections among various systems of oppression', but it still fails to 'explicitly address the systematic oppression of nature' (1987: 17). Warren concludes by arguing for a 'transformative feminism' which will supersede earlier feminisms (including the earlier *eco*feminism of biological essentialism) and is to be equated with ecofeminism (1987: 17–20).

Plumwood also examines the relationship of ecofeminism to

feminism. Using the familiar wave metaphor to explain recent feminist history, Plumwood sees 'critical ecological feminism' as 'a third wave or stage of ecofeminism moving beyond the conventional divisions in feminist theory ... It is prefigured in and builds on work not only in ecofeminism but in radical feminism, cultural feminism and socialist feminism'; even so, 'this critical ecological feminism conflicts with various other feminisms, by making an account of the connection to nature central in its understanding of feminism' (1993a: 39).

Ecofeminism's claim to constitute a feminism in its own right seems beyond dispute. We must wait to see whether it also has the capacity to supersede earlier feminist formulations in accordance with the visions of Warren and Plumwood.

AN ETHIC OF CARE

There are other uniquely ecofeminist contributions to environmental thought. Given the actionist orientation of ecofeminist thought, it comes as no surprise to find that much of this thought is developed within an *issue* framework. Ecofeminists have focused particularly upon the politics of women's bodies and on new reproductive technologies (Diamond and Kuppler 1990; Razak 1990), on the psycho-social aspects of birth and mothering, on issues of peace and disarmament (Birkeland 1993; K.J. Warren 1994), on questions of population and female infanticide (Cuomo 1994), on the economic and social conditions of women in developing countries (Shiva 1989), on women's work generally (Merchant 1996; Salleh 1990; 1994; 1997), and on the nature of western science/technology.

We will consider just two aspects of ecofeminist thought: one that encapsulates the distinctiveness of ecofeminist politics, and one that does likewise for ecofeminist ethics.

We have seen that ecofeminism views the structures of patriarchal power as not primarily institutional but lodged within time-hardened mindsets — mindsets so entrenched, that their components have been given the status of 'unchangeable nature'. It follows that the focus of effective action for social change has to include spheres of human interaction not conventionally thought to have political significance. All human interactions are characterised by relations of dominance, and all levels of human relationship must therefore be the focus of any realistic strategy for political change. *The person*, the aphorism runs, *is the political.* Hence the stress that is placed, within ecofeminist thought, on the political importance of human relations at the micro-level.

Of course, this focus is also present in other feminisms, though a stress upon micro-level politics, upon psycho-sexual factors, and upon the political importance of emotional needs and motivations is largely absent from non-feminist theory. Within ecofeminism these factors are used both at the level of the ordinary individual, to explain pathological behaviour within human relationships *and* between individuals and their sustaining environment, and at the level of national and global political power to explain the irresponsible, irrationally ecocidal activities routinely indulged in by proud men of power. Charlene Spretnak writes: 'identifying the dynamics — largely fear and resentment — behind the dominance of male over female is the key to comprehending every expression of patriarchal culture with its hierarchical, militaristic, mechanistic, industrial forms' (1988:1). Other pathological doubts and urges stemming from oppressive power relationships have been identified as 'sexual identity, the fear of death, the link between personal worth and power, the repressed need to belong, and other expressions of personal insecurity' (Birkeland 1993: 19). Pathologies will continue to govern human relationships, at all levels, and continue to manifest in 'ecocidal' behaviour, says Birkeland, until we surmount 'the presumed necessity' of relationships based upon power (1993: 19). Nothing less ambitious is the political project of ecofeminism.

And so to the question of an ecofeminist ethic. Rather than the rights-based ethics of traditional western ethical discourse, Karen Warren posits a *relational* ethics based upon contexts within which interactions occur, and this relational ethics is characterised by compassion and humility in treating with what is 'other than I':

> an ecofeminist perspective about women and nature involves this shift in attitude from 'arrogant perception' to 'loving perception' of the nonhuman world ... Any environmental movement or ethic based on arrogant perception builds a moral hierarchy of beings. In contrast, 'loving perception' presupposes and maintains *difference* — a distinction between the self and other, between human and at least some nonhumans — in such a way that perception of the other as other is an expression of love for one who/which is recognized at the outset as independent, dissimilar, different (1990: 137).

This unselfconscious reference to a loving relationship with what is 'other than I' occurs frequently within ecofeminist writing. Chaia Heller, for example, writes that 'authentic love is a celebration of the distinctiveness of the other', insisting that 'knowledge of other people and of nature must be gleaned from actual labour or "caring for" the beloved. It cannot be "acquired" by meditating in isolation' (1993: 233). An ethic of care cannot, in other words, be derived from a process of intellectual abstraction. This 'loving perception' is the base in which Warren grounds an ecofeminist ethic of *care*, described thus: 'ecofeminism makes a central place for values of care, love, friendship, trust, and appropriate reciprocity', wherein 'one is doing something in relationship with an "other", an "other" whom one can come to care about and treat respectfully' (1990: 143).

There is also a psychological basis for relational ethics. Drawing on the object-relations school of feminist psychology (for example, Chodorow 1978), this position maintains that the liberal view of the ego as radically apart is a quintessentially male view that stems from the male child's anger at his first dismaying realisation that he is different to his mother, a sense of difference experienced as betrayal, and the source of deep-seated, gender-focused resentment. The growing male comes to define himself as 'not female' and seeks, in the venting of his anger, to dominate the female — which includes nature. And, because the undifferentiated, pre-separation reality for boys was female, it is nature that subsequently becomes the undifferentiated natural backdrop against which individuals must struggle for identity. Girls 'maintain their sense of identity with their mothers for a much longer time than do boys', and thus their sense of self 'is bound up with relationship' (Zimmerman 1987: 31). This conduces to a relational sense of self and, hence, a caring, relational ethics (on this, see Gilligan 1982; Zimmerman 1987: 31–32), whilst the male preoccupation with 'rights' can be related to the male need to erect barricades around his separateness across which other individuals may not intrude (Zimmerman 1987: 32).

Though object-relations theory is strongly congruent with relational ethics, it is differentially employed within ecofeminist ethics, and is less frequently invoked now than was the case in the 1980s. Warren, for example, seems to draw upon it only minimally.

Warren does draw fruitfully upon an earlier essay by Jim Cheney (1987), in which Cheney promoted the virtues of an ecofeminist ethic of care over rights-based ecophilosophies. He describes a 'moral community' based upon relations of care as:

> a gift community in which selves are not atomistic entities protected by bundles of rights derived from, or tied to, bundles of properties or interests internal to the individuals. It is a community in which individuals are what they are in virtue of the trust, love, care, and friendships that bind the community together, not as an organism but as a community of individuals (1987: 129).

Cheney's criticism of rights-based ecophilosophy concentrates on deep ecology, a poorly chosen target because, of all the radical ecophilosophies based upon 'atomistic entities protected by bundles of rights' that he *could* have attacked, in choosing deep ecology he picked the one ecophilosophy which is conspicuously *not* rights-based. That Cheney's case *against* deep ecology is flawed is, however, not to say the same about his case *for* the relational ethic of care. It has attracted criticism, however, and it is to this and other objections to ecofeminism, or aspects thereof, that we now turn.

SOME PROBLEMS, INCLUDING THE TENACITY OF ESSENTIALISM

In a wide-ranging paper Janis Birkeland identifies several 'misconceptions about ecofeminism': that 'it is dualistic, partial, anti-rational, and essentialist' (1993: 21). None of these assessments, she argues, withstand analysis.

The charge of 'dualism' stems from a perception that the privileging of a female standpoint maintains masculine/feminine dualism whilst merely reversing the ascription of superior value. Birkeland is probably right to label this a misconception. She reacts too strongly, however, when she insists that the charge is beneath contempt, for the claim that 'the concept of gender itself helps maintain the dualistic hierarchy' has been made from *within* ecofeminism (by Christine Cuomo 1992: 361). And ecofeminism is, of course, gender-based theory.

The notion of ecofeminism as 'partial' or 'incomplete' is also difficult to dispatch. Ecofeminism can be held to be 'partial' or 'incomplete' if it does not adequately theorise environmental degradation. It is in something of a cleft stick here. Its most prominent proponents have been at pains to eschew reductionist explanation. Rightly so. But this makes for a degree of ambiguity over the causal status to be accorded 'patriarchy' in ecofeminist analysis. If it is not the base social pathology, with all other axes of domination secondary manifestations of it, upon what basis can it be claimed to be the most persistently significant axis of domination? What *is* its relationship to other forms of exploitation, if these are not causally linked to it, but run alongside patriarchal oppression through time? Despite the best efforts of ecofeminism's key theorists to move away from a patriarchy-reductionism, these uncertainties continue to exert a pull back towards it.

The most important context in which this becomes a problem is the linked ascription of woman/nature exploitation to patriarchy. It can be argued that the exploitation of nature, at least, is not reducible to patriarchal social relations — more specifically, not reducible to the patriarchal identification of female and nature. This position has been well put by Robyn Eckersley:

> how, then, do we explain the existence of patriarchy in traditional societies that have lived in harmony with the natural world? How do we explain Engels' vision of 'scientific socialism', according to which the possibility of egalitarian social/sexual relations is premised on the instrumental manipulation and domination of the nonhuman world? Clearly, patriarchy and the domination of nonhuman nature can each be the product of quite different conceptual and historical developments. It follows that the emancipation of women need not necessarily lead to the emancipation of the nonhuman world, and vice versa (1992: 68; a similar point is made by W. Fox 1989).

Just as suspicions of reductionism have not been entirely dispelled, so, too, has the charge of essentialism not been answered to everyone's satisfaction. For Salleh the charge of essentialism is mere 'insult' (1997: xi). Birkeland is similarly emphatic that essentialism has no place within ecofeminism: the notion 'that women possess an essential nature — a biological connection or a spiritual affinity with nature that men do not ... would be inconsistent with the logic of ecofeminism'. Birkeland subscribes to the 'cultural experience' position: 'the assertion of "difference" is based on the historical socialization and oppression of women, not biologism' (1993: 22). But it can be argued that this merely transfers a biologically based essentialism to an essentialism that is culturally grounded. Each position can be said to constitute 'special pleading' based upon gender. Each position involves a claim to 'epistemic privilege, of superior ecological insight and awareness' (Judy Evans 1993: 181; see also Wittbecker 1986: 181). Cuomo also rejects the validity of a 'special claim' even when it is based in culture rather than in biology. In her view — an ecofeminist view, as the conclusion of her paper makes clear — the 'veneration' of 'feminine values' 'promotes, rather than dismantles, a logic of domination' (1992: 352).

Thus, the situation remains unsettled. The biological essentialism of early ecofeminism has been rejected in favour of the privileging of a 'female standpoint' based upon a shared history of cultural oppression; this may, itself, merely transfer essentialism from one ground to another. But the question can be asked: if there is no historical or cultural basis upon which a special female empathy with the natural world can be claimed, what is left of ecofeminism? This chain of thought is followed by Judy Evans, and at the end of it she returns to liberal or first-wave feminism, arguing:

> while there are many causes for which women may, and some of us will think, should work, by only one — a demand for equal treatment — is our cause as women advanced. And thus it makes sense, I suggest, to speak ... not of ecofeminists, but of ecologists who are feminists too. It follows ... that women have no especial, innate tendency towards, or interest in, ecological concerns; and that while ecology will be one of the causes for which we may work, if we do so as feminists it must be on a basis of equality with men, or rather, while striving for that (1993: 187).

For Evans, as with most liberal feminists, the stumbling block is nature. She concludes with a warning against 'celebrating the natural', for 'that could entrench more or less every aspect of the female condition many of us have sought to renounce. Having fought to emerge from "nature", we must not go back' (1993: 187). Nervousness on this score is likely to preclude realisation of Warren's and Plumwood's desire for ecofeminism to attain a hegemonic position within feminism — to *become* feminism.

Within the environment movement others may reach a position

similar to Evans's but for different reasons. If women have no especial insights into nature, and if, as ecofeminists sometimes observe (for example, M. Daly 1987: 21), men need not expect women to let them off lightly by resolving the ecological crisis those same men have created, activists (of either sex) for whom the prime goal *is* the resolution of that ecological crisis may conclude that ecofeminism offers them little. Of course, this is not a problem if one's ecofeminism is essentialist (a similar point is made by Biehl 1993: 63). An essentialist comeback thus seems distinctly possible, and some ecofeminists have begun to argue for a more flexible line on the matter. For Noel Sturgeon there are strengths and weaknesses in an essentialist position, and she sets herself the task of finding 'precisely when and where both essentialism and anti-essentialism are useful, and where and to whom they may prove disabling' (1997: 8). Essentialism, for example, can sharpen the sense of oppositionality needed for effective activism (1997: 17). Cuomo, too, insists that 'strategic essentialisms ... can effectively mobilize women' (1998: 124), and she pleads for ecofeminists to again 'take seriously work that has been rejected as essentialist' (1998: 125).

The marginalisation of primal spirituality within ecofeminism has also been resisted. Karen Warren, noted earlier as someone who has detailed ecofeminist criticisms of earth-based spirituality, nevertheless argues for a place at the table for ecofeminist spiritualities, because such spiritualities can be 'life-affirming' and 'personally empowering' (1993: 130). Carol Adams, whose 1993 book, *Ecofeminism and the Sacred*, was important in rehabilitating earth-spirituality within ecofeminism, sees ecofeminism's celebration of diversity as providing in-principle support for spiritualist groundings:

> we reject an either-or approach; we do not believe that we must decide between working to help human beings or working to stop environmental abuses, between politics and spirituality, between humans and the rest of nature. We recognize that addressing issues related to the sacred furthers ecofeminist goals. We do not dematerialize the sacred or despiritualize matter. The idea of diversity amidst relationship does not erase differences among us; nor does it deny our commonalities (1993: 4).

And Ruether not only defends the retention of a spiritualist perspective within ecofeminism, but even finds a place for a capital-G God:

> God, in ecofeminist spirituality, is the immanent source of life that sustains the whole planetary community. God is neither male nor anthropomorphic. God is the font from which the variety of plants and animals well up in each new generation, the matrix that sustains their life-giving interdependency with one another (1993: 21).

The problem with the Birkeland position, then, is that it seems to assume a single entity when there is patently scope for considerable disagreement within ecofeminism. Responding to Janet Biehl's claim

(1991) that ecofeminist politics evince massive internal contradiction, Salleh observes that 'because ecofeminist politics grows out of a plurality of social contexts, it will have many contradictions' (1993b: 94; see also 1991: 30; Thrupp 1989). There are already ecofeminists seeking an alternative signifier to 'ecofeminism' or who affix a qualifying prefix to their ecofeminism. Thus, Carolyn Merchant (1990) and Mary Mellor (1992a; 1992b) describe themselves as 'socialist ecofeminists'. 'While radical eco-feminist philosophy embraces intuition, an ethic of caring, and weblike human relationships, socialist eco-feminism would seek to give both production and reproduction a central place in materialist analysis', writes Mellor (1992b: 44). It is, in fact, the socialist tradition within environmental thought that has most trenchantly critiqued ecofeminism. That rich possibilities for a fusion between the two have not hitherto been realised is lamented by Mellor (1992b: 45), whilst Merchant sets out the terms such a linkage might take:

> socialist feminism incorporates many of the insights of radical feminism, but views both nature and human nature as historically and socially constructed. Human nature is seen as the product of historically changing interactions between humans and nature, men and women, classes and races. Any meaningful analysis must be grounded in an understanding of power not only in the personal but also in the political sphere (1990: 103).

Other ecosocialists, looking askance at the 'intuition, ethic of caring, and weblike human-nature relationships' within ecofeminism, see less prospect for fusion, finding ecofeminism to be incurably neo-romantic. Faber and O'Connor (1989a) took this position in *Capitalism, Nature, Socialism* and sparked considerable debate. Thrupp (1989), Salleh (1991), Martin O'Connor (1991) and Mellor (1992b) contested the issue, but Faber and O'Connor gave only marginal ground:

> we would argue that radical ecofeminism is 'neo-romantic' ... 'Romantic' is associated with 'intuition'. It has often been 'anti-science and technology' and it inverts Enlightenment by privileging body ('human biology') over mind. Romanticism also is associated with organic theories emphasizing emotional ties to the community ('caring'). ... In short, we would argue that radical ecofeminism is infused with romantic values (1989b: 177–78; see also 1991; Biehl 1991).

Faber and O'Connor include the care ethic in their criticism and ecosocialists are not the only movement theorists to be sceptical of ecofeminism's care ethic. Cuomo finds in it precisely that self-sacrifice and self-effacement that feminism has struggled against for years:

> female caring and compassion for oppressors are cornerstones of patriarchal systems. Women have forgiven oppressors, stayed with abusive husbands and partners, and sacrificed their own desires because of their great ability to care for others ... the care ethic actually causes moral damage in

> women and, therefore, caring is not always a healthy and ethical choice for a moral agent (1992: 355; see also Card 1990).

Rather than a non-specific attitude of care, Cuomo wants to know whether a potential object of care merits that care: 'to talk of caring and compassion in the abstract, without naming the object of the caring and the context in which the caring occurs, is ethically uninformative' (1992: 354), because 'the meanings and ethical relevance of acts of caring and compassion are determined by their contexts and their objects' (1992: 356). Cuomo's criticism does not rescue the care ethic; it entirely demolishes it. Once care is rendered situation-dependent, it is no longer an ethic, no longer a *principle* for shaping action, because it has been rendered potentially subservient to other considerations. And as the greatest monsters our species has produced have usually seen some form or item of life as a meritorious object of care — their cat, their mother, the *herrenvolk* — Cuomo's notion of contextualised care is *also* 'ethically uninformative'. Yet, there *is* force to her argument; for how is an undifferentiated ethic of care ever to serve as a guide for real-world action if it does *not* provide a mechanism for judging between rival claims upon our care? All political action, after all, proceeds from the assumption that some ends and interests are more careworthy than others. If that assumption is not made the only alternative is terminal political paralysis.

Ecofeminism's actionist credentials have also been questioned in another context. A more general query is raised — again from an empathetic insider's perspective — by Joni Seager (1993: 247–52), who identifies an 'apolitical undercurrent' within ecofeminism, a 'think-your-way-out-of-oppression' politics 'that does not necessarily suggest a political direction for a feminist re-thinking of the ecology question':

> despite the many ecofeminists who do not fall into this trap, there is a strong apolitical, acultural, and ahistorical undercurrent to ecofeminism that is especially limiting. Environmental destruction takes place in a political and politicized context. Environmentalism must remain a political movement. Such a movement, while it should be concerned with the psychic well-being of its supporters should not exist primarily to minister to their personal needs (1993: 251–52).

A charge that ecofeminism uses against deep ecology — that it is ahistorical and apolitical — is thus used by Seager against ecofeminism. But this is surely a less formidable problem for ecofeminism. Seager mistakes an ultimately minor, peculiarly hairy-chested North American variant of deep ecology *for* deep ecology. She similarly mistakes a particular, historically dated North American ecofeminist essentialism *for* ecofeminism — and this nothwithstanding her acknowledgement that 'the ecofeminist umbrella is a big one, and there are many important shades of difference among people who self-identify as ecofeminists' (1993: 252).

Finally, from a non-partisan perspective within environmentalism, ecofeminists may be seen to hold their feminism as primary, with their ecologism — the 'eco' bit — very much subordinate; a mere label of orientation rather than an identification. 'Ecological feminists take the commitments of feminism ... as far less problematic than the histories of environmental thought and science created by men', writes Cuomo, approvingly, 'even when feminism neglects to take seriously the interests of human and nonhuman beings' (1998: 53). Put bluntly then, ecofeminists may be seen to identify with a sisterhood of feminism, but not with a family of environmentalism. Evidence adduced in support of this view would especially instance the hostility evinced in ecofeminism to deep ecology. As we have seen, for a time in the early 1990s ecofeminist papers seemed almost to be written to a formula, with the inclusion of an almost ritualised denunciation of wrongheaded deep ecology being a required component of that formula. More specifically targeted attacks upon deep ecology also abound; some of these (for example, Doubiago 1989; Slicer 1995) are passionately voiced. We have also seen that several ecofeminists have been concerned to explicate the especialness of *eco*feminism, and to critique the perceived inadequacies of earlier feminisms in the process. But in few cases has this critique achieved the prominence or passion of the critique of deep ecology. Thus, Slicer — having noted that 'ecofeminism, we should remember, is a critique not only of androcentric environmental philosophy but of some feminist theory as well' (1994: 35) — proceeds to ignore her own stricture as she devotes the rest of her paper to a critique of deep ecology and other 'errant' environmentalisms. By contrast, I know of no purpose-built ecofeminist repudiations of liberal feminism, and few have shown the willingness of Ynestra King (1981; 1989), Val Plumwood (1993a) and Ariel Salleh (1997) to decisively distance their ecofeminism from 'masculinising' first-wave feminism. For Salleh, whilst environmentalist paradigms are 'pre-feminist', feminist paradigms are 'pre-ecological', and so 'ecofeminism interrogates the very foundations of mainstream feminism, by pointing to its complicity with the Western androcentric colonisation of the life world by instrumental reason' (1997: 13). Plumwood is also perceptive on the point:

> women, in this strategy [the liberal feminism of 'uncritical equality'], are to join elite men in participation in areas which especially exhibit human freedom, such as science and technology, from which they have been especially strongly excluded ... But the approach of liberal feminism fails to notice not only the implicit masculinity of the conception of the individual subject in the public sphere (and indeed the subject of post-enlightenment rational discourse generally), but also its other exclusionary biases, and fails to challenge the resulting bias of the dominant model of the human and of human culture as oppositional to nature (1993a: 27–28).

Thus, she insists that ecofeminism reject 'approaches to women's liberation which endorse or fail to challenge the dualistic definition of women and nature and/or the inferior status of nature' (1993a: 39). But she proceeds, nevertheless, to posit a family of all feminism:

> critical ecological feminism would also draw strength and integrate key insights from other forms of feminism, and hence have a basis for partial agreement with each. From early and liberal feminism it would take the original impulse to integrate women fully into human culture (1993a: 39).

As noted, it is difficult to find evidence of a similarly perceived family relationship to other ecologisms. But it would not be difficult to make a case for the proposition that there is considerably more commonality between the projects of deep ecology (say) and ecofeminism than there is between liberal feminism and ecofeminism. Indeed, ecofeminism remains significantly more marginalised within broader feminist circles, wherein it is the object of considerable misgiving, than it is within environmentalism, wherein (I have claimed) it is now the predominant paradigm. Such an assessment is, however, one with which few ecofeminists are likely to agree.

4

RELIGION, SPIRITUALITY AND THE GREEN MOVEMENT

RELIGION, SPIRITUALITY, GREEN AMBIGUITY

Few questions better reveal the complex nature of green thought than those to do with religion and spirituality. For environmentalists who reach a commitment from the evidence of 'orthodox' science, the paths of religion and spirituality represent irrelevant, if harmless folly at best; wacky, cause-discrediting voodooism at worst. Certainly, the hostility many environmentalists hold towards science and rationality often takes the form of a search for a new cosmological/spiritual/religious basis for human life and inter-species relationships. And some look to a differently founded science, one seen to have much in common with certain religious ideas and promising an end to the long divorce between scientific and religious ways of appertaining truth. Fritjof Capra, in particular, stresses affinities between the new cosmological constructs emerging from quantum physics and chaos theory — constructs now largely accepted within theoretical physics — and the worldviews long held within certain eastern religions. 'In the eastern view ... as in the view of modern physics', he writes (1992: 321), 'everything in the universe is connected to everything else and no part of it is fundamental.'

BUDDHISM, TAOISM AND NEO-PAGANISM

Given Capra's observation, it is not surprising that there is, within the green movement, considerable interest in the promise proffered by eastern religious traditions for a refurbishment of the human-in-world relationship. Though Hinduism and Islam have been promoted as religious systems with the capacity to underwrite a new environmental enlightenment, most attention has focused upon Buddhism and

Taoism. (For Hinduism see, for example, Capra 1993: 97–104; Deutsch 1986; Dwivedi 1990; Gupta 1993; Jacobsen 1994; Palmer 1990: 52–54. For Islam see, for example, Izzi Dien 1990; 1997; Khalid and O'Brien 1992; Marshall 1992: 127–36; Palmer 1990: 60–61; Sardar 1989; 1990.)

Buddhism is one of those religions characterised in the west as 'quietist': its injunctions lead to contemplation and worldly inaction rather than activity and interventionism. As Peter Marshall explains it, as long as a person's focus is on material acquisition:

> he or she will be chained to a cycle of reincarnation ... The only way to escape from the cycle of rebirth (samsara) is thus to overcome one's ego and dissolve all desire. One must recognize that the world of appearances, of permanence and diversity, is maya, an illusion. At the same time, one should recognize that one is part of the universal whole. If this is achieved, then individual consciousness will pass into the state of nirvana, nothingness or nonexistence (1992: 42).

At first acquaintance it is difficult to see how such a disengaged view can speak to today's environment movement, concerned as this movement is with real world tragedies and real world action. How can a program for ecologically informed social change emerge from a tradition that sees the ultimate human goal as the attainment of pure non-being? For Gary Snyder, who has probably done more to make these connections explicit than anyone else, the relevance is straightforward: 'Buddhism holds that the universe and all creatures in it are intrinsically in a state of complete wisdom, love and compassion, acting in natural response and mutual interdependence' (1985: 251). All beings — not only humans — come from and seek to return to nirvana. They are bonded in cosmic purpose. Buddhism, thus, professes an appropriate inter-species empathy, and it also rejects the mind–body, sacred–profane, man–nature dualities that many people see as key pillars of nature-alienated, western value systems. Joanna Macy, for example, places emphasis on 'the centerpiece of the Buddha's teaching', which is 'the dependent co-arising of all phenomena' (Macy 1993: 53) — a conception of a fundamentally democratic cosmos, one devoid of inter-species hierarchy, and certainly devoid of any species-specific claim upon privileged interests. This insight is enormously empowering for green activists, Macy maintains (against the charge of quietism); for one takes strength to act from, and through those beings with whom one is in such intimate relationship and on whose behalf one acts. (For other explications of Buddhist-green congruences, see Badiner 1990; Bilimoria 1998; Byers 1992; de Silva 1998; Inada 1987; K.H. Jones 1993; Kalupahana 1986; Kaza 1993; Kornfield 1993; Macy 1991; 1993; Marshall 1992: 41–53; Nash 1990: 113–16; Palmer 1990: 54–56; Spretnak 1991: 33–78.)

Taoism, which draws from several classical Chinese writings, of which the key text is *Tao Te Ching* of Lao Tzu from the sixth century BC, has made an even more substantial impact upon current environmental thought. Numerous papers and books have sought to explicate the relevance of the Tao to an age of ecological crisis (for example, Ames 1986; Capra 1992; 1993; Cheng 1986; Devall and Sessions 1985; Jeremy Evans 1973; Goodman 1980; Hall 1987; L.E. Johnson 1991: 269–72; La Chapelle 1988: 90; Marshall 1992: 11–23; Novak 1993; Palmer 1998; Peerenboom 1991; H. Smith 1972; Sylvan and Bennett 1992). Fritjof Capra has incorporated the Tao into the title of his widely read book on emerging paradigms within science — now regarded as one of the classic texts of green literature — *The Tao of Physics* (1992, first published in 1976). Taoism, he writes, 'offers one of the most profound and most beautiful expressions of ecological wisdom' (1983: 412). We have seen that deep ecologists also draw from the Taoist well, seeing in particular their notion of the enlarged 'Self' as resonant within Taoism. Devall and Sessions write: 'the Taoist way of life is based on compassion, respect, and love for all things. This compassion arises from self-love, but as part of the larger Self, not egotistical self-love' (1985: 11). Sylvan and Bennett, on the other hand, are highly critical of deep ecology, purport to demonstrate the points at which Taoism and deep ecology fail to mesh, and attempt to construct a green ethics ('Deep Green Theory') that is a more valid outgrowth of Taoist thought:

> Taoism is throughout decidedly ecologically oriented; a high level of ecological consciousness is built into it, a main recipe amounting to Follow Nature as the basis of practice ... The richness of Taoism, the philosophical topics traversed to which nothing corresponds in Deep Ecology, will mean that Deep Ecology is often left far behind (1992: 21).

Yet, it is not easy to come to grips with Taoism — not easy, in any case, for people accustomed to the rational, material ways of the industrialised west. It is subtle, elusive, as intangible as a lick of smoke. Sylvan and Bennett, for instance, see the Tao as 'the One simple and undifferentiated, the unity behind all multiplicity ... before the differentiating and perhaps distorting conferring of names'. The Tao is pre-naming then, or perhaps nameless: certainly 'it is not an existent concrete thing' (1992: 25). If more substantiality than this is required, we could perhaps say that the Tao, which is usually translated as 'the Way', posits all that is, was, and is yet to come as a single flow of becoming, though that flow, or way, gathers together an infinity of constituent ways, constituent taos. The essence or motive power of the Tao is the *te*, and virtue consists in respecting the free play of the particular *te* of each tao. Lawrence Johnson explains the significance of this:

> what we must not do is depart from the natural rhythms and harmonies and balances of the world. Rather, we act in accordance with the nature of things. Such action does not go against the grain of things and causes it no disruption. While such action does not, in that sense, do anything, it does what there is to be done (L.E. Johnson 1991: 271).

Western environmentalism has seen in this philosophy of 'letting be' a view of living in nature that is respectful and non-exploitative, and thus in sharp contrast with traditional western views of nature (and also with the other great Chinese philosophical tradition, Confucianism, which enjoins mastery over nature; Palmer 1998). Taoism implies a mode of action which minimally disrupts the natural flow. At the same time, the notion of individuated taos within a larger flow of the Tao preserves the individual within the greater whole (what Sylvan and Bennett call the 'grand', 'capital' or 'overarching' Tao; 1992: 29), though no particular tao has priority over any other, and no individual tao can ever take precedence over the Tao. Marshall sees in this 'a kind of ecocentric impartiality' (Marshall 1992: 17), whilst for Sylvan and Bennett, in relation to humans' dealings with each other and with other life, this involves 'deep love — which can also be taken to involve compassion, pity, commiseration, care and respect' (Sylvan and Bennett 1992: 72).

There is clearly a high degree of congruence between Taoism and the values of the current environment movement. But does Taoism provide appropriate principles for problem-solving in the real world? It might seem, on the face of it, to lead to a way of being in the world that is as non-engaged as Hinduism, Buddhism, and other eastern traditions are often thought to be. Its adherents, though, are at pains to insist that this is not so. Sylvan and Bennett argue that the central Taoist principle of *wu wei* ('non-action') is best described as 'acting naturally'. Whilst this enjoins 'letting things take their own natural course; letting Beings be' (1992: 35), *wu wei* is not inactivity; it means, rather, that 'real work should therefore go with and not against the grain of things' (Marshall 1992: 19).

The subtle, somewhat ethereal philosophy of the Tao can, then, be seen to provide those precepts for light living and a non-aggressive relationship with other life that are currently sought within the green movement. It does so, moreover, without — according to its champions — having to bear the cross of contemplative withdrawal which is seen to encumber the claims of most eastern religions to provide an appropriate basis for the current environment movement.

The other non-Christian religious 'tradition' (it is questionable whether many distinct, geographically bounded traditions can be conjoined into a single 'tradition') to make impressive appeal within current environmentalism is neo-paganism. This is drawn largely from the spiritual heritage of native peoples, especially the displaced inhabitants of the so-called 'new world(s)' — the native peoples of North America, in particular (A.L. Booth 1997: 267–70; A.L. Booth 1999; Booth and Jacobs 1990; Callicott 1983; 1994; L.E. Johnson 1991: 267–69; Knudtson and Suzuki 1992; Mander 1991; Marshall 1992: 137–48; Nash 1990: 117–19; Spretnak 1991: 89–109; Swanson 1991), but also the Australian Aborigines (D.H. Bennett 1985; Callicott 1994: Wright 1990) and the Maoris of New Zealand (Patterson 1994), as well as other remnant, usually embattled native societies (Callicott 1994; Knudtson and Suzuki 1992; La Chapelle 1985; Norberg-Hodge 1991; Skolimowski 1993) — and from the pre-Christian peoples of Europe (Callicott 1994; Eisler 1987; McDonagh 1986: 204–09; Marshall 1992: 89–96; Metzger 1989; Sale 1985b; 12–21; Sheldrake 1990: 10–22).

The native North American, it is argued, had such a deep conception of the interrelatedness of living things that all life was considered to constitute a community, and the individual components of that community were anthropomorphised. As Nash puts it, 'Indians regarded bears, for example, as bear *people*. Plants were also people. Salmon constituted a nation comparable in stature and rights to human nations' (1990: 117); whilst Lawrence Johnson observes: 'when killing game, Indians would frequently pray to the spirit of the game animal — or of its species — giving the spirit their reverence and explaining the necessity. The natural world was seen as a continuing community to be lived in' (1991: 268).

Western philosophy frowns upon this: to attribute human qualities to non-human entities is to succumb to the 'pathetic fallacy'. Present-day ecologists are likely to frown upon it too, on the ground that it denies to another lifeform its entitlement to respect on the basis of its essence-in-itself. There is, however, a great deal here — and in the relationship of other pre-modern peoples to the land and its lifeforms — that is attractive to ecologically attuned people. The notion of a community of living things chimes in with philosophical ecocentrism and the ecological impulse to species humility within the larger matrix of life. The appeal this makes is reinforced by traditional society's use of ritual to sustain the concept of a community of all life; or, in societies where the conception of inter-species relationships stops short of that, to regulate a relationship between species that preserves ecosystem integrity and guarantees a degree of humility in trans-species interactions. The social role of ritual within ecologically wise living is argued by Dolores La Chapelle:

> ritual is essential because it is truly the pattern that connects. It provides communication at all levels — communication among all the systems within the human organism; between people within groups; between one group and another in a city and throughout all these levels between the human and the non-human in the natural environment (1985: 250).

And the construction of appropriate inter-species relationships is not all that is involved here. A revitalisation of ritual, it is thought, can reconstitute a sacred dimension to living that has long been discarded within the dominant Christian and secular traditions. This is a major theme, for example, in the best known investigation of ecologically informed native cosmologies, that of Knudtson and Suzuki (1992).

Most writers drawing upon pre-modern traditions in the search for environmentally and socially benign ways of living in today's world do counsel a degree of caution. There are exceptions, but few recommend the wholesale relevance of the traditions with which they are concerned. The most wholehearted calls for a new nature-based spirituality tend to draw from various traditions to advocate an 'a-cultural' neo-paganism (Eisler 1987; La Chapelle 1978; 1988). By 'a-cultural' I do not mean 'a-historical'. The writers I have in mind not only draw insights from widely disparate cultural traditions; they draw, as well, upon evidence put forward by certain prehistorians to argue that the very earliest cosmologies were earth-reverencing, female pantheisms which were later overthrown by male warrior sky-god cosmologies (Eisler 1987). The earth-spirituality wing within ecofeminism figures prominently here (for example, Adler 1989; Allen 1990; Eisler 1987: 59-77; Metzger 1989; Sjoo and Mor 1987; Spretnak 1986; 1989; 1991: 114–55; Starhawk 1979; 1989; 1990), though other greens are also drawn to this position. What they envision is a conglomerate paganism, one making extensive use of ritual, incantation, shamanism (Knudtson and Suzuki 1992: 10; Roszak 1971) and the Goddess tradition to rediscover the rhythms of natural processes.

The revival of the Goddess is fundamental. The Goddess roughly equates with a spiritualised Mother Earth; but such an observation leaves unsaid the cosmic symbolism of the Goddess as distinct from the male Godhead of the Judeo-Christian tradition. Spretnak explains it thus:

> no one is interested in revering a 'Yahweh with a skirt', a distant, judgmental, manipulative figure of power who holds us all in a state of terror. The revival of the Goddess has resonated with so many people because She symbolises the way things really are: All forms of being are One, continually renewed in cyclic rhythms of birth, maturation, death. That is the meaning of Her triple aspect — the waxing, full, and waning moon; the maiden, mother and wise crone. The Goddess honours *union and process*, the cosmic dance, the eternally vibrating flux of matter/energy: She expresses the dynamic, rather than the static model of the universe. She is *immanent* in our lives and the world. She contains both female and male, in Her womb, as a male deity cannot; all beings are *part of Her*, not distant creations (1989: 128; see also Sjoo and Mor 1987: 45–86).

A particular manifestation of the Goddess revival centres around Gaia, the Great Mother of Greek mythology. Gaian worship was given a huge fillip by its use as a metaphor for a scientific theory that postulated life on earth as constituting a single, self-correcting organic system. The scientific progenitors of this theory — James Lovelock, and others — insist that Gaia is a central mechanical process involved in sustaining life on earth: not a Goddess, not endowed with moral significance, not teleological (Allaby 1989: 112). For many, though, the Gaia hypothesis dovetails neatly with their spiritual concerns, and it is precisely the mythological resonances that have been taken up (for example, Roszak 1981: 63–64; Sale 1985b: 3–18).

Earlier, it was argued that the current environment movement is not, in essence, backward-looking — not concerned to re-establish the conditions of a pre-industrial 'golden age' when people merged harmoniously with biophysical processes rather than contended with them. Clearly, this generalisation does not apply to those who seek a resurrection of earth-based paganism. Marshall, who intelligently surveys the cosmologies of several American Indian tribal groups, and who believes that 'their interweaving of humanity and animals and their reverence for nature stand as a model of a harmonious way of relating to the earth' (1992: 148), nevertheless argues that the desire to emulate the 'old ways' of the Indians represents 'a revival of Romantic "primitivism", of a belief that many aspects of existing civilization are evil and that people lived a better life in earlier, more "primitive" times' (1992: 147).

The impact of eastern religious and neo-pagan traditions within the green movement highlights the perceived failure of mainstream Christianity to provide a relevant basis for an ecologically benign society. Christianity is often seen as unhelpful at best, hand in glove with the forces of rape and pillage at worst. Why is this the case?

CHRISTIANITY AND ECOLOGICAL VALUES: LYNN WHITE AND HIS CRITICS

The extent to which Christian teachings are implicated in the environmental crisis was placed on the agenda early in the history of modern environmentalism, when Lynn White Jr published his now classic article, 'The Historical Roots of Our Ecologic Crisis', in the March 1967 issue of *Science.*

White maintained that the crisis of the earth's living systems, as we face it now, can be sourced directly to 'the Christian dogma of creation'; particularly to the view that 'man' is 'made in God's image', and as such 'shares in God's transcendence of nature' (1967: 1205). The rest of living creation was placed on earth solely for mankind to dispose of to his (and therefore God's) greater earthly glory as he sees fit. 'In the more recent past', writes Attfield of the White position, 'the roots of the crisis' may be attributable to technological imperatives and the scientific ethos that sustains those imperatives', but 'the beliefs implicit in Genesis' underwrite the destructive marriage of science and modern technology (Attfield 1991: 21; White 1967: 1206).

It is usually argued that the comparatively recent triumph of the western scientific paradigm provided the technological wherewithal to apply unprecedented stress to biophysical systems. But, White argued, 'the leadership of the West, both in technology and science, is far older than the Scientific Revolution of the 17th century, or the so-called Industrial Revolution of the 18th century' (1967: 1204). The harnessing of nature to human ends was already proceeding apace, a trend that dates from Christianity's triumph over paganism. The guardian spirits of trees, rocks, streams and other natural phenomena no longer commanded the protective respect that was accorded them in pagan times. 'By destroying pagan animism Christianity made it possible to exploit nature in a mood of indifference to the feelings of natural objects', wrote White. 'The spirits in natural objects, which formerly had protected nature from men, evaporated ... and the old inhibitions to the exploitation of nature crumbled' (1967: 1205).

This point is developed by Rupert Sheldrake, though he softens the rigidity of White's pagan/Christian bifurcation. Sheldrake looks at the impact of Christianity on the old religions, arguing that in some cases, in the Celtic Church in Ireland and Britain for example, Christianity constituted a development or *culmination* of the old religions (1990: 15–24). The spread of the cult of Mary was a grafting of the Mother Goddess onto Christianity. But robust Protestantism ended all that. Christianity had long since adopted a patriarchal ecclesiology, but now the masculinisation of God became complete. This was particularly so within zealous Protestantism, which denounced as pagan the cult of Mary and the reverencing of the ancient sacred places. Sheldrake, then, brings the disastrous day of Christianity's divorce from nature forward to a time much closer to our own: to the Protestant Reformation of the sixteenth century (see also Bratton 1993: 232–39). Protestantism, he says, joined hands with the emerging belief in science and reason and their attack on superstition and blind faith:

> the material world was governed by God's laws, and incapable of responding to human ceremonies, invocations or rituals; it was spiritually neutral or indifferent, and could not transmit any spiritual power in and of itself. To believe otherwise was to fall into idolatry, transferring God's glory to his creation ... The domains of science and religion could now be separated: science taking the whole of nature for its province (Sheldrake 1990: 20–21).

The irony was that, having let the genie of secular humanism out of the bottle, the Protestant ascendancy was itself overthrown. 'For the secular humanist', says Sheldrake, 'nothing is sacred except human life' (1990: 23). There is no reason, then, why the purified God of

Protestantism should be accorded any more authority than the deposed God of Catholicism. 'Those whose religion is founded on a protest against other people's unreasoning', Sheldrake observes, 'have been ill-equipped to defend an unreasoning faith of their own' (1990: 23).

White and Sheldrake trace the development of the dominant, nature-imperialising character of western science to an underlying base in Christianity, though for Sheldrake this was a particular, comparatively recent form of Christianity. If core Christian beliefs constitute the motor that has driven science to the dismaying but logically consequent conclusions of the mechanistic paradigm, then it is clear, says White, that science and technology cannot resolve our problems — because these are merely secondary manifestations of a more fundamental malaise. Instead, we must change the religious assumptions that have given rise to, and sustain that 'holy' alliance of science and technology. A new religion is needed. Zen perhaps, though White has misgivings about this. More promising, he thinks, is a resurrection of a minority Christian tradition, that of St Francis of Assisi — 'the greatest radical in Christian history since Christ' (1967: 1206; see also Bratton 1993: 217–29; Hooper and Palmer 1992; Hughes 1996). St Francis, says White, 'tried to depose man from his monarchy over creation and set up a democracy of all God's creatures', for all creatures have souls, and in all is reflected the glory of the Creator (1967: 1206). Sheldrake, too, considers various non-Christian religious traditions before opting for a nature re-enchanting Christian mysticism (1990: 153–71) within which 'God is not just immanent in nature, as in pantheist philosophies, and not just transcendent, as in deist philosophies, but both immanent and transcendent, a philosophy known as panentheism' (1990: 167).

White's seminal paper had considerable influence, and evoked a powerful reaction, though this response has largely been confined to environmentally concerned Christians. Non-Christian greens have tended to accept the White case at face value. Among green Christians, too, a division is apparent. Christians supportive of the White position tend to hold somewhat revisionist views of Christianity generally, usually urging a shift in the direction of a nature-based Christian mysticism similar to Sheldrake's. Those who hold a commitment to the mainstream denominations tend to defend Christianity against White's charges. Such people are more likely to support the Christian stewardship position, a position that deep ecologists, in particular, have been at pains to reject (for example, Devall and Sessions 1985: 120–26). Among mainstream Christians the position of the Rev. Andrew Dutney is, thus, somewhat unusual:

> White was not the first to accuse Christianity of complicity in the ecological crime, nor the last to present the evidence for the prosecution, but his 1967 essay was decisive in the case of 'The Earth v. Christianity'. There is no need to go over the evidence again. While some theologians

> would like to do some 'plea bargaining', many accept the charge as it stands. In any case, as I was recently reminded at a gathering of environmentalists, the lay jury has already returned a verdict of 'guilty as charged' (1992: 77).

The first notable riposte to White was Lewis Moncrief's 1970 response in *Science*, the same prestigious journal in which White had published. White's charge that Christianity was *primarily* responsible for the environmental crisis demonstrated, wrote Moncrief, a lack of understanding of the real motor of historical change. Religious beliefs *background* the changes taking place through time; they define 'the ball park in which the game is played', but 'the kind of game that ultimately evolves is not itself defined by the ball park' (1970: 58). Other social processes such as technological change, democratisation and urbanisation are more potent and more complicit in environmental degradation, and are not consequent upon religion; they proceed under their own dynamism, though 'the Judeo-Christian tradition has probably influenced the character of each of these forces' (1970: 514).

The broad context of Moncrief's anti-White critique has endured. This is the extent of White's reliance on ideas to explain social phenomena. In arguing thus, White is up against the weight of Marxist historical explanation, which has it that economic forces ('relations of production') determine the tides of history, and that the dominant ideas within any space-time locus are part of the 'superstructure' of social existence — the visible, but secondary manifestations of the less easily recognised, but more primary factors determining the terms of existence below the surface. The dominant ideas will not be what drives a society along; these are just the means of legitimising structural social relations which are not dependent on ideas, but which cause those ideas to be generated. This point is made — outside an overtly Marxist framework — by F.B. Welbourn, who argues against White that 'it is difficult to give so much primacy to the causal efficacy of ideas' (1975: 571), on the ground that the social function of ideas selected for pre-eminence is to justify changes occurring in the material basis of society.

This is not the place to discuss Marx's doctrine of historical materialism, but the claim that belief systems have no social-shaping capacity seems not so ironclad as to have the status of universal law. There is the problem of explaining, for instance, why the growth of advanced, interventionist technology has been exclusively a feature of European (or Europeanised) history. The first steps to technological take-off occurred elsewhere as well, but either failed to proceed or were channelled to other ends. Without reference to cultural factors then, it is difficult to entirely explain patterns of technological development. A less reductionist explanation than the classical Marxist view of the relationships between economics, technology and worldviews thus seems necessary, and it may be that ideas play an important and quasi-independent role in the explanation of social and historical developments, even if they seldom constitute sufficient explanation by themselves. White himself has shifted ground on this matter — or perhaps he has merely clarified an earlier misreading — explaining that his choice of the word 'roots' was carefully made ('the historical *roots* of our ecologic crisis') to avoid imputing a single cause for ecological crisis: 'no sensible person could maintain that all ecologic damage is, or has been, rooted in religious attitudes', though 'in the end one returns to value structures' (1977: 57–58).

White has also been charged with misinterpreting the Old Testament — in particular, with misinterpreting the Old Testament view of the relationship between humankind and nature. In Genesis (1:26), God created man to have 'dominion over the fish of the sea, and over the fowl of the air, ... and over all the earth, and over every creeping thing that creepeth upon the earth', and man was commanded, in Genesis 1:28, to 'be fruitful, and multiply, and replenish the earth, and subdue it'. White interprets this to mean that 'God planned all of this [creation] explicitly for man's benefit and rule: no item in the physical creation has any purpose save to bear man's purposes' (1967: 1205). Support for this reading can be found elsewhere in Genesis. After the expulsion from Eden ('the Fall'), the initial state of harmony between humans and other animals was no longer tenable. God tells Noah 'And the fear of you and the dread of you shall be upon every beast of the earth, and upon every fowl of the air, upon all that moveth upon the earth and upon all the fishes of the sea; into your hand are they delivered' (Genesis 9:2).

Nevertheless, there has been dispute over whether White gets this right. It is widely held that, on the Old Testament view, nature is not sacred.

The Protestant theologian, Karl Barth, makes a sharp distinction between 'upper creation' (heaven) and 'lower creation' (earth), insisting that the latter 'is not to be feared and worshipped as a thing divine' (Barth 1959: 62). God is wholly other: He cannot be revealed through nature. Creator and creation are 'radically distinct ... it is idolatrous to worship creation, and so there is certainly nothing sacrilegious about treating nature as resource for human benefit' (Attfield 1991: 25).

Nature is humankind's to use, then, but this does not signify an end to human responsibility. John Passmore was the first to subject White's interpretation of the Old Testament to close scrutiny. He concludes: 'Genesis, and after it the Old Testament generally, certainly tells man that he is, or has the right to be, master of the earth and all it contains. But at the same time it insists that the world was good before man was created, and that it exists to serve God rather than to serve man' (1974: 27). For Passmore, the villain of the piece is not Genesis, nor even the Old Testament generally, for notions of unlimited human despotism over nature did not become current within Christianity until the third century AD, by which time the Christian position had been substantially modified by Stoic and other Greek teachings:

> there is a strong Western tradition that man is free to deal with nature as he pleases, since it exists only for his sake. But they are incorrect in tracing this attitude back to Genesis ... It is only as a result of Greek influence that Christian theology was led to think of nature as nothing but a system of resources, man's relationships with which are in no respect subject to moral censure (1974: 27; for Passmore's full case see pp. 13–17).

Other writers have claimed that White erred in staying exclusively with Genesis, maintaining that a reading of the other Old Testament scriptures indicates that his interpretation of Genesis is flawed. Thus, Attfield points out that the laws of Leviticus and Deuteronomy 'set considerable limits to human dealings with nature', whilst Psalm 104 is a remarkable, almost ecological, account of God's providence for all creation's creatures, with humankind enjoined to respect its place in God's scheme of things (1991: 25–26).

For Attfield it is fundamentally wrong to attribute support for a human despotism over nature on the basis of Old Testament teaching. The 'dominion of man' accorded by the Old Testament is not despotism, maintains Attfield, the original biblical concept of 'dominion' being somewhat different to our own: 'if Genesis authorizes us to rule nature, it authorizes only the kind of rule compatible with the Hebrew concept of monarchy' (Attfield 1991: 27). This, he argues, was never a concept of absolute monarchy, for the monarch was responsible to God for the health of the realm, and faced replacement if that responsibility was not satisfactorily discharged; responsibility entailed a steward's care for the flora and fauna of God's creation.

Arguments of this type — claiming misunderstanding based upon shifts of meaning from the original Hebrew — are not uncommon. Most concern themselves with 'dominion', but rival interpretations of 'subdue' ('be fruitful, and multiply, and replenish the earth, *and subdue it*') are also advanced (for example, Innes 1987: 14). To my mind, the implications of 'and subdue it' are of even greater import for Christian ecology than 'dominion', and without the specialist insights of the linguist it is difficult to assess the validity of these 'rival meanings' arguments.

The New Testament is thought to be somewhat less ambiguous. Though he finds the record not entirely straightforward, John Austin Baker comments: 'in seeking for any kind of theology of humanity and nature, the Christian cannot but be grateful that his or her Bible does not consist merely of the New Testament' (1990: 24), whilst 'the New Testament ... places man at the centre of universe', observes historian John Young: 'the anthropocentric basis of Christianity was made more explicit in the writing of St. Paul, which represents a fusion of Hebrew thought with that of the Greek Stoics' (1991: 56).

White offended people mightily, and the reaction was lively and massive. But there has since been a reaction against the reaction. Christian ecology thrives, its writings often acknowledging a debt to White — not necessarily agreement, but certainly a debt — for placing the question of the ecological credentials of the Christian religion at the forefront of debate.

CHRISTIAN STEWARDSHIP

Of the various forms of Christian ecology, the most hostile to the White thesis is Christian stewardship. Attfield (1991) provides its most detailed exposition.

Attfield's interpretation of 'dominion over nature' is that humankind is to be a 'steward' over creation, 'charged by God with responsibility for its care' (1991: 27). The task is to ensure that God's handiwork is maintained in good health, drawing sustenance and even profit from it whilst managing it sustainably and looking to the interests of its living components. This act of stewardship is the highest of the responsibilities God has placed upon his 'designated manager' (O'Riordan 1981: 204) — humankind. To wantonly destroy nature may not be sacrilege, but it is a slight upon God's craftsmanship and a renunciation of the most important responsibility with which we have been charged. And the ecological crisis has come about because there has been a gradual weakening of the stewardship ethic, with protective obligations to nature becoming devalued.

Present-day proponents of Christian stewardship are at odds over the precise nature of the relationship between humans and other life appropriate to a benign reading of 'dominion over'. Attfield's concept of stewardship accords considerable status to the handiwork of God:

> most adherents to the stewardship view have implicitly accepted that intrinsic value is to be found among nonhumans as well as humans; this granted, stewards of the earth should be seen not only as managers of resources but equally as curators of treasures or as trustees of the biosphere ... on the stewardship view nature has characteristically also been regarded as of value in itself (1983: 216–17).

And, he suggests, this reading of the human–other life relationship would have been, historically, a significant position within the Christian tradition (1983: 211). Perhaps. But it is my strong impression that it is certainly a minority position today. A more characteristic position is that advanced by John Black.

Black rejects an ecocentric relationship based in the work and teachings of St Francis (such as that envisaged by White) since, he argues (1970: 205), the adoption of a Franciscan ethic would so sanctify nature as to render all science and civilisation impossible. (Similar criticisms have been made of a return to pagan conceptions of nature:

paganism too, it is argued, so sanctifies nature as to doom our species to such a non-interventionist stance toward nature that civilisation could be maintained at only the most rudimentary level.) Black argues instead that nature is best accorded the degree of protection a proper reading of the scriptures makes appropriate by enlarging the scope of ethical consideration to include the good of generations of humans yet unborn: 'by redefining "mankind" in terms of the whole of humanity, dead, living, and as yet unborn, we may perhaps be able to assess what we do in terms of the good of mankind, regardless of any individual's position along the time axis' (1970: 123; see also Elsdon 1981: 155). To Black, only a Christian stewardship ethic, which accords other life mere instrumental value but also sees it as the heritage owed by the present to the future, offers practical hope. Others agree. This from a Jesuit Father:

> we can summarize the Old Testament presentation of man and nature. Man is the high point of creation. He stands in a special relationship to God. There is something of God in him ... Man is limited, but free. His destiny is to unlock the mysteries of nature, to initiate and find his own way in the created universe in which God has set him. In doing so he can and does overstep his limits ... Man is not just superior to the animals. He is quite different. He is responsible and must act responsibly towards the animal world (Scullion 1973: 28).

How to sort this out? Much depends on the status of Genesis 9:2, quoted earlier: 'And the fear of you and the dread of you shall be upon every beast of the earth ... into your hand they are delivered'. This seems to be the most unrestrained observation in Genesis on the human–other life relationship (and given that this new, post-Fall arrangement is supposed to be a key part of God's punishment of *humankind*, it does seem strange that the larger weight of the punishment falls less upon humans than it does upon the rest of God's living creation!). However, this observation is followed by one of the least ambiguous affirmations of humankind's stewardship responsibilities, for the covenant to which Noah was bound is a covenant with all of living creation: 'this is the token of the covenant which I make between me and you and every living creature that is with you, for perpetual generations' (Genesis 9:12; on this, see McDonagh 1986: 123).

Much also depends upon how the key Christian interpreters have dealt with this relationship through the ages. The Augustinian focus upon salvation at the expense of creation gave overwhelming priority to the human–God relationship. It is humans, after all, who are fashioned in God's image. As the post-Fall compact excluded creatures from a share of God's infinite grace, and as '[human] salvation could only be communicated through the Church', all that was non-sacred, 'defined as every being/thing outside the Church *or in nature*' was 'deprived of grace and incurably corrupt' (Watson and Sharpe 1993: 220).

So the weight of historical scholarship seems less than promising for the sort of 'intrinsic value in nature' stewardship ethic envisaged by Attfield. There is some indication of headway. At a major ecumenical meeting held under the auspices of the World Council of Churches (WCC) as part of the 1992 Rio Earth Summit, one of the WCC's groups, noting that 'anthropocentric, hierarchical and patriarchal understandings of creation lead to the alienation of human beings from each other, from nature, and from God', recommended as follows:

> the current ecological crisis calls us to move towards an eco-centred theology of creation which emphasizes God's Spirit in creation (Genesis 1:2, Ps. 104), and human beings as an integral part of nature ... A hierarchical understanding of the Imago Dei, putting human beings infinitely above all creation, must be replaced by a more relational view. Human beings are created for the purpose of communion with God and all the living and non-living things (Granberg-Michaelson 1992: 78–79).

But, in the main, current theological authority is as indisposed today to shift from the Augustinian position as it has ever been. Karl Barth juxtaposes the realm of nature ('brute') with the realm of history ('mind'), drawing such rigid barriers between the two that it seems impossible to derive a concept of stewardship therefrom — certainly not an Attfieldian ethic, at any rate: 'man is the creature on the boundary between heaven and earth; he is on earth and under heaven. He is the being that conceives his environment, who can see, hear, understand and dominate it' (Barth 1959: 63). No joy, either, from Paul Tillich's interpretation of the biblical God as 'the God of change', in contrast to the nature gods of Israel's neighbours, which were 'gods of stability' (Santmire 1977: 86): 'to the birth of man out of nature and against nature corresponds the birth of prophetism out of paganism and against paganism ... This means, first of all, that He is the God who acts in history toward a final goal' (cited in Santmire 1977: 86). In such a conception God is no longer interested in creation. What counts is the unfolding of human destiny *away* from a base nature, a nature increasingly less relevant, towards a future measured as progress in terms of the ever-growing distance between humankind and nature.

PANENTHEISM AND THE REINTERPRETATION OF 'CREATION'

A second significant form of Christian ecology is 'process theology' — an approach borrowing heavily from the metaphysics of Alfred North Whitehead. Whitehead posited a world of flux, a world whose organising principles were to be found within ongoing *process* rather than a stop-start world of first causes and consequences. From 'the Newtonian point of view there is no reason to be found in the essence of a material body — in its mass, motion, and shape — for the law of gravitation, or for the existence of any forces or stresses between bodies' (Mathews 1991: 33). In Whitehead's process theory such interactions are not secondary properties explaining linkages between more primary categories; they constitute, in themselves, primary existence (Merchant 1992: 127). Such views have had a profound impact upon certain ecologically minded theologians.

It is easier to define process theology than to establish its content. It is a notion of Christianity as unfolding tradition: that is, not absolute, not eternal, not changeless, not anchored in an inflexible dogma, and open to cross-fertilisation with other religious traditions (Cobb 1982 and McDaniel 1990a; 1990b are particularly interested in Christian–Buddhist intersections). Jay McDaniel describes Christianity as:

> a continuing historical movement within which, in each new age, new generations with new ideas rightly participate. Through their participation, the religion can itself take new forms. A new generation's faithfulness to Christ does not lie in absolutizing the Christian past ... Rather, faithfulness lies in being open to a God whose call perpetually comes from the future rather than the past, beckoning for openness to truth, goodness and beauty, whatever their sources ... It is because Christianity is a process, initiated but not completed by Jesus, that it can become more ecological despite its ambiguous past (1986: 37–38).

For orthodox Christians the Bible has the status of Received Truth; but for process theologians there is no immutable truth. Process theology is organic in conception and argues a return to thinking about the universe as a 'cosmos'; a single, integrated system united by universal principles, which portray all things in the world — human, natural and divine — as related components of a greater whole. It is a project which aims to re-establish a 'natural theology'.

Note how this flows into a vision of science. Since the disintegration of holistic knowledge the only recognised facts about the universe as a whole have been reached by isolated aggregations of data within various disciplines, and there has been no imperative for the integration of these facts into a transdisciplinary cosmology — a cosmology that would re-integrate humanity and nature.

Process theology seeks a rapprochement with science then, though it does insist upon a humble science, one that eschews the Promethean hubris that has characterised it in its mainstream tradition, and one that acknowledges a mystery within life that scientific endeavour can reach only in part (McDaniel 1986: 34). Given its base in Whiteheadian metaphysics, it is understandable that process theology has some scientifically prominent supporters, of whom the most eminent is probably the distinguished Australian biologist, Charles Birch. Birch's collaboration with theologian John Cobb has given green thought some of its most influential writings — most notably their jointly authored book *The Liberation of Life* (1981).

Birch and Cobb — and indeed, process theologians generally — are at pains to establish theirs as pre-eminently an *ecological* theology. All beings are viewed as democratic participants within larger value-infused wholes. This applies to plants and even sub-atomic particles (Birch and Cobb 1981: 152–53). 'An ecological spirituality will be open to, celebrative of, and transformed by, plurality — the sheer diversity of different forms of life, human and non-human', writes process theologian Jay McDaniel (1990a: 31). Unlike Christian stewardship then, where ecocentrists such as Attfield must contest the issue with proponents of stewardship who oppose ecocentrism (such as Black), ecocentric principles would seem to inhere within process theology.

But where stands God? Is God radically apart from the created world (as Christian opponents of an ecological Christianity and some proponents of Christian stewardship would have it), or is God synonymous with the evolving cosmos (as pantheists would have it)? For process theologians the answer stands somewhere between these extremes. For Sallie McFague creation is 'the body of God ... bodied forth in the eons of evolutionary time', though 'God is not reduced to the world if the world is God's body' (1990: 213). Process theology is, thus, classically 'panentheist': God has both a separate and an immanent existence. Birch and Cobb define it thus: 'God is not the world, and the world is not God. But God includes the world, and the world includes God. God perfects the world and the world perfects God. There is no world apart from God, and there is no God apart from some world' (1981: 196–97). The 'some world' is important: 'God can exist without *this* world' but, because relationality is a first principle of all life, including God's, 'God's life depends on there being some world to include' (Birch and Cobb 1981: 197).

Somewhat similar to process theology, though it seems to have followed an independent evolution, is the so-called 'creation theology' of certain Roman Catholic mystics. Like process theology, creation theology opposes the dominant soul/body, spirit/nature and mind/body dualisms, and it, too, is concerned to reconstitute the Christian tradition, not merely defend it. Its key notion is the idea of continuing creation. Creation is not to be identified with a remote event at the beginning of time, but with God's continuing involvement in the world. God sustains the world and is with it, creatively, in every age. Thus, the world is continually being 'recreated': it is a dynamic and interdependent whole in which new forms emerge. Neither nature nor humanity are immutable.

The wellsprings of this ecological theology tend not to be in Whiteheadian science. Recourse is had instead to the spirituality of the medieval 'Rhineland mystics' — Hildegard of Bingen, Meister Eckhart, Mechtild of Magdeburg, Nicholas of Cusa, and Julian of

Norwich (on these sources, see M. Fox 1984; 1989) — and to the tradition of St Francis. Hildegard (1098–1179) is held to stand at the head of an alternative, non-dualistic tradition that joyously celebrates creation itself. Creator and creation relate as man and woman (Uhlein 1983: 56), God infusing all that lives with godly spirit. Interestingly, the term Hildegard chooses for this process whereby God infuses the world is 'greening' (*viriditas*): 'God's Word is living, being, spirit, all verdant greening, all creativity' (Uhlien 1983: 49). This leads to quite a different conception of sin. 'If creation is a cosmic blessing, then true sin represents a rupture in the cosmos', writes Matthew Fox: 'the ultimate sin, Hildegard recognizes, is ecological: a sin against the earth, against the air, against the waters, against God's creation' (1983: 10).

God is, thus, a god of flux and imagination; a creative influence within an organic unity, or — to jump to ecological language whilst saying the same thing — a creative influence within an interdependent web of life. In *The Coming of the Cosmic Christ* (1989), Matthew Fox argues that since the Enlightenment Jesus Christ has been stripped of his cosmological context. The 'Cosmic Christ'? Simply Christ as the all-pervading spirit in the universe — Hildegard's 'greening'. Fox refers to this as 'the pattern that connects all the atoms and galaxies of the universe, a pattern of divine love and justice that all creatures and all humans bear within them' (1989: 7). The sacred is everything — all of creation. Here is a view of the God–nature relationship that seems rather more pantheist than panentheist: God is immanent, but not separate. It is hardly surprising that Fox has incurred the Vatican's wrath, particularly as he has enthusiastically embraced native American spirituality as a fecund source of spiritual inspiration (as have other creation theologians; for example, Joranson and Butigan 1984: 5–6). 'The churches need to beg the forgiveness of native peoples', writes Fox (1989: 6), 'whose sense of the Cosmic Christ was never lost.'

'Creation theology' is primarily concerned with cosmological interpretation then, and the 'high' cosmologists within this tradition are Thomas Berry and Brian Swimme (T. Berry 1988; 1990; T. Berry with Clarke 1991; Swimme 1988; Swimme and Berry 1992; see also the laudatory appraisal by J. Berry 1993). Thomas Berry, like Fox, distances himself from the Christian tradition wherein 'earth worship was the ultimate idolatry' and in which the 'personal savior orientation ... easily dispenses with earth except as a convenient support for life' (1990: 152). The earth, writes Berry, is infused with spiritual energy that flows through all life (1990: 153), an informing swirl of spirit that is genuinely cosmic in dimension. Having lamented the attitude that 'caused or permitted humans to attack the earth with such savagery', and the fact that this was perpetrated 'by a Christian-derived society, and even with the belief that this was the truly human and Christian task' (1990: 154), he spells out the alternative:

> we need a spirituality that emerges out of a reality deeper than ourselves, even deeper than life, a spirituality that is as deep as the earth process itself, a spirituality born out of the solar system and even out of the heavens beyond the solar system. There in the stars is where the primordial aspects take shape in both their physical and psychical aspects. Out of these elements the solar system and the earth took shape, and out of the earth, ourselves (1990: 155).

SOME PROBLEMS: 'FLAKY SPIRITUALITY' IN A MATERIALIST AGE

Prominent around my small city is a car entirely plastered over in Christian slogans. Greens feature prominently within this motorised collage, and not positively. 'Greenies No, God Yes' is one eye-catching exhortation. The medium may be somewhat eccentric, but the message is familiar. Despite the 'Rio theology' of the World Council of Churches, despite the promotion of a long-standing alternative Christian tradition by panentheists and proponents of Christian stewardship, the ponderous weight of mainstream Christian theology, Protestant and Catholic, remains a formidable obstacle to a green refurbishing of Christianity.

The most robust form of western Christianity today — Protestant fundamentalism — is, in the main, implacably opposed to environmentalist aspirations, and its political power reached all the way to the White House in the presidencies of Reagan and the elder Bush, when legislative gains in environmental protection, incrementally won over many years, were swept away. Unlike the populism of the late nineteenth century and early twentieth century, the Protestant fundamentalism that underpins the New Right today does not target the threat to established values contained in programs of rapid industrialisation. Rather, industrialisation is itself seen as a pivotal component of national greatness and the heritage of 'traditional values' which stands as bulwark against Godless 'humanism'. Fundamentalism's theology focuses on the relationship between God and the unique soul, to the exclusion of any relationship between humans and God's creation. God Himself is immutable and unchanging and uninvolved in the world in any ongoing sense, whilst the focus of 'personal being' is to attain one's *super*-natural destiny through God's redemption.

The latter stance is particularly significant. The status of the tangible world is dramatically downgraded; it is the seat of sin and imperfection; the mire from which the soul — all that is worth saving — struggles to escape. To devote one's life to reaching a paradise far superior to the world of earthly substance readily leads to apathy concerning the fate and condition of the physical world — especially when it is destined for the destruction of Armageddon in any case. President Reagan's fundamentalist (and hard-line anti-environmentalist)

Secretary of the Interior, James Watt, took this view. His public utterances contained regular references to the transience of material phenomena and to the imminence of the apocalypse. It is worth noting Isaiah 24: 4–6 in this context:

> The earth mourneth and fadeth away, the world languisheth and fadeth away, the haughty people of the earth do languish. The earth is also defiled under the inhabitants thereof; because they have transgressed the laws, changed the ordinance, broken the everlasting covenant. Therefore hath the curse devoured the earth, and they that dwell therein are desolate: therefore the inhabitants of the earth are burned, and few men left.

It is possible to read such verses as a warning of the consequences of too-cavalier a treatment of God's creation. Some do. Green religious fundamentalists do exist; they occasionally turn up in my postgraduate classes, as often as not with an interesting and plausible position to defend. God may be distinct from, and above creation, but the universe remains 'the product of God's creative mind and energy' (Harris 1989: 5) and therefore merits respect. But the visitation of devastation upon the earth can also be taken as cause for joy, for the wicked are getting their come-uppance, a fate which those favoured of God, those destined to escape the earth, can expect to avoid.

It should come as no surprise, then, to find that, for every Attfield, every Cobb, every Berry, there exists a host of Christian denouncers of 'green' theology. The writers cited by Stephen Fox articulate a more 'typical' Christian perspective:

> 'we too want to clean up pollution in nature', said an editorial in *Christianity Today*, 'but not by polluting men's souls with a revived paganism.' A Jesuit writing in the Catholic journal *America* described ecology as 'an American heresy' that challenged those faiths concerned primarily with man: 'For Christians, nature cannot be fully nature except through man.' R.V. Young, Jr., declared in *National Review*: 'There is no "democracy" of all creatures; a single human soul is worth more than the entire material creation. Just as man is here to serve God, so nature is here to serve man.' Another writer warned in the *Christian Century* about 'a sort of mindless ecological imperative ... ultimately reactionary and anti-human, as well as anti-Christian' (1981: 373).

Here is a catalogue of comments made, by and large, in passing. But among the more substantial anti-green writings Christian perspectives also figure. Early on the attack was Lutheran pastor Richard Neuhaus (1971), and the vehemence of his denunciation has probably not since been surpassed. He described the 'nature worship' of the fledgling environment movement as 'classically fascist', its devotees exclusively to be found among upper- and middle-class people seeking to preserve elite privileges against the claims to betterment of the lower orders. And he was especially scathing of the notion that God's presence is to be sought in the works of nature rather than those of human industry.

All very well, then, for Attfield and the panentheists to point to a genuine pedigree for an alternative Christian tradition, one that fixes attention upon God immanent in creation rather than upon a personal human salvation that is not for this earth. That no less a figure than Albert Schweitzer can be adduced in support of this tradition (Birch and Cobb 1981: 148–49; L.E. Johnson 1991: 134–41), and the fact that the World Council of Churches has conceded its appropriateness, certainly accords it a foot in the theological door. But the weight of inertia will take some shifting. Those I have loosely lumped together as panentheists and Christian mystics face, perhaps, the most formidable task. Even the World Council of Churches remains ambivalent on the matter. At its 1991 Convention a Korean delegate, Chung Hyun-Kyung, argued that nature was sacred and purposeful, and that it was time to put all life rather than a single, privileged species at the centre of creation. She received sustained criticism for succumbing to 'syncretism' — the practice of incorporating beliefs and practices from other religions within Christianity — a tendency against which the orthodox are ever vigilant. For panentheists the tripwire will always be the status of non-human life within Christian dogma.

A case in point is provided by an interesting little book by Tony Campolo — the same evangelist to whom President Clinton turned for guidance during his travails of 1998. Campolo argues *for* a Christian responsibility of stewardship; but he will have no truck with the ecocentrism of, say, Attfield. For Campolo 'creation care' is 'a Christian mission' (1992: 131), but the *sin* of according value to nature in, and of itself is to be resisted: 'to treat nature as some kind of spiritual personality with whom we can merge through meditation and psychic surrender is a serious mistake. God breathed His own breath only into the human race. Only we are created in God's image' (1992: 177); and 'while we sense something sacred in all of life, because we claim that all life is sustained by God, we nevertheless believe that there is a hierarchy in God's creation' (1992: 179).

Campolo also observes the 'seductive' influence upon Christianity of flirtations with native American spirituality, owning to having been present 'at Christian gatherings where there were extensive readings from Black Elk'; moreover 'these readings were treated as though they possessed more truth than the Bible' (1992: 175). It is, thus, easy 'to cross the line from Christianity into paganism', a line Campolo is sure that Matthew Fox has well and truly crossed (1992: 175). Birch and Cobb are at pains to point out that they 'are committed to values and insights which we interpret to be Christian' (1981: 8); but Campolo might also have instanced the Christian-Buddhist fusions explicitly sought by Cobb (1982) and McDaniel (1990a; 1990b), or the creation cosmology of Swimme and Berry.

The creation cosmologists have attracted criticism from other quarters, too. It has been argued, for instance, that the division of Christian thought into two distinct traditions — one based in personal redemption for Original Sin, one creation-based — renders too simplistically a great deal of theological complexity, including substantial differences between the medieval mystics upon whom the cosmologists draw, and the extent to which matters of sin and redemption remained central questions for them (Ruether 1992: 242; J. Young 1991: 61). Ecofeminists have also found creation cosmology wanting, on the same denial-of-plurality ground that has been charged against deep ecology. 'A view of the whole as spiritually or ecologically significant', writes Plumwood, 'is in no way a substitute for a recognition of the great plurality of particular beings in nature as capable of their *own* autonomy, agency and ecological or spiritual meaning' (1993a: 128).

Criticisms can also be made of the capacity of Buddhism, Taoism and native religions to provide a basis for a new ecologically wise society.

Apart from the not-inconsequential objection that the robust character of orthodox Christianity presents a formidable obstacle to the movement of Buddhist, Taoist and neo-pagan spiritualities into the western values mainstream — despite the impressive inroads all have made within the green movement — there are also more specific problems. Each of these traditions, it is held, has been simplified, distorted and misused in its green applications (for examples, see Curtin 1994; Peerenboom 1991; Rolston 1990) — though observers making this claim often continue to urge the relevance for the tradition, at least in some sense, once the mistakes have been rectified.

One observer who notes the somewhat indiscriminate nature of these adaptations of 'eastern cults, philosophies and religions', the Canadian naturalist John Livingston, argues that it is a development that is essentially *anti*-ecological: 'it was contemplative, to be sure, but it was inward', and it has little in common with 'what I feel to be the spiritual experience of the naturalist'. This is because 'as it extended beyond the individual, it extended to other (human) individuals on a parallel "trip" ... it had little to do with nonhuman nature. Its "environmental" aspect was social and cultural, not ecologic' (1981: 60). For many of those who believe the central green project to be the establishment of a new, less destructive relationship with other life, then, a human-centred spirituality is unhelpful; a blind alley.

Perhaps the most substantial objection to the construction of an environmental ethic based upon religious traditions that are textually enigmatic and only ambiguously prescriptive is that, in conditions of environmental crisis, subtlety is precisely not the commodity upon which one would want to rely. An example. The small Australian state of Tasmania has long been enmeshed in a 'to preserve or to clear-fell' imbroglio over the state's native forests. One of the stoutest champions of the clear-fell position is a former Liberal (conservative) Premier and Minister for Forests. 'Clear-felling', as the name suggests, involves razing the *entire* forest — not merely the logging of selected trees. No 'letting Beings be', no 'flowing in line with the overall Tao' here; on the contrary, this is public policy that *does* 'interfere deeply with the natural course of things'. Yet, the ex-premier has revealed in interview that Lao Tzu sits upon his bedside table, and therein he delves each night after a hard day's work on behalf of the clear-fellers. Surely this can only be so because the Taoist perspective is not overt, and those who would see what they wish to see are able to do so.

In a similar vein, it might be asked why it is that the so-called 'Asian Tigers', precisely those countries in which the influence of the ecologically benign religions of the east is strongest (not just Buddhism and Taoism, but Shinto and Jainism), are the world's most aggressive plunderers of their own and others' environments. Buddhism, in particular, remains bedevilled by its tradition of quietism, and visitors to Buddhist countries frequently comment upon the complacent acceptance which greets the perpetration of the most appalling environmental destruction. It is difficult to avoid the conclusion that, when it comes to determining the public and private choice of ecologically relevant behaviour, the non-directive nature of eastern religious traditions renders them unsuited to serve as determinants of, or even moderators upon action.

Several voices have expressed concern over the uses made of native peoples' spiritual traditions. A recurrent argument is that it is impossible to generalise single themes from such vast cultural diversity. Even between native North American nations there is such diversity that it is not possible to talk sensibly of 'native American society'. The denial of cultural particularity can even be seen as insulting. Additionally, Calvin Martin has argued that white interpretations of 'the' native North American relationship with the land have customarily avoided coming to terms with the mythic world of the native peoples, adopting instead 'the tendency to interpret another culture using the norms and values of one's own culture as a point of reference' (1981: 153; see also Gaard 1993: 308; A. Smith 1994). People locked into western ways of seeing cannot even approach the deep immersion in nature and the nature-based mythic cosmology that so characterised native North American societies. These societies were 'saturated with the primordial Power of Nature which seemed to pulsate throughout all creation' (C. Martin 1981: 155). Martin (1978) also insists that, despite their 'saturation' within 'the primordial Power of Nature', native North Americans did not, in their cosmologies, provide the explicitly ecological ethic that many 'Indian-romanticisers' attribute to them. A similar position is taken by the animal rights philosopher, Tom Regan, who argues that it was fear of well-recognised environmental constraints rather than a benign ecological cosmology that determined native American behaviour patterns (1982: 226–30). It is possible to go further: to argue that, in fact, aboriginal Americans were less than perfect ecological housekeepers. In support of such a stance one might adduce the evidence of buffalo kill sites, where animals far in excess of need were driven to their death and left to rot (Simmons 1969: 52, 54). It is possible that this practice was crucially implicated in the extinction of the North American camel and a species of bison.

Whilst this debate was in full swing along came the sensational revelation that Chief Seattle's famous 1854 letter to 'the great white chief' in Washington was the 1972 fabrication of a Texan television scriptwriter. As Chief Seattle's letter was the most celebrated and best known 'green' North American Indian text, and a classic text within the corpus of green writing generally, that, it would seem, was game, set and match to the critics. I am not so sure. Baird Callicott has persuasively defended the relevance of native American ethics to the modern environment movement against Martin and Regan, conceding that in all cultures there will be individuals not constrained by the general ethos, and that, on this score, the ecological record of the 'Indians' is bound to be flawed, but that 'the American Indian cultures provided their members with an environmental ideal, however much it may have been, from time to time or from person to person, avoided, ignored, violated or, for that matter, grudgingly honoured because of fear of punishment' (Callicott 1983: 257). As for the damaging revelation concerning Chief Seattle's 'letter', this really seems to be of little ultimate consequence, for there are similar ecologically informed writings by American Indians that are as moving and insightful as the Seattle letter — and that are of undoubted authenticity (for example, Black Elk 1932; Lame Deer 1972; Momaday 1970; Standing Bear 1933; Storm 1972).

One final point. Much discussion about religion and environmentalism is less concerned with the relationship between particular religious traditions and ecological precepts than it is with environmentalism as *a religion in its own right*. Thus, Richard Lowry talks of 'the new "religecology"' wherein 'the collective religious commitment to cleaning up the environment creates a kind of therapeutic community in which all can purge themselves of personal guilt by simple and immediate acts of penitence' (Lowry 1971: 354), whilst Anna Bramwell has it that ecologism 'is itself a new quasi-religion that has values, a creed, a way of life and a priesthood' (1994: 16). Dixy Lee Ray, wishing to pour scorn upon the notorious 'environmentalist extremist', Al Gore, former vice-president of the United States, entitles her paper 'The Gospel According to Gore' (1993: 188). And Richard North writes: 'ecologism is a faith, and it provides us with many reminders of a medieval and post-medieval world, in which faith mattered. One could see Greenpeacers as crusaders, with the industrialist cast as the infidel' (North 1995a: 40), whilst 'antilogging campaigners feel no need to compromise in their defence of the spotted owl. Indeed, they cannot, because they are activists of the green religion of suburban America, in which the purity of wilderness is sacred' (North 1995a: 41; see also 1995b).

Can environmentalism legitimately be termed a religion? William Alston lists nine essential characteristics of a religion: belief in supernatural beings; a distinction between sacred and profane objects; ritual acts focused upon sacred objects; a god-sanctioned moral code; feelings of awe, mystery, guilt; adoration in the presence of sacred objects and during rituals; prayer and other forms of communication with the supernatural; a worldview which includes a notion of where the individual fits; and a cohesive social group of the like-minded (Alston 1967: 141–42). Some people in the environment movement score positively on several, perhaps most, of these indices; others hardly at all.

How this question is resolved, however, is not really the point. What is significant is that the claim that environmentalism is a new religion is made with a rhetorical passion that scarcely obscures the absence of hard reasoning behind it. The charge is never analytically defended; indeed, it is invariably employed as smear and insult. It is, then, *denigrative* to charge as 'religious' the basis of an issue commitment. And it is assumed by those making the charge that most witnesses to it will concur in that assessment. Though Oelschlaeger argues that 'the legitimating narrative' of the west is 'a biblical culture', which 'conditions the basic narrative that guides the process of discussion' (1994: 63), I would contend that, as religious commitment seems to be so negatively valued in public discourse, there is no point — nothing to be gained — from seeking to base a movement for an ecologically responsible society in religion.

There is, not surprisingly, much ambivalence over the promotion of religious and spiritual perspectives within the green movement. Many would like to see such developments disappear. We have seen that there have been attempts to downplay Mother Earth spirituality within ecofeminism. Similarly, Robert Goodin's formulation of a green praxis in which effort is concentrated upon foundational values explicitly rejects religious or spiritual orientations. These 'crazier views', he writes, 'are not connected to the deeper logic underlying the distinctiveness of green political theory'. They are 'green heresies' and cost the green movement 'invaluable allies' (Goodin 1992: 17). Given that theological questions 'are contentious matters, on which opinion is deeply divided,' it follows, argues Goodin, that 'linking the green case to spiritual values in this way thus seems to borrow an awful lot of trouble — and unnecessarily so as it happens' (1992: 40). It is 'unnecessarily so' because there are impeccable rational/secular justifications for green imperatives. John Dryzek is another who insists on the need to build the green movement upon a sound basis in rationality. He acknowledges the ecological function of myth and tabu in traditional societies, but maintains that there is no going back:

> contemporary societies have, quite simply, lost their innocence. For better or for worse, selfconsciousness about individual and collective choice is the norm. To return to the unselfconsciousness of traditional or preliterate societies is simply impossible ... [Furthermore] pre-literate societies may possess excellent feedback devices in terms of instantaneous reaction damped by tradition, but they lack the resilience necessary to cope with any severe disequilibrium conditions in the relationship between human and natural systems ... Tradition and myth may have been functional for the survival of preliterate anarchies ... But myth, tradition, religion, and charisma clearly destroy the 'openness' required for resilience (Dryzek 1987: 220).

Even Jonathon Porritt, having extolled the virtues of a new Gaian spirituality, suddenly veers away: 'I hope this doesn't sound too mystical, for we need to reassert the unity of humankind and nature without necessarily relying on quasi-religious concepts' (Porritt 1984: 208). And Morris Berman, in his critique of science's obsession with empiricism and rationality to the exclusion of 'enchantment', makes it clear that he does not advocate a return to 'mystical or occult philosophies', for these 'wind up dispensing with thought altogether' (1981: 194).

The most damning criticism in this vein has come from the eco-anarchist, Murray Bookchin. Bookchin sees recourse to spirituality as craven, for it sanctions avoidance of the tough political activity demanded of a green commitment. 'The real cancer that afflicts the planet is capitalism and hierarchy', he writes: 'I don't think we can count on prayers, rituals, and good vibes to remove this cancer; I think we have to fight it *actively* and with *all the power we have*' (1988: 7). Elsewhere he writes that green spirituality is 'flaky'. His own 'social ecology', by contrast, 'does not fall back on incantations, sutras, flow diagrams or spiritual vagaries. It is avowedly *rational.* It does not try to regale metaphorical forms of spiritual mechanism and crude biologism with Taoist, Buddhist, Christian, or shamanistic "Eco-la-la"' (1987a: 19).

Some, such as Porritt and Winner, find this distressing, seeing it as a manifestation of the deep-seated, politically damaging, and apparently irreconcilable conflict between those who have come to a green commitment from the old hard-nosed Left, and those coming from a 'New Age' perspective, wherein pre- or non-rational modes of understanding are favoured. Porritt and Winner's sympathies are with the latter, though they concede this to be a vexed question, and one likely to increasingly strain the green movement's ecumenical tolerance (1988: 240). They are almost certainly right in that.

5

GREEN CRITIQUES OF SCIENCE AND KNOWLEDGE

THE GREEN CRITIQUE OF MECHANISTIC SCIENCE

We have seen that much environmentalist attraction to religion and spirituality stems from a reaction to the social role of modern science. This can take the form of a complete rejection of the central claims and principles of science (as is the case, for instance, with neo-pagan spirituality) or it can take the form of a project aimed at ending the centuries-long divorce between science and religion (as is the case with process theology).

It comes as no surprise, then, to find that there is, within the environment movement, a full range of assessments of science as a way of knowing the world, and of its capacity to generate resolutions to chronic ecological and social crisis. At one extreme are those people, prominent within 'problem-solving' government bureaucracies and academic departments of environmental science and sub-departments of environmental economics, for whom the privileged status of science is not problematic. For such people environmental problems are technical in nature and to be resolved by appropriate applications of environmental science — and they *will* be resolved, for there is no holding back the boundless problem-solving might of science. Passmore sums up this position well: 'western science is still fecund, still capable of contributing to the solution of the problems which beset human beings, even when they are problems of the scientist's own making' (1974: 177). Others see no problem with science in its essence, but are critical of the priorities it has set itself. 'Many people', notes John Young, 'are concerned not about science itself, but merely the direction of much of modern scientific enquiry' (1991: 77). For these people the priorities of science are set for it — they are set within the power structure of science. This power structure need not only include scientists, and it works to discriminate against certain scientific foci, perhaps particularly the environmental focus of scientists who make nuisances of themselves by drawing attention to the ecological consequences of other scientists' science (on this, see B. Martin 1981).

Still others retain an attachment to the idea and ideals of science, but urge a reconstituted science, one based upon principles that are radically different from the reductionism and instrumentalism of the dominant tradition. This position is characteristic of many ecophilosophers. 'At an elevation still to be attained', writes Holmes Rolston, 'science can help provide a clearer vision', though such a science stands in need of 'a revised environmental ethic' (1990: 71–72). Warwick Fox also argues that science and environmental ethics 'should be — and need to be — allies' (1994: 212). Noting that science has become fixated upon its 'technical agenda', Fox pleads for a re-orientation of scientific priorities towards what he calls science's 'interpretative agenda', its social responsibility to provide public education on matters of scientific complexity, so that we ordinary people can make sense of 'the complex issues that confront us'. This information, furthermore, needs to be ecologically relevant above all else, and to be couched within a worldview that provides science with a visionary underpinning appropriate to a less-technologically exuberant time (1995b: 114). Fritjof Capra, too, noting that 'most of what scientists are doing today is not life-furthering and life-preserving but life-destroying' (1988: 151), urges scientists to 'adopt a holistic framework' that 'is in agreement with the most advanced scientific theories of physical reality' (1993: 71). Capra's exhortation has support in other quarters, too. Physicist Paul Davies's unhappiness with the anti-science stance of large sections of the environment movement is well-known. Less well-known is that, in Davies's view, the accepted precepts of theoretical science are already pretty much in line with the worldview of the environment movement:

> the images which are emerging from these new developments are very close to what we might think of as an ecological view of the world. They emphasise things like information, organisation, complexity. They are really more concerned with what we might think of as the cosmic network rather than the cosmic clockwork, they link human beings in, not as overlords of the universe, but as an integral and important part of the outworking of the processes of nature ... I think that the new physics and the new biology and the new way of looking at the world will make it possible for human beings to have a greater sense of dignity, to feel that they have a place in nature, to be integrated within nature. The new developments rest easily with the ecological view of the world in which human beings are fully integrated into the processes of nature instead of standing apart looking out on nature, or manipulating it (Davies 1992a: 29–30).

For Davies the shifts sought by Rolston and Fox are already in place. We will return to the insights of the 'new physics'.

And finally, there is a segment of the environment movement, a not-inconsiderable segment, that sees science and the hubris of scientists as *primarily* responsible for advanced ecological degradation. Within the sociological literature on environmentalism much is made of a starkly contrasted 'environmental values set' and a 'dominant values set'. These tabulations almost always characterise the 'ecological paradigm' as highly sceptical of the scientific method and the central claims of science. In Cotgrove and Duff's table, for instance, the dominant social paradigm's 'confidence in science and technology', 'rationality of means', and 'separation of fact/value, thought/feeling' is set against the environmental paradigm's 'limits to science', 'rationality of ends', and 'integration of fact/value, thought/feeling' (1980: 341). It can be argued that too much complexity and nuance is overlooked in these tabulations, but it remains true that for many environmentalists the scientific project has been sullied beyond redemption. Given the primary complicity of science in environmental crisis, it is foolish, the argument runs, to look to science for the answers. To take that path is to opt for a never-ending sequence of technological fixes, each striving to resolve the unforeseen damage wrought by the one before.

Despite the variety in green evaluations of science, most standpoints rest upon a common historical account of how science came to be what it is. Critiques of dominant scientific assumptions point the finger at Francis Bacon and Rene Descartes, though it is clear that the elevation of our species to the point where it is thought not to be subject to nature's laws can be traced as far back as the Ancient Greeks.

We have noted the radical separation of humans from nature in the realm of ethics, and this is a bifurcation that also has important scientific implications, for the distinction between subject (human) and object (non-human) provides the ethical justification for experimentation on other life, upon which our corpus of scientific knowledge has been based.

Belief in the right of our species to manipulate other life through scientific experiment was firmly in place by the time of the Renaissance, but there remained a certain uneasiness on the matter up to the seventeenth century, when the triumph of scientific rationality over biblical revelation was won. This incalculably important epistemological revolution paved the way for the explosion of technological advance and European mercantile expansion in the following century — the century of 'Enlightenment' — that in turn fed directly into the industrial revolution.

It fell to Francis Bacon to define the 'science-into-technology' project. He argued that the scientific pursuit of knowledge was not an end in itself, but the means whereby humankind could finally bring nature to heel, establishing dominance over 'her' processes and re-ordering the world to suit the interests of just one of its species. Humankind's purpose, Bacon insisted in 1626 in *The New Atlantis*, is 'the Knowledge of Causes, and Secrett Motions of Things; and the Enlarging of the bounds of Humane Empire, to the Effecting of all Things possible' (1990: 34–35). Knowledge was to be sought because it was the means, 'expressed and applied in technology', whereby 'humans assume power over the material world' (A. Jones 1987: 236). For Bacon, this involved a complete break with past ways of knowing, with past ways of acquiring knowledge, and with past ways of assessing the status of knowledge. Morris Berman terms this 'a violent shift in perspective' (1981: 29), a massive switch from faith in the unerringness of thought to faith in the empirical massing of data.

Green critiques have most picked up on the enthusiastically violent character of humankind's 'interrogation' of nature. In Bacon's view, Alwyn Jones observes, knowledge had to be forcibly wrung from nature, for 'she' would not willingly give over 'her' secrets (1987: 236). As Berman notes, for the first time:

> knowledge of nature comes about under artificial conditions. Vex nature, disturb it, alter it, anything — but do not leave it alone. Then, and only then, will you know it. The elevation of technology to the level of a philosophy had its concrete embodiment in the concept of the experiment, an artifical situation in which nature's secrets are extracted, as it were, under duress (1981: 31).

The case for an interventionist science, based on the contrived experiment, is thus made most explicit by Bacon. Perhaps, though, the contribution of Descartes is of greater significance, because it was he who first made philosophically coherent the subject/object bifurcation that was to become an essential principle of both western philosophy and the scientific method. According to Capra:

> the birth of modern science was preceded by a development of philosophical thought which led to an extreme formulation of the spirit/matter dualism. This formulation appeared in the seventeenth century in the philosophy of Rene Descartes who based his view of nature on a fundamental division into two separate and independent realms; that of mind (*res cogitans*) and that of nature (*res extensa*). The 'Cartesian' division allowed scientists to treat matter as dead and completely separate from themselves, and to see the material world as a multitude of different objects assembled into a huge machine (1992: 27).

This had massive implications — for the morality of animal experimentation in particular. We have seen that the interventionist tradition in science was well and truly in place before Bacon and Descartes happened along, but that there were misgivings on the matter. This unease stemmed from the ambiguous moral status of the animal kingdom up until the seventeenth century. In one of his fine early papers John Rodman explores the historically concurrent traditions of *theriophilia* and *theriophobia*. Theriophilia? The view 'that (some) beasts and humans are closely related; are equal in some basic sense' (1974: 19). Its opposite, theriophobia, is 'the fear and hatred of beasts as wholly or predominantly irrational, insatiable, violent and vicious beings whom man strangely resembles when he is being wicked' (1974: 20; to 'wicked' we might add 'libidinous' and 'gluttonous'). Both outlooks, says Rodman, have a basis in the Aristotelian notion of the 'great chain of being'. This was characterised by gradings, or levels, of 'soul'. In Aristotle's view all living entities participate in the most basic form of soul, that concerned with the reproductive and nourishing processes that he called the 'nutritive' form of soul. Animals have, in addition, a capacity for appetite; for pleasure and pain. This is 'sensitive' soul. And humans have, too, the capacity for abstract thought — the 'rational' form of soul. Only the rational soul survives bodily death, 'but all forms of life (vegetable, bestial and human)' strive for a share of immortality 'by reproducing themselves'.

Unfortunately, writes Rodman, this gesture in the direction of a graded cosmic democracy occurs only within Aristotle's natural philosophy. When he discusses *human* institutions, as in the *Politics* and the *Ethics*, he adopts the position of 'regarding the world from the standpoint of the "good-for-man", so that "happiness" and "well-being" was restricted to man' (1974: 20):

> whereas in his writings on the overall universe of living beings Aristotle stressed continuities, parallels, differences of degree, the difficulty of drawing sharp lines, and the value of all beings as participants in the system of life, in his practical writings he took the part of man, sharpened the differences between man and beast [and] relied upon theriophobic patterns of metaphor (1974: 21).

Stemming from this confusion (and running alongside it through time) are many full-fledged theriophiliac — as well as theriophobic — writings. In antiquity the Pythagoreans and Plutarch were theriophiles. Later came St Francis of Assisi. In the late Renaissance is Montaigne, who suggested that beasts were 'closer to God and virtue' than were people; whilst, later still, David Hume would maintain that non-human creatures not only experienced the full range of 'human' emotions but also had a capacity for reason. As recently as the eighteenth century, European courts sat to try pigs for murder — and ordered their subsequent execution — on the assumption that such beasts had souls and the capacity for rational moral choice (Rodman 1974: 18).

But the dualism of Descartes represents the most enduring attempt to tip the scales in the struggle between theriophile and theriophobe in favour of the latter. Cartesian dualism was not simply of subject and object (that is, of mind and matter), but a basic principle of being. This dualism governs peoples' own natures, wherein the rational is divided from the bestial, and the external world, wherein humankind is fundamentally separate from the beasts. The subject/mind — and that alone — was really alive. Gone was the nutritive and sensitive soul. The familiar dictum of Descartes, 'cogito ergo sum' ('I think, therefore I am') brilliantly freed the rational mind from any need for, or dependence upon corporeality. The non-rational part of the human, the body, was now relegated to the status of inert matter — matter that was governed by eternal physical laws set in motion by God.

Similarly, because they did not partake in mind, or soul, beasts were also mere inert matter — matter organised, machine-like, to function in accordance with the laws of the universe. Moreover, as only the soul, the living part of existence, can experience pain — mere machines cannot do that — it follows that animals can feel no pain. Pain is experienced 'only by conscious beings capable of understanding their bodily sensations. Beasts might go through the external motions that were, in men, the symptoms of pain, but they did so without experiencing pain as a sensation in the mind' (Rodman 1974: 22; see also Sheldrake 1990: 40–41). The apparent screams of pain of which the more complex animals are capable, therefore, really have the status of the discordant noises that emanate from any faulty or damaged machine.

All this is to be found in Descartes' Fifth Discourse. Having 'desensitised' the natural environment, Descartes spells out its implications in the Sixth Discourse. Here, he rejects the notion that the pursuit of knowledge is its own end, joining hands with Bacon in the view that the point of knowledge is the control and exploitation of nature via science and technology. It is, in Descartes' own phrase, to make us 'masters and possessors of nature' (the import of these 'five dramatic words' is considered by Easlea 1973: 253–58; see also Rodman 1974; Sheldrake 1990: 37–40). For this to occur any lingering doubts about the morality of the experimental method had to be banished — and the theriophiliac tradition clearly established such doubts. Nature — and animal nature in particular — had thus to be shown to be devoid of powers of sensation; to be, in fact, nothing more than mere 'mechanism'. As Rodman points out, most philosophical discussion of Cartesian dualism has focused on mind/body separation; but the more significant application of this dualism is the sharp separation it makes of humankind from the non-human world. This 'was the real Cartesian revolution' (1974: 23).

The Sixth Discourse includes a justification of experimentation: this 'involves coercing, torturing, operating upon the body of Nature so as to transform it — unless Nature's body is an unfeeling, soulless mechanism, in which case torture is not torture'. Thus, 'the initial "progress" of modern science came covered with blood from the dissecting room. It required, for its continuation, the termination of "superstitions" inhibiting both the dissection of the human cadaver and experimentation upon dead and live beasts' (Rodman 1974: 23). Discussions of the relative contributions of Bacon and Descartes frequently portray them as strongly contrasting: with Bacon, the first great prophet of empiricism, urging the need to amass data, and Descartes urging the rigorous application of pure, methodical, mathematically based thinking as the path to knowledge (for example, M. Berman 1981: 33–34). But Descartes, as Rodman points out, was himself involved in experiments on animals, and even began to prepare 'a definitive treatise on animals ("the whole architecture of their structure, and the causes of their movement")' (1974: 23). Descartes' view of beasts as unfeeling machines was taken up with enthusiasm by the new physiologists. Rodman cites a disapproving contemporary description of a physiological seminary of the late seventeenth century:

> they administered beatings to dogs with perfect indifference, and made fun of those who pitied the creatures as if they had felt pain. They said that the animals were clocks; that the cries they emitted when struck were only the noise of a little spring which had been touched, but that the whole body was without feeling. They nailed poor animals upon boards by their forepaws to vivisect them and see the circulation of blood, which was a great subject of conversation (cited in Rodman 1974: 23).

Experimental science was, thus, 'founded on acts of violence against non-human nature'. The 'historical function of Cartesian philosophy' was to make this legitimate and clear the way for action 'uninhibited by intellectual doubt or moral guilt' (Rodman 1974: 23).

This, then, is the *historical* case against Cartesian science, a case advanced today primarily within the environmentalist paradigm. But such a critique does not need to be overtly historical. Extracted from this context, the critique takes the following terms: science has become the only acceptable form of knowledge in modern industrial society. But its approach, variously labelled as 'atomistic' or 'reductionist' (because it 'reduces' reality to an understanding of the smallest identifiable component parts), 'mechanistic' (because it sees all existence, including animated existence, as behaving in the fashion of automata, according to unvarying laws and principles), and 'instrumentalist' (because knowledge is not sought for its own sake, nor even in the interests of the thing under investigation, but for the ultimate good that knowledge might prove to be for humans), has led to a fragmented view of the world, with an understanding of aspects of reality being gained at the expense of the whole. This epistemology of fragmentation, the critique holds, lies at the heart of the ecological crisis confronting us today. Let us look at this case in more detail.

We have seen that the distancing of the thinking subject from an objective reality 'out there' is the framework within which western scientific thought has developed. 'It is a presupposition of science', writes Jones, 'that the mind, as an independent entity, is able to experience and grasp objective reality' (1987: 236). Corroboration of individual findings is made possible through the application of universally accepted procedures in controlled experiments — the 'scientific method'. In this way, the scientist seeks to understand the parts and elements of reality, assuming that an understanding of the parts provides all that essentially needs to be known about the whole, and that knowledge of the linkages between the parts is at a subordinate level to knowledge of the parts themselves. The whole is thus seen as less important than the parts, the assumption being that the essence of any phenomenon, no matter how complex, is to be found in an understanding of its constituents.

This approach, Jones notes, fragments reality 'into discrete elements' in order to make 'separate analysis of the parts possible. But by acting in this way there is a tendency to perceive reality as though it *is* fragmented, with interactions between the parts operating mechanistically rather than being dynamically integrated' (1987: 237) within synergies. The reductionist organisation of knowledge, it is generally conceded even by its staunchest critics, has its uses (for example, Capra 1982: 93). The problem with the reductionist paradigm lies in its not knowing its place; in its claim to the status of 'master' paradigm. But its explanation is partial, and this is 'compounded by the development of highly specialised disciplines in both the natural and social sciences' — disciplines that, although isolated from each other, have sought to explain the whole 'through theories specific' to their individual perspectives (1987: 237). This organisation of knowledge is designed to give detailed understanding of the part of reality on which any discipline is focused. Such fragmentation also leads to an 'instrumentalist worldview, in which means, based especially on criteria of efficiency and utility, become redefined as ends' (1987: 239). This is because ends are defined by Cartesian science as technical. Any other sort of end requires the intrusion of values, and such is deemed anathema to the acquisition of 'real' knowledge of the world.

For many environmentalists such a science can only lead to the insensitive and widespread exploitation of other life, none of which is accorded any form of ethical protection, nor even any intrinsic scientific importance.

In its *reductio ad absurdum* this applies to human beings as well; the most technocratically enchanted dreamers of progress envisage the heroic human project eventually dispensing with its imperfect animal mediary altogether. Teilhard de Chardin's (1959) vision of a 'noosphere' has the human body ceasing to exist, having been succeeded by pure disembodied souls and their technological contrivances, whilst the science-fiction writer, Arthur C. Clarke, 'appears to view the prospect of "the obsolescence of man" with undisguised relish' (Rodman 1975: 136) — our bit-part in history over once we have been made redundant by our own creations. For Clarke, 'the tool we have invented is our successor. Biological evolution has given way to a far more rapid process — technological evolution. To put it bluntly and brutally, the machine is going to take over' (Clarke 1962: 216); though we can take some comfort, for 'it will have been no mean thing for man to have served as a "bridge" (as Nietzsche once put it) to a species higher than man' (Rodman 1975: 136).

At a less rarefied level, the science of value-free instrumentalism can also lead — and has led — to pathologies of modern life that contrast starkly with the high-mindedness that is claimed for the scientific project. It is argued that science has so subverted democracy that only the formal democratic shell remains, for real knowledge has been rendered accessible only to a priesthood of the scientifically literate, whilst the products of its research provide the means for the few to hold political power against the many. And it is argued that science has allowed the larger part of its effort to be directed away from life-enhancing research to the production of weapons of almost inconceivable destructive power. And science, it is held, has so sold its soul to the devil of commerce that it focuses upon the product to the virtual exclusion of the consequence. The result of this is that the planet has been plunged into environmental crisis as the 'negative externalities' of applied scientific 'achievement' have reached and surpassed the assimilative thresholds of the earth's natural systems. And all this under the amoral guise of a 'scientific objectivity' which was thought to be 'politically neutral'. As John Young puts it, 'science, it is claimed, is value-free, objective; the use which society chooses to make of scientific discovery is its own business' (1991: 79). Phrases such as 'self-serving', 'mealy-mouthed', 'criminally irresponsible', even 'cowardly' might be urged as more apt descriptions for this stance, a stance that retains currency among scientific practitioners to this day. Young (1991: 77–95) gives it quite a going-over.

This, then, is the environmentalist critique of modern science. It is not a critique exclusive to the green movement; other points of critical vantage have provided similar analyses. It is also not a critique that leads to any consensus concerning alternatives. As we have seen, some critics wish simply for the same processes and assumptions of modern science to be redirected to more socially and environmentally benign ends. At the other extreme are those who would have us abandon the enterprise of western science entirely. Between these extremes stands the more common green response: that science needs to be reconstituted upon a different basis, drawing upon different core assumptions, and conceiving differently the fundamental nature of existence. It is to these critiques we now turn.

THE 'NEW PHYSICS' IN ENVIRONMENTAL THOUGHT

Most people assume *ecology* to be the science of the environment movement. But increasingly ecological thought is drawing upon developments in other sciences, especially the so-called 'new physics'. We have seen that process theologians such as John Cobb and Charles Birch are close to the 'new physics', their thought drawing upon Whitehead's process metaphysics. Whitehead rejected the scientific orthodoxy of a reality reducible to disembodied physical laws. Writing in the first decades of the twentieth century, he was an early populariser of, and speculator upon the significance of Einstein's theory of relativity and of the new insights of quantum physics, particularly the latter's revolutionary insistence upon universal unpredictability and uncertainty as primary physical categories. He rejected the aridity of mechanistic science, seeking in its stead a new science based upon process, creativity, and what Marshall has termed 'a vision of vital relatedness' (1992: 359). Nature, Whitehead claimed, was to be understood holistically, not atomistically, and humanity was lodged inextricably within nature, not radically apart from it.

The key developments upon which Whitehead so fruitfully drew took place in the first years of the twentieth century, when relativity theory and sub-atomic physics revolutionised ideas about the material basis of the universe, its principles of causation, and the nature of space-time. One of the most far-reaching insights was that matter is not all that there is to the universe. Newtonian physics, and its central idea of indestructible particles of matter moving in accordance with the universal mathematical laws of gravity, underwent substantial revision:

> the stuff of the universe isn't the hard lumps of bits and pieces that we had thought, and even if you consider things which are undeniably hard lumps, like concrete or rock or tables and chairs, when you look at these things at the atomic level they dissolve away into vibrating patterns of ghostly energy ... At the level of quantum physics, atomic physics, the nature of reality, the atomic realm, is very, very different in form from our daily experience, and that difference can best be encapsulated by two key words. One is 'indeterminism'. The behaviour of an atomic system is not something that we can predict from moment to moment. It's inherently unpredictable and uncertain, and hence indeterministic. And the other word that I think captures quite well what is going on is 'fuzzy'. Qualities or propertfies that seem to be well defined on the everyday scale, such as having a location in space, or having a definite movement, are smeared out at the atomic level ... So basically the idea that the whole universe of matter is just a gigantically elaborate machine is simply wrong (Davies 1992a: 24).

Quantum theory, then, demonstrated that sub-atomic units of matter do not behave with the mathematical predictability posited by

Newtonian physics. As Capra describes it, not only are particles of matter changeable and 'nothing like the solid objects of classical physics', but they move within 'vast regions of space':

> at the sub-atomic level, matter does not exist with certainty at definite places, but rather shows 'tendencies to exist', and atomic events do not occur with certainty at definite times and in definite ways, but rather show 'tendencies to occur'. In the formalism of quantum theory, these tendencies are expressed as probabilities, and are associated with mathematical quantities which take the form of waves (1992: 77–78).

Together with Einstein's theory of relativity, which seemed to demonstrate that space and time are not separate but exist on a single continuum, the effect of these developments was to move the emphasis from matter to *motion*; to sudden shifts 'from one "quantum state" to another', and to the significance of fluid relations that ultimately gather in all reality (Capra 1992: 82). In such fluidity, the greatest epistemological casualty is scientific *certainty*. The foremost principle of the new conception of reality is indeterminism, which is a genuine universal principle, not merely a gap in current human comprehension. 'The universe really is indeterministic', Paul Davies states, 'at its most basic level' (1992b: 31).

More recently, chaos theory and its notion of self-organising systems has taken the principles of unpredictability, uncertainty and random shift still further, applying them to systems of far greater complexity than the level of sub-atomic interaction, such that 'the notion of predictability or of a repeatable experiment, long taken to be the basis of scientific method, is now in extreme doubt' (Marshall 1992: 381). It is now maintained that chaos waits just below the surface gloss of order, ready to pull apart any apparent state of equilibrium. Yet, within this condition of flux, of random shift, a more profound and intangible order is still to be discerned. Chaos and order are, then, not incompatible, but coexistential within a larger whole; as Davies and Gribbin note, 'chaos evidently provides us with a bridge between the laws of physics and the laws of chance' (1991: 35; see also Prigogine 1994). The environment movement has its favourites among the theorists of the new physics, and one such is David Bohm. Bohm's notion of an 'implicate order' in which 'the totality of existence is enfolded within each region of space (and time)' so that 'wholeness permeates all' (1983: 172; see also 1994) also stresses the underlying structure of the larger whole against which the flux of chaotic self-organisation is played (Bohm and Peat 1989: 137–40).

Though the implications of these developments have taken long to penetrate the popular consciousness — and still longer to exert any major impact on the theory, practice and technological application of science in its other branches — the *potential* consequences are hard to understate. Their cumulative effect is to undermine reductionist science, for it can no longer be credibly held that knowledge resides in an understanding of the constituent parts of biophysical systems. The apparent support the new physics lends to green ways of seeing has been made explicit by a number of observers (for example, Birch and Cobb 1981; Capra 1983; 1992; Devall and Sessions 1985: 88–89). In essence, this lies in the reinforcement it provides for the non-reductionist paradigm also found within ecology, including the latter's stress upon the unforeseen system-wide consequences of technological interventions in biological systems.

But it is a legacy complicated by the different conceptions of 'ecology' held by those who constitute the green movement and who draw upon the extra-scientific conception of ecological holism that inspires the ethics and social thought of the green critique, and the smaller college of scientific practitioners of ecology. It is to the science of ecology and insights derived therefrom that we now turn.

ECOLOGY: THE 'SUBVERSIVE SCIENCE'?

How does the science of ecology fit the dominant scientific paradigm, as outlined above? It would seem not to fit very well at all. With its stress on the importance of relational dynamics, and the challenge thereby implied to the reductionism of the dominant scientific paradigm, ecology would seem to constitute quite a departure from standard understandings of the nature of scientific knowledge. And at this point of departure stands the green movement. The green movement is the 'ecology movement'; its action-shaping principles are held to be ecologically *sourced*. Bryan Norton (1991) maintains that the principles of ecological science — relatedness, dynamism, systematicity, creativity and differential fragility — constitute the point of shared agreement uniting an otherwise disparate movement. These are principles, moreover, of intrinsically radical social import. Ecology, Paul Sears maintained as early as 1964, is 'a subversive subject', whilst for Paul Shepard, 'the ideological status of ecology is that of a resistance movement' (1969: 9). An ecologically informed worldview is, thus, seen to prescribe ways of understanding and being in the world that are at odds with entrenched ways. 'Ecological awareness', Capra writes, 'recognizes the fundamental interdependence of all phenomena and the embeddedness of individuals and societies in the cyclical processes of nature' (1988: 145). Certain social principles are held to be consequent upon this insight.

A science which ignores ecology, basing itself rather upon *fragments* as the locus of real knowledge, is a science which is little concerned with ends, for the world of 'ends' is the world of subjectivity. Non-ecological science is concerned with means, and problem resolution is purely a *technical* endeavour. But 'a crucial issue emerges as to whose interests are served' by a science concerned with means, with fragmentation, and with merely technical ends. As the means-orientation of contemporary science/rationality 'effectively silences public debate over the ends or purposes of life', such a science presupposes, 'not just control over nature, but ever more effective control and domination of society itself' (Jones 1987: 239). Because a science of holism threatens a science of reductionism and fragmentation, it also threatens the embargo on ends-based thinking that is implicit in a science of fragmentation, and opens a line to radical possibility.

So runs one line of argument. But there is another, one which draws no radical *social* principles from ecology, for it draws no radical *scientific* principles from it. As is the way with discipline-channelled knowledge, ecology, despite its synthesising insights and methodologies, grew from the biologies, wherein it still tends in practice to be located. Though the term 'ecology' dates from the mid-1800s, it is only with the growth of environmental concern in the 1960s that it became a *socially* significant science. This is, perhaps not coincidentally, also the period which marks the beginning of widespread disillusionment with modern science and technology. Part of that disillusionment arises from the perception that science is deeply implicated in many current environmental problems. Yet, most academic ecologists have not seen themselves in an 'alternative science' role. They have instead seen themselves as mainstream scientists, pursuing knowledge parallel with, in pursuit of the same ultimate ends as, and complementing the work of other scientists. As Neil Evernden puts it:

> it is a source of irritation to some ecologists that their discipline, which they endeavour to make as scientific and objective as possible, has become linked in the public mind with a rag-tag collection of naturalists, poets, small-scale farmers, and birdwatchers who constitute a visible part of the environment movement. Their grievance has semantic roots, in that 'ecology' is now being used to connote something quite different from its academic namesake. But these ecologists are also dismayed at seeing their work lumped together with the pseudo-science ... of what the media calls 'the ecology movement'. The connection between these and ecology proper is, they would say, extremely tenuous (1985a: 5).

But some *do* argue that ecology embodies principles that set it apart from, and in opposition to mainstream science. 'Orthodox' science selects 'aspects of nature (simple, mechanical, repetitive) and those cognitive processes (counting, measuring) that are amenable to manipulation and control'. It 'systematically ignores' aspects that are not; yet, 'these are likely to be most important for achieving ecological understanding'. Exploitation is, thus, 'embodied in the sort of knowledge modern science yields'; it is knowledge selected for 'exploitative purposes' (Biggins 1978: 220–21). Concern about 'the mechanistic-reductionist character of modern science has been a recurring theme' in ecological writings (Biggins 1978: 222). In this way of thinking ecology contains the possibility of a different science around which environmentally benign technology and environmentally benign social relations can be constructed. Roszak argues this: 'its sensibility — wholistic, receptive, trustful, largely non-tampering, deeply grounded in aesthetic intuition — is a radical deviation from traditional science' (1972: 400). Ruether also lays stress upon the social and ethical engagement *inherent* within ecological science:

> ecology is the biological science of biotic communities that demonstrates the laws by which nature, unaided by humans, has generated and sustained life. In addition, its study also suggests guidelines for how humans must learn to live as a sustaining, rather than destructive, member of such biotic communities. Thus, unlike those modern physical and biological sciences that have claimed to be only descriptive, ecology suggests some restoration of the classical role of science as normative or as ethically prescriptive (1992: 47).

We will look at this ecology-into-ethics prescription shortly. First, though, the challenge ecology is said to present to the mainstream scientific tradition.

Realisation that the precepts of ecology are not entirely compatible with the orthodox scientific view of nature grew only slowly through the 1960s and 1970s. Many ecologists, anxious as they were to remain within the collegial shade of an approving scientific community, resisted this view, arguing that ecology was a discipline for 'integrating' holism and reductionism. David Biggins rejects the timidity of this comfort-first position, arguing that ecology is intrinsically holistic and that it cannot be made compatible with mechanistic-reductionist science (1978: 220–23; 1979: 56). He maintains, instead, that science must be *transformed within* a holistic framework, not merely integrated with it. 'The field of ecology is a most likely area in which such a transformation might be worked out', he says, because modern science's mechanistic strait-jacket provides inadequate *knowledge* of reality, and it is, as well, *ethically* inadequate (1979: 56). Ecology, says Biggins, has gone some way towards overcoming both deficiencies.

The literature of the green critique contains many similar assessments. Capra argues that scientists must not be 'reluctant to adopt a holistic framework, as they often are today, for fear of being unscientific' (Capra 1993: 71). Rolston sees science as consisting of 'two parts': an 'evolutionary ecoscience', which 'describes the way nature operates'; and a 'technological science', which involves interventions in nature to generate products for human use. These two sciences are at odds: 'the same science that, theoretically and descriptively, has revealed the extent of biological diversity has, practically and prescriptively, often pronounced nature valueless, except in so far as it can be used instrumentally as a human resource' (1990: 70). In Rolston's view, accelerated decline in biodiversity began 'when science-based models were exported to traditional societies', and these cultures were 'opened up for development'. Science, to be 'part of the solution' must be based upon values that are wider than the reflexes of human economic preferences (1990: 71). Unfortunately, the 'logic of home' that is inherent within ecology has been replaced, in western science, with a 'logic of resource use' that is drawn from economics. Birch and Cobb also urge the scientific mainstream to accept the logic of ecology:

> instead of looking for mechanistic explanations of phenomena and viewing structural and ecological explanations as provisional and unsatisfactory, scientists can clarify the conceptuality and methodology appropriate to a world which is far more complexly interconnected than mechanical models have suggested (1981: 84).

Whilst conceding a place for mechanistic thinking (1981: 89), they argue for a shift from Newtonian 'substance thinking' to the 'event thinking' of the new physics, which they draw on to inform their 'living systems' ecological model. This is a model of entities 'which are what they are because of the environment in which they are found. The objection to the mechanical model is that it leads to the study of living entities in abstraction from their environments' (1981: 94).

Thus, the ecological and dominant models of science promote quite different conceptions of nature. Biggins notes that the mechanistic model 'provides an attitude that is appropriate to the manipulation of nature' (1979: 56). Our earlier discussion of the Cartesian foundations of modern science demonstrates why this should be so. An ecological viewpoint, by contrast, sees inter-species relationships quite differently. An ecological model of nature is not one of hierarchy, with the strong exercising dominion over the weak, but one of interdependency — of biotic *community*. Ecology, then, suggests a softer, more co-operative ethics than Tennyson's characterisation (in 'In Memoriam') of 'Nature, red in tooth and claw' or the Darwinist principle of 'survival of the fittest' that has been used to promote a society based upon competitive individualism (Ruether 1992: 57). Bookchin has raised this view of a benign ecology to a central place within his ecologically inspired social anarchism. For Bookchin, the core anarchist principle of mutualism is drawn straight from ecology, which is an 'artful science', a 'form of poetry that ... interprets all interdependencies (social and psychological as well as natural) non-hierarchically'. If nature consists of non-hierarchical interdependencies, ecology 'knows no "king of beasts"; all life forms have their place in a biosphere that becomes more and more diversified in the course of biological evolution' (1980: 271; a stress upon the need for a science based upon the insights of *evolutionary* ecology is provided by W. Fox 1995b).

There are exceptions. Andrew Brennan, as we have seen, does not believe a holistic ethic can be derived from scientific ecology, for ecology is no less reductionist than any other science, and properly so. For Brennan the answer to the question, 'what kind of explanation does ecology offer as a science?', is that 'it offers just the same range of explanation found elsewhere in the sciences' (1988: 28). It will nevertheless be apparent why ecology is thought to be a science wherefrom certain social and ethical prescriptions inevitably flow. To observe thus is to make no particularly startling observation. All scientific paradigms bring with them subsidiary political and ethical interpretations. Darwinism, for example, informed the brutal social Darwinism of Herbert Spencer and is used today to underpin ideologies of the capitalist market. Yet, a great many practitioners of mechanistic science seem blind to this simple and readily verifiable fact; hence their absurd claim that science is, uniquely, value-free. Western civilisation's basic stance towards non-human nature stems from the *value* position of mechanistic science that untrammelled intervention in biophysical processes is appropriate and justifiable. On the other hand, as Biggins notes, an 'organic' model implicitly enjoins minimal or non-interference: 'the replacement of that attitude by the modern ones of manipulation and control was essential to the project to conquer nature. Thus scientific knowledge is laden with values and via such values it motivates certain lines of activity in the world' (1979: 56).

For its part, ecology also suggests 'not only how the world *is*, but how we *ought* to act in the world', and this is how it should be, says Biggins, because 'natural' knowledge is the soundest foundation for values: 'where else can we base values? How else would we know how we ought to act in the world except by knowing the world?' (1979: 56; for a view opposing such 'biological determinism', see Ross 1994: 1–20). These values are often said to inhere within the best known summation of the principles of ecology, Barry Commoner's 'four laws of ecology': 'everything is connected to everything else', 'everything must go somewhere', 'nature knows best', 'there is no such thing as a free lunch' (Commoner 1972: 33–46). Indeed, these 'laws' are mostly taken as precepts to guide action. Once granted the ecological insight that consequences of actions ramify throughout systems, Ruether argues, 'the knower must take responsibility for shaping reality in ways that can be benign or destructive' (1992: 39). Thus, the interrelationship of all things is a scientific insight imbued with ethical consequence: from the science of how nature works *as a system* an ethic naturally arises. As Birch and Cobb note: 'the human treatment of other living things has heretofore been directed in large part by assumption of utter discontinuity. If there is continuity ... an ethic is needed that takes into account the subjective as well as the objective reality of other living things' (1981: 140).

GAIA

Within the science of the green movement there is a theory of super-ecology; a theory revolutionary in its scientific consequences. This is the *Gaia hypothesis*, the proposal that life on earth co-ordinates, regulates, and self-corrects in such a way that it is maintained even through substantial alterations to the geological and chemical conditions that sustain it.

Its formulator, James Lovelock, argued that the only feasible explanation for the Earth's highly improbable atmosphere is that it is continuously fine-tuned, and that the manipulator is life itself. The atmosphere, in other words, is a dynamic extension of the biosphere. Here is a hypothesis which proposes the entire world as one vast self-regulating organism: 'the entire range of living matter on Earth, from whales to viruses, and from oaks to algae, could be regarded', writes Lovelock, 'as constituting a single living entity' (1979: 9; see also Lovelock 1988a: 35: 'the tradition that sees the Earth as a living organism' is a tradition 'to which I subscribe, and I believe it to have a firm scientific basis'). Moreover, this single living entity 'is capable of manipulating the Earth's atmosphere to suit its overall needs, and is endowed with faculties and powers far beyond those of its constituent parts' (1979: 9).

This is a theory that puzzles some people. Uneasy with the apparent anthropomorphising of a regulatory mechanism, they assume that it is intended as metaphor. However, as Stephen Yearley points out (1992: 145), the Gaia hypothesis really does propose that the earth is regulated — and regulating. It is also a theory that excites considerable controversy. Some commentators attempt to minimise controversy by insisting that it be treated purely as a scientific hypothesis, shorn of the green movement holism that is seen to discredit it. Thus, the microbiologist Lynn Margulis, who collaborated with Lovelock in the initial development of the hypothesis, has since disavowed Lovelock's insistence on seeing the earth as a living organism, for such a conceptualisation 'alienates precisely those scientists who should be working in a Gaian context'. Margulis prefers to characterise Gaia as 'an extremely complex system with identifiable regulatory properties which are very specific to the lower atmosphere' (1988: 50). Lovelock was initially at pains to distance himself from 'the many small fringe groups, mostly anarchist in flavour, who would hasten our doom by dismantling and destroying all technology' (1979: 123; see also Allaby 1989), though the period of dialogue he subsequently entered into with the environment movement, a period that coincided with a degree of ostracism from the scientific establishment, saw a softening of some of these positions in *The Ages of Gaia* (1988b), the sequel to his 1979 book.

Edward Goldsmith, the long-serving editor of the *Ecologist*, is an enthusiastic publicist for the Gaia hypothesis, believing that both Gaian science *and* the philosophical holism it has inspired are of unsurpassed importance. Goldsmith is one of those unimpressed by the argument that ecology is inherently subversive. He argues that ecology has come to 'conform closely to the paradigm of reductionistic and mechanistic science', to the extent that it has been 'perverted in the interests of making it acceptable to the scientific establishment and to the politicians and industrialists who sponsor it' (1988a: 65). He exempts the Gaia hypothesis from this stricture, however, arguing that Gaia renders the 'perverted' concept of ecology unsupportable, and that it can serve to rescue and re-found the discipline. But he also endorses the ethic of holism and the spiritual and political projects inspired by Gaia. In a 1988 paper he lists sixty-seven principles of a Gaian worldview, many of which affirm social rather than scientific principles (1988b). The readiness with which principles of social, ethical, and even spiritual import can be drawn from the Gaia hypothesis explains the enthusiasm with which it has been taken up by parts of the environment movement that are not scientifically credentialled, and may even be antagonistic to science.

Number 22 of Goldsmith's sixty-seven principles is 'Gaian processes are teleological' (1988b: 169). Noting that the idea of a teleological world is anathema to reductionist science, for which 'only man because of his "intelligence", his "consciousness" and his "reason" is seen to be capable of purposive behaviour' (1988b: 169), Goldsmith then explains why Gaia must be 'purposive':

> programming is not a random process. Who did the programming and why? If living things are endowed with instructions, it is because these instructions were developed over hundreds of years, along with all the other adaptive features of Gaia, as part of a teleological strategy for achieving Gaia's overriding goal of maximizing her stability (1988b: 169).

Lovelock himself has gone to great lengths to stress the *non*-teleological nature of Gaia, his conceptualisation of Gaia as a living, self-regulating entity notwithstanding. He has used computer models to demonstrate that the earth's regulatory processes do not require goal-directedness, but occur in a mechanical stimulus–response fashion (Lovelock 1988b: 42–63). In vain. Goldsmith's position has carried far more persuasive power than Lovelock's. Rupert Sheldrake, for example, maintains that the attribution of 'livingness' necessarily entails a teleological standpoint, for the 'organising principle with its own ends and purposes' that is at the heart of the Gaia hypothesis 'is itself a teleological attribution' (1990: 134). This is a position with which it is hard to take issue. As Sheldrake observes, it is not necessary to impute consciousness on the earth's part to impute the goal-directedness of a teleological process.

Sheldrake links the Gaia hypothesis to his own controversial theory of 'morphic resonance'. This theory argues against the existence of timeless biophysical laws. In their place, it posits mere habits which, through space-time resonance in the form of 'morphic fields', become engrained as 'laws' but are subject to variation since random happenings set in train modifying or contradictory morphic fields of their own. For Sheldrake, 'the purposive organizing field of Gaia can be thought of as her morphic field', an integrating force that ensures 'the integrity of the system, and enables it to regenerate after damage' (1990: 135). As with all organisms, says Sheldrake, 'Gaia builds up habits through repetitive patterns of activity. The more often those patterns are repeated, the greater the likelihood of them happening again' (1990: 136).

For Lovelock and Margulis here was a hypothesis to excite the curiosity of the scientific mind. What it became was the ideological basis of a new and sometimes counter-scientific vision, a new ethic, and a new spirituality for a dynamic social movement.

But it did suggest, too, a new 'way of knowing'; a new scientific epistemology. 'Gaia forces upon us a concern for the planet and its state of health and offers an alternative to our near obsessive concern with the state of humanity', writes Lovelock (1988a: 44), and David Abram has developed this into a claim for a new (or rather *re*newed, for he traces a long scientific pedigree for this stance) dialectical science, one based upon 'a sensuous participation between ourselves and the living world that encompasses us' (1988: 126). This is appropriate because the Gaian view of a world that is 'continually creating itself', rather than 'a machine — a fixed and finished object' that 'cannot respond to our attention' — forces upon us 'quite literally, *a return to the senses.* We become aware, once again, of our sensing bodies, and of the bodily world that surrounds us' (1988: 126–27). In a science that seeks communication with nature rather than control, experimentation 'would come to be recognised once again as a discipline or art of communication between the scientist and that which he or she studies' (1988: 127).

Some take a contrary view. Matthias Finger, for example, argues that Lovelock's insistence on seeing the scientific and technological activities of modern civilisation as an essential aspect of Gaia in its current phase of development precludes 'thinking critically about science and technology' (1988: 203), whilst the exclusion from relevance of other social factors perpetuates the disastrous culture–nature epistemological rift (1988: 208). Nevertheless, the epistemological principles Abram draws from the Gaia hypothesis seem largely congruent with the epistemological principles drawn from the ecofeminist critique of science; and Abram, in common with ecofeminist science, takes Nobel prize-winning geneticist, Barbara McClintock, as the exemplar of a resurrected tradition of co-participatory science (Abram 1988: 128).

ALTERNATIVE 'WAYS OF KNOWING': THE ECOFEMINIST CRITIQUE OF SCIENCE

A critique of science figures prominently within ecofeminism, and this critique takes place at different levels. At one level it is a critique of gender power relations and gender-biased research priorities within scientific institions, the institutional structure of science being seen as one of western society's most powerful and entrenched centres of patriarchy. In Bowling and Martin's summation:

> science is based on the professional creation and certification of knowledge which is tied to powerful interest groups, notably the state, corporations and the scientific profession itself. Patriarchy is based on male control of dominant social structures and the exclusion of women from positions of power through means such as direct discrimination, socialisation and the gender division of labour. Patriarchy within the scientific community is manifested through male control of elite positions and various exclusionary devices (1985: 308).

Most people at the post-technical level in science — though, significantly, not at the technical level — are men, and the concerns of science represent male concerns, perhaps most noticeably in the establishment of health research and biotechnology priorities, where the claim is persistently made that male-determined research funding systematically discriminates against the interests of women, and, beyond that, works to further perpetuate gender-based relations of dominance and subordination.

At this level the critique is concerned with the practice of science rather than its essence. But ecofeminism also provides a deeper critique. Here the argument holds that there is a scientifically relevant women's perspective that western science ignores or devalues in generating what are essentially masculine paradigms 'built on assumptions of competition and hierarchy' (Bowling and Martin 1985: 308). Ecofeminist writings occasionally instance scientists whose conception and conduct of science is exemplary. Barbara McClintock is foremost among these.

McClintock's scientific method was characterised by a deep regard on the part of the potential knower for the known, an approach which renders both knower and known 'subject' and casts aside the subject/object dualism that is central to the mechanistic paradigm. Eschewing scientific detachment, McClintock held that scientific investigation must be based upon 'a feeling for the organism' (Keller 1983: 198), a 'deep reverence for nature, a capacity for union with that which is known' (Keller 1983: 201), for 'good science cannot proceed without a deep emotional investment on the part of the scientist' (Keller 1983: 198). Such an approach significantly qualifies the conventional view of science as an exclusively rational process. In Keller's words, McClintock's science reflects:

> a different image of science from that of a purely rational enterprise ... We are familiar with the idea that a form of mysticism — a commitment to the unity of experience, the oneness of nature, the fundamental mystery underlying the laws of nature — plays an essential role in the process of scientific discovery. Einstein called it 'cosmic religiosity'. In turn, the experience of creative insight reinforces these commitments, fostering a sense of the limitations of the scientific method, and an appreciation of other ways of knowing ... In McClintock's mind, what we call the scientific method cannot by itself give us 'real understanding' (1983: 201).

It will be apparent how well such an engaged, 'empathetic' science fits ecofeminist precepts. A science that permits 'intimacy without the annihilation of difference' (Keller 1985: 164) chimes with the ecofeminist concern for a caring engagement with nature in a way that does not subsume individuality. Here is a science that 'allows us to see the profound kinship between us and nature', a return to the diminished scientific tradition in which 'the goal of science is not the power to manipulate but *empowerment*, the power to appreciate, the power to be humble' (Hallen 1992: 58–59). Hallen contrasts this with the ideological project of science set in place by Bacon's injunction to '"put nature on the rack and torture her" so that nature — the ultimate "other" — will reveal her secrets' (1992: 58). Much of the ecofeminist critique of mechanistic science proceeds from such historical comparison. Mies and Shiva see reductionist science as 'a specific projection of western man that originated during the much acclaimed Scientific Revolution' (1993: 22), and, as such, 'a source of violence against nature and women', dispossessing each of 'their full productivity, power and potential' (1993: 24). Such analysis points up the perceived incompatibility between the key status-of-nature and procedural assumptions of mechanistic science and the relational ethics of ecofeminism. It also dovetails with the ecofeminist critique of hierarchical dualism, the latter being seen to inhere within the structure of science. Indeed, the mechanistic paradigm not only leads directly to ecological catastrophe, but it also carries much responsibility for the creation of a structure of political and technological power that systematically manipulates, devalues and oppresses women. Mechanistic science is perhaps the *primary* embodiment of that women/nature identification that hierarchical dualism renders subordinate to the masculine/cultural.

The pathfinding ecofeminist history of science is that of Carolyn Merchant, whose *The Death of Nature* (1980) advanced a gendered interpretation of the Cartesian revolution and the close relationship between European mercantilism and the subjugation of women and nature. She notes the demise of organic cosmology and its displacement by the metaphor of the world-machine; she notes, too, the practical consequences of that change — the removal of moral constraints upon mining, for example (1980: 3–41). She examines the political factors behind the witch trials, noting the place of witch persecutions within a wider, gender-based struggle over medical paradigms (particularly concerning reproduction) that was probably *the* key site at which the proponents of mechanistic science staked their hegemonic claims (1980: 132–63). And she makes much of the crudely violent, sexual imagery employed by the early champions of the new science, particularly Bacon (1980: 168; on the metaphoric universe of the fathers of mechanistic thought, see Keller 1992: 56–72). The equation of nature and female is put in the context of the sixteenth century's prevailing categorisation of women as lewd, licentious and libidinous, and requiring strict regulation on that account. As with women, so with nature. Merchant describes Bacon's project thus:

> the new man of science must not think that the 'inquisition of nature is in any part interdicted or forbidden'. Nature must be 'bound into service' and made a 'slave', put 'in constraint' and 'molded' by the mechanical arts. The 'searchers and spies of nature' are to discover her plots and secrets. This method, so readily applicable when nature is denoted by the female gender, degraded and made possible the exploitation of the natural environment. As woman's womb had symbolically yielded to the forceps, so nature's womb harbored secrets that through technology could be wrested from her grasp (1980: 169).

The project of selling nature/woman into bondage is thus made most explicit by Bacon, who set in place a science that 'implicitly embodies the view of a male knower who manipulates, dominates and exploits the object of his knowledge' (Okruhlik 1992: 65). In contrast to the empathetic relationality of McClintock's science, the mechanistic paradigm is inherently masculine. It defines a new masculinity, one which heroically transcends all constraints, including the constraint of being woman-born, woman-nurtured:

> the characterization of both the scientific mind and its modes of access to knowledge as masculine is indeed significant. Masculine here connotes, as it often does, autonomy, separation, and distance. It connotes a radical rejection of any commingling of subject and object, which are, it now appears, quite consistently identified as male and female (Keller 1985: 79).

Psycho-sexual interpretation of this sort is prominent within the ecofeminist critique of science. We have seen that Brian Easlea maintains that economic factors alone cannot explain the savagery of the witch persecutions. There were, he argues, less tangible but equally potent motivations in play. And it is the same today. In *Fathering the Unthinkable* (1983) he argues that current science has little to do with the promotion or affirmation of life, but is fuelled by a 'hyper-aggressive' will to power, a masculine desire to circumvent the sense of exclusion from birth and creation processes. Hence the drive to conquer the earth, to shape it in accordance with the heroic male project of transcending all constraints, to do away with nature and the *natural* processes of birth and death. And for this the tool is science: 'a kind of surrogate sexual activity in which scientists would penetrate to the hidden secrets of an essentially female nature, thereby proving their mankind and virility without necessarily running the risk of attempting the same with real, live and perhaps far from passive women' (Easlea 1981: 89; see also Hallen 1992, Keller 1992: 39–55).

Ecofeminist science lays emphasis upon 'holism, harmony and complexity rather than reductionism, domination and linearity' (Rose 1986: 72). It conjoins, says Hallen, the stress to be found in both ecology and feminism on 'creative activity over inert matter', on 'relation over substance', and on 'objects as subjects over subjects as objects' (1992: 56). It brings to bear upon science the feminist stress upon lived experience. At these points ecofeminist science nudges the somewhat older epistemological tradition of phenomenology, and it is to that we now turn.

ALTERNATIVE 'WAYS OF KNOWING': GREEN PHENOMENOLOGY

'Science' is a mode of knowing: an epistemology. The green critique of science is, thus, a critique of the dominant western epistemology. Capra claims a key role for an epistemological dimension within green thought:

> in the Cartesian paradigm, scientific descriptions are believed to be objective, i.e. independent of the human observer and the process of knowing. The new paradigm implies that epistemology — understanding the process of knowing — has to be included explicitly in the description of natural phenomena (1997: 39).

Though western environmental thought strongly critiques the epistemological framework that it deems inadequate or wrong-headed, the development of an epistemology appropriate to the environment movement's knowledge aspirations has been less in evidence. Claims have been made, however, on behalf of a dissident epistemology, one which, it is held, can serve as a basis for the new knowing that environmentally benign modes of living will mandate. This is *phenomenology*.

In philosophy, anthropology, environmental psychology and environmental geography, phenomenological explications have long been commonplace, and through the late 1970s and the 1980s several writers began to promote phenomonology as epistemologically appropriate for green scholarship (for example, Marietta 1982; Relph 1976; Seamon 1982; 1984a) — and in sharp contrast to the established operating norms of science.

I want to come at this by consideration of what phenomenology is not. Whilst phenomenology is a dissident, or alternative epistemology, the dominant mode of investigation in all the social and physical sciences — even in the disciplines and sub-disciplines noted above as having accorded credibility to the phenomenological tradition — is *positivism*. Positivism is the philosophical and pedagogical stance which argues that the only knowledge that can be held to be 'true' or 'valid' or 'worthwhile' is obtained empirically. It is associated with such terms as 'objectivity', 'quantification', 'prediction', and 'repeatability' (Seamon 1982: 120). At its extreme it argues that no statement is meaningful unless it can be verified in concrete terms. The positivist dominance is the natural concomitant of the Cartesian revolution, and its pre-eminence has been contested over the years by the minority phenomenological tradition (as well as by Marxist structuralism and various postmodernisms).

Phenomenology is the study and description of phenomena in terms of their essential and particular qualities, as these reveal themselves through authentic human experiencing. It approaches reality through qualitative data assembled and described via *experiential* modes of explanation: as opposed to the stress positivism places on quantitative information, amassed and assessed in accordance with pre-existing laws and theories.

It is not hard to see why phenomenology is regarded with suspicion within mainstream academia. When one argues a need to separate oneself from all preconceptions, common sense understandings, scientific laws and blueprint-ish theories in order to experience a thing in its essence, to get inside it, to *see it as it would see itself*, one is arguing for an apparent subjectivity that is anathema to most research paradigms. 'Potentially', writes David Seamon, 'there are as many phenomenologies as there are things, events and experiences in the world' (1982: 121). Phenomenological disclosure, then, remains open, recognising that understanding is of the moment, a merely temporary gestalt. No articulation can entirely enclose an essence; no explanation can fully and immutably exhaust meaning.

Against the charge that this can only result in the end of systematised knowledge, the phenomenologist will set his/her counter-charge against positivism. The application of theory to the explanation of substance, says the phenomenologist, is less an explanation of the phenomenon that theory purports to explain than an exercise in 'forcing real-world processes and events into a set of imposed and arbitrary cerebral constructs' (Seamon 1982: 121), and in fact collapsing the distinction between reality itself and those rigidly maintained 'cerebral constructs'. In the words of phenomenology's foremost proponent, Maurice Merleau-Ponty:

> all my knowledge of the world, even my scientific knowledge, is gained from my own particular point of view, or from some experience of the world without which the symbols of science would be meaningless. The whole universe of science is built upon the world as directly experienced, and if we want to subject science itself to rigorous scrutiny and arrive at

> a precise assessment of its meaning and scope, we must begin by reawakening the basic experience of the world of which science is the second-order expression (1962: viii; on Merleau-Ponty's relevance to ecological thought, see Abram 1995).

The intellect is not neutral then, but manipulative; and conventional methods, concepts and theories do not so much discover as conjure their effects. By contrast, the phenomenologist avoids theoretical laws and abstract concepts in organising the content of whatever is being studied. The student approaches the thing-to-be-known via 'empathetic looking and seeing', and attempting accurate *qualitative* descriptions. This does not mean that phenomenological explanation must remain at the level of the particular. It does seek synthesising insights:

> although phenomenology begins with descriptive accounts of concrete things, events and experiences, these specific idiosyncratic descriptions are in the end secondary. Rather, phenomenology is primarily eidetic, i.e. a major goal is to seek out within the uniqueness of concrete phenomena more general experiential structures, patterns and essences ... Generalisation is an aim of both positivist and phenomenological research, but the difference is that phenomenological generalization ... does not establish a guiding theoretical framework beforehand, but rather works to allow general patterns to appear in their own time and fashion through the various specific individual instances (Seamon 1982: 121–22).

Positivist and phenomenological approaches are also at odds over the nature of the relationship between subject (researcher) and object ('thing' researched). Seamon sets out the crucial differences. In both theory and methodology, positivist research establishes an apartness 'by which the researcher supposedly sees the thing impartially and objectively'; that is, as something untouched by the researcher's subjectivity. By contrast, 'the phenomenological approach seeks to maintain ties of meaning between researcher and phenomenon'. Generalisation must arise from the 'specific descriptive accounts, not beforehand', and to this end the personal idiosyncracies of the researcher constitute 'an asset rather than a liability'. This is because different people are sensitive to different aspects of a phenomenon; thus each uniquely individual perspective adds to, and enriches understanding (1982: 122). Against the remote and disengaged gaze of positivist science, the phenomenologist sets an approach characterised by wonder and respect for the inviolate otherness of the other. A pronounced expression of this attitude — though perhaps excessively epiphanous — is supplied by David Abram, writing in an overtly phenomenological context, of an encounter with a bison in a Javanese nature reserve:

> I found myself caught in a non-verbal conversation with this Other, a gestural duet with which my conscious awareness had very little to do. It was as if my body in its actions was suddenly being motivated by a wis-

> dom older than my thinking mind, as though it was held and moved by a logos, deeper than words, spoken by the Other's body, the trees, and the stony ground on which we stood (1996: 21).

We have seen that the subject–object dualism of the Cartesian/positivist tradition has particular relevance for the human–environment relationship. Phenomenology rejects the dichotomised person–world conceptualisations of positivism, arguing for a seamless unity between person and world. This 'immersion-in-world' is a prime focus of phenomenological investigation. It makes for an emphasis on holism that contrasts with a reductionist stance; it is here, in its 'in-worldliness' and the epistemological alternative it presents to body/mind, subject/object dualism, that its attraction to many environment movement theorists is to be found. Phenomenology, Evernden writes, 'implies that it is misleading to speak of an isolated self surveying a world, for the person is from the start *in* the world, and consciousness is *of* the world' (1985a: 59). The search for a green epistemology is in the first instance a search for an alternative to the dualisms of mainstream science and the destruction that is consequent upon them; this is just what phenomenology supplies. Rejecting as arrogant humankind's claim to a monopoly upon subjectivity, phenomenology acknowledges subjectivity at large in the world. It thereby collapses the most potent of the Cartesian dualisms, and in so doing lines up with the central epistemological project of ecological thought.

SOME PROBLEMS: OLD OR NEW SCIENCE?

We have seen that the green critique encompasses virtually all possible variations of attitude towards science — from uncritical exuberance to emphatic rejection with, in between, various visions of a reconstituted science appropriate to green ways of seeing. Each position is itself a critique of other positions within the environment movement; nevertheless, some recurring claims have slipped through our net, and should be gathered up here.

Defenders of science often express bitterness about what they see as a serious double standard on the part of anti-science greens: that those who condemn science nevertheless accept and make political capital from scientific findings when they happen to suit their own purposes (Paul Davies refers to this as 'New Age plunder'; 1992a: 29). Environmental problems, furthermore (defenders of scientific orthodoxy point out), are *identified* by science, and such identification is extremely important given the insidious nature of many global pathologies. Thus, John O'Neill argues (1993: 145) that many environmental problems would never be debated were it not for science setting out the terms in which they can be construed as problems: global warming and the extent of biodiversity decline are good cases in point. Science is, therefore, necessary. It is, moreover, reliable, at least up to the point that expectations of it are realistic. It is not *sufficient* because its pronouncements will never be so complete that ethical and political judgements are rendered unnecessary (O'Neill 1993: 145–48). But nothing 'new' is needed; science does not need to be differently constituted, into 'holistic' or feminist science, say, or to be 're-moralised'.

Social and ethical claims that are drawn from the new physics have attracted specific criticisms. A movement that seeks to retrieve corporeality/embodiment/grounding from disembodied abstraction cannot be wholly comfortable with the metaphysical speculation of the new physics. So it is with Devall and Sessions, for whom the computer-based modellings on which much theoretical physics and scientific ecology (including the Gaia hypothesis) is based 'distort the living reality'. This is because 'the *sensuousness* of the natural world is left out of their formal, computerized or mathematical abstractions' (1985: 89). It can also be argued that the micro focus of the new physics cushions its practitioners against real world implications. Thus, David Abram argues:

> to participate mystically with subatomic quanta or even with the ultimate origin of the universe does not really force science to alter its assumptions regarding the inert or mechanical nature of the *everyday* sensible world, and so does not really threaten our right to dominate and manipulate the immediate world around us. Biologists, who study this very world — the world that we can directly perceive, often with our unaided senses — are in a much more precarious position politically (1988: 128–29).

On the other hand — though still not to the good of the new physics — it can be argued that, instead of avoiding the real-world consequences of their ideas, scientists and commentators on science tend to *over*state the extent to which science determines social reality. Scientific paradigms, the argument runs, have modest relevance in this regard: in fact, it is social imperatives that shape science. Such a view insists that the linear science-into-technology equation is misleading; that technological envisioning, and the funding decisions that ensue, in large part determine the science that we get. Even in the purest of theoretical enterprises scientists are as likely to borrow from social models, or from idiosyncratic interpretations of the social world around them, as society is to mould its social ideas in accordance with scientific models. Ted Benton puts this position well. 'Scientific thinking', he maintains, 'is dependent upon the availability, within the wider culture of the scientists, of social and cultural practices which can serve as sources for metaphorical thinking', and so 'scientific knowledge will be to some extent at least shaped by its specific and possibly localized cultural context, and by the constellation of social, political and economic interests which fund research and govern its institutional form'

(1994: 35). Brian Martin agrees: 'looking for inspiration from modern physics can be a deceptive process', for 'what is found in these quests may simply be an exotic version, a distorted reflection, of our familiar, banal, everyday experiences' (1993: 39).

If society chooses its science, rather than being itself chosen by science, then it is likely that a science seen to throw up unpalatable social models (and let us concede, for the purposes of the argument, that the new physics does that) will be marginalised for political reasons. Capital, bureaucracy, and the knowledge-industry infrastructure like the science we now have. This is what has made possible and sanctioned advanced industrial civilisation, the context wherein capitalists, bureaucrats and knowledge-industry gatekeepers have attained power and prestige. Such centres of strategic power have their own potent stopping mechanisms, and can be expected to resist the progress of new models of science if and when certain unacceptable implications deemed to inhere within them become broadly apparent.

It might be argued that this has already happened, in part at least, with ecology. Does ecology provide a framework which can require a fundamental revision of existing patterns of scientific thought? For many, perhaps most, green-leaning people it clearly does. But the environment movement's faith in ecology also has its in-the-movement detractors.

Thus, Neil Evernden — and he is writing from a *radical* position within environmental thought — is critical of the 'wishful ecology' that is intended to 'help us "feel" our way into a healthier relationship with the world', a version of ecology that would not be recognised 'as any kin to academic ecology' (1985b: 16). The environment movement's 'wishful ecology' selectively emphasises such ecological virtues as stability, diversity and health, whilst ignoring competition, exclusion and survival, even though the latter concepts are every bit as vital to a proper understanding of ecology as the former set. We choose, in other words, to ignore 'the "dark" side of ecology' (1985b: 16).

Evernden also stresses the *social* ambivalence of ecological science. It is more frequently used to support the *status quo* than the cause of social reform, for it can readily be employed as 'an informant of technological and bureaucratic intervention' in order to 'facilitate continued growth with a minimum of environmental backlash ... Indeed, one could cynically conclude that the role of the ecologist is to identify niches for humans to expropriate' (1985b: 16). We have already noted that many professional ecologists resent the political uses to which 'ecology' has been put, and they also resist defining 'ecology' in opposition to the mechanistic paradigm. In Evernden's view such people are not evincing intellectual or moral timidity; their concept of ecology is at least as valid as the more radical interpretation upon which the green movement draws. 'Contrasting viewpoints' can legitimately be drawn from ecology then:

> ecology cannot be presumed to be the exclusive ally of the environment movement, for it provides information that could just as well be used to manipulate nature as defend it. Nor is there any reason to assume that the findings of ecology provide credibility for environmentalists' social prescriptions. They could just as easily suggest a refined continuance of the status quo (1985b: 17).

An additional problem with the project of extracting ecological social ethics from ecological science comes from formal logic. This is the 'is/ought' dichotomy, first formulated by the eighteenth century philosopher, David Hume, and later recast as the 'naturalistic fallacy', which holds that no prescriptions as to what *ought* to be can be logically derived from imputing any sort of status to what *is*. Radical environmental philosophers saw this as a problem early in the piece, though they tended not to bend their efforts to disputing the fallacy itself. In large part they tried instead to demonstrate that environmentalist ethical formulations — and particularly Leopold's pathfinding derivation of the 'land ethic' from ecological science — do not fall foul of the is/ought impasse (Biggins 1979; Callicott 1982; Marietta 1979; Rolston 1975).

This debate was largely a product of the late 1970s and early 1980s. But it has since been revisited by Warwick Fox, who, unlike the earlier discussants, does dispute the naturalistic fallacy on its own terms. He argues that 'every formal system of reasoning' necessarily proceeds from certain assumptions ('axioms') that are held to be so self-evident that they are 'proofless'; that is, they cannot themselves 'be proved or disproved within the system to which they attach'. For intrinsic value theorists the notion that moral considerability must inhere in any entity that has 'a good of its own' is assumed — taken as axiomatic. It is not derived by logical inference and, hence, falls outside the strictures of the 'naturalistic fallacy' (1990: 193; for a refinement of Fox's position, see 1994: 211).

However this problem is ultimately resolved, it is clear that, for the present, social and ethical extrapolation from principles of ecological science is not a simple and uncomplicated process.

What of the Gaia hypothesis? Needless to say, it has been attacked as yet another manifestation of loopy New Age pseudo-science. Yearley, for instance, quotes a *Guardian* article in which Lovelock — despite his record as a supporter of, and participant in some of the more Promethean undertakings of current science — is dismissed as 'a scientist of sorts', whose ideas could 'appeal only to "scientifically illiterate greens"' (1992: 146). Paul Davies, certainly not a 'scientifically illiterate green', finds much of interest in Gaia (1992a: 27), on the other hand; so, the critical traffic is not all one way.

Within the environment movement uneasiness takes a different tack. Some of the misgivings about ecology and the new physics also apply to the Gaia hypothesis. And much has been made of the notion that, if life on earth has a self-correcting capacity that is so effective that it has succeeded in maintaining itself through aeons of geophysical change, there is no need to be concerned about the environmental 'crisis' of today. Gaia has coped with worse before, and we can be confident that 'she' will cope again. Lovelock fuelled the misgivings this interpretation has generated by dismissing the seriousness of some of the issues that engage environmentalist concern. CFC depletion of the ozone layer is of little import (1979: 115–16). Nuclear power is endorsed as 'clean' energy. Industrial pollution in the developed countries is of little concern: 'the evidence for accepting that industrial activities either at their present level or in the future may endanger the life of Gaia as a whole, is very weak indeed' (1979: 107–08). In fact, little that takes place in the temperate zones is of major importance. This is not because human activities are no cause for worry *per se*; it is because 'we have been led to look for trouble in the wrong places' (1979: 111). The 'real' environmental crises are the consequences of rapid destruction of tropical forests, the increase in carbon dioxide levels in the upper atmosphere, and ignorance of the crucial, system-wide importance of certain biophysical processes specific to wetlands, estuaries and the continental shelves (1979: 111–15; 118–21). We should thus 'take special care not to disturb too drastically those regions where planetary control may be sited' (1979: 114), and these 'core regions' are to be found between latitudes 45° north and south (1979: 120).

It is not germane to our present purposes to set out Lovelock's reasons for these assessments. He has, in any case, since modified the complacent acceptance of industrial pollution that was so noticeable in his 1979 book; he has also modified his position on the nuclear industry (1988b: 173–75), and on ozone layer depletion (1988b: 170). But his confidence in the capacity of industrial society to find technical solutions to its environmental problems, and his insistence on considering Gaian problems from a perspective that concerns itself with life in the abstract, rather than in terms of threats to its specific contemporary manifestations, continues to disturb many within the environment movement. So, too, does the ambiguity that surrounds the place of humankind within Gaia. We are told, for example, that humans are not possessors of the planet; that even the 'tenant' metaphor is inappropriate, for 'the stable state of our planet includes man as a part of, or partner in, a very democratic entity' (Lovelock 1979: 145). Less than two pages on, however, Lovelock enthuses over the question, 'do we as a species constitute a Gaian nervous system and a brain which can consciously anticipate environmental changes?' (1979: 147). An affirmative response to this question — and this is anticipated by Lovelock — assuredly requires us, and certainly entitles us, to intervene in the world at a level that, in terms of the relationships it involves with other forms of life, is more accurately described as 'despotic' (albeit benevolently so) than 'democratic'.

And so briefly to the ecofeminist critique of mechanistic science, and here I would make just one point. It can be argued that ecofeminism's insistence that women are not *naturally* closer to nature than are men, and its rejection of the woman/nature conjunction on the ground that this has been a historical site of women's oppression, provides a ground for rejecting as specifically feminist any concern with mechanistic science. Kathleen Okruhlik moves towards this position:

> the suggestion ... is that our concepts of rationality and objectivity are so deeply masculinist in their seventeenth-century origins that any truly feminist account of science must be premised upon their rejection. The dangers of this response, however, are very great. We surely do not wish to accept the patriarchal identification of women with nature, thereby supporting the inference that if the mechanical philosophy was bad for nature then it must have been bad for women too (1992: 65–66).

The inference I take from this passage is that, within feminism, the criterion against which 'the mechanical philosophy' is to be judged is solely its impact upon women, and if it can be demonstrated that it has been 'bad' for nature but *not* for women, then feminism has no cause to reject it. Later, this position is put with even less scope for ambiguity:

> one can see very well how ecological concerns might have been better served had nature continued to be viewed metaphorically as an organism rather than a machine. The situation, however, is not so clear with respect to women. We do not want to say that whatever is bad for nature is bad for women because that is to buy into the very identification of women and nature that feminist theory *ought* to be resisting (1992: 76).

The ecofeminist critique of science may not, on this evidence, command overwhelming support within wider feminist circles.

And so to phenomenology which may, on the face of it, not be compatible with *any* sort of scientific method. But there *is* a tradition of phenomenological science and it traces as far back as the eighteenth century poet-scientist, Johann von Goethe (1987). In Goethe's science sensual, non-analytical observation of nature, using touch and smell as well as sight, is privileged. There is a marked revival of interest in Goethean science (for example, Bortoft 1996; Brook 1998; Hoffman 1994; 1996; Seamon 1976; Seamon and Zajonc 1998; Steiner 1988), and it may yet attain a prominence within ecological thought which has hitherto eluded it.

Phenomenological epistemology has been much criticised, though not, as far as I know, by green-leaning writers attempting to minimise its influence within green thought. In any case, it would be a mistake to assume that the green movement has embraced, in wholesale fashion, phenomenological epistemology as an appropriate off-the-shelf alternative to the Cartesian scientific tradition. Ecology movement theorists are only intermittently preoccupied with questions of the status and basis of knowledge. Phenomenology has thus made an uneven impact within the environmental critique. But most people trying to interpret the nature of 'place', and to relate place to human experience, draw heavily upon phenomenology, and make explicit that debt. We will shortly turn to this often overlooked aspect of environmental thought.

Thomas Kuhn (1970) popularised the concept of 'paradigm' within epistemological discourse; the notion that scientific knowledge is organised within tightly consistent knowledge frameworks, which can be abruptly undermined when their key assumptions are successfully challenged by rival frameworks proceeding from different assumptions and first principles. There is evidence that we are now at a time when the dominant paradigm of reductionist-mechanistic science is threatened with just such a Kuhnian paradigm shift — and the green critique of science is inextricably caught up in this struggle for epistemological primacy. Both the new physics, with its stress on self-organising, spontaneous systems, and ecology, with its insistence on the primacy of relationality, are at least potential rivals to the mechanistic paradigm.

A further contender for the status of 'master scientific discourse' is the neo-Darwinism of Richard Dawkins — specifically his claim that the unit of evolutionary selection is the gene and not the organism, and certainly not the species (or the planet) (1976; 1986). This science disputes a first principle of the new physics: randomness and spontaneity but with an underlying order. Neo-Darwinist science insists that the notion of an underlying order is unsustainable; that, cosmologically speaking, there is no organising principle; that all is purposeless. And, in its insistence that it is the gene rather than the larger system that is the key determinant in the unfolding of life, it would also seem to sit uncomfortably with some of the paradigmatic assumptions of ecology. It is argued that, with its affirmation of self-interest, competition, greed, and perpetual disequilibrium, the 'new biology' undermines the central claims of ecology within science, and lends scientific support to the ideological claims of economic rationalism within social science (for example, Worster 1995; Zimmerman 1996). Nevertheless, the significance of these developments for the green project is not entirely clear. Some interpret the new biology as lending credence to environmentalist priorities in a way that was distinctly not the case with mechanistic science. The noted conservation biologist, Michael Soule, accepts the key assumptions of the new biology whilst remaining committed to the goals of environmental activism (1995). And the Dawkinsian, Daniel Dennett, mounts a spirited defence of biodiversity, concluding:

> is this Tree of Life a God one could worship? Pray to? Fear? Probably not. But it *did* make the ivy twine and the sky so blue, so perhaps the song I

> love tells a truth after all ... it is surely a being that is greater than anything any of us will ever conceive of in any detail worthy of its detail. Is something sacred? Yes, say I with Nietzsche. I could not pray to it, but I can stand in affirmation of its magnificence. This world is sacred (1995: 520).

The foundations of modern science may currently be under more rigorous challenge — from within science — than at any time in the past 250 years. But the point I would make here is that, despite misgivings within the environment movement about the various paradigmatic rivals, a case can be made that they all potentially provide environmentalist aspirations with a degree of authority that has been hard to win from mechanistic science. How to predict the future of the green dialogue with science? The situation is fluid, the picture confused, and much will depend upon the resolution of debates within science itself. But a less conflictual relationship between green aspirations and the authority of science is at least conceivable.

6

RECLAIMING PLACE: SEEKING AN AUTHENTIC GROUND FOR BEING

A LITERATURE OF PLACE

Poets, novelists and essayists have long made the idiosyncratic especialness of place a prominent literary theme, but the elevation of place-writing to 'genre' status is predominantly the achievement of a robust North American tradition of nature writing. Such writing — the writing of the *experiencers* of place — adopts an approach to the comprehension and interpretation of place that could serve as paradigmatic of the phenomenological method. In his modern masterpiece, *Arctic Dreams*, Barry Lopez writes:

> the land retains an identity of its own, still deeper and more subtle than we can know. Our obligation toward it then becomes simple: to approach with an uncalculating mind, with an attitude of regard. To try to sense the range and variety of its expression — its weather and colours and animals. To intend from the beginning to preserve some of the mystery within it as a kind of wisdom to be experienced, not questioned (1987: 228).

These literary constructions of place are, respecting their phenomenology, largely unselfconscious and a-theoretical; though Lopez does note that his writing 'focuses mostly on what logical positivists sweep aside' (1997: 24). Lopez best suits our purpose here, because he, almost uniquely, stands back and reflects upon what he is doing (Romand Coles writes of Lopez having 'a world that is *both inside and outside* of the conversation; 1993: 242). He does this in *Arctic Dreams*, and he does it, too, in his many short essays. There is even a certain tension between the Lopez who is immersed unselfconsciously in place and the Lopez who draws back to examine his project with a degree of detachment. In his essay 'The Stone Horse', he notes how a critical interrogation of an ancient horse of stones on the desert floor impeded his appreciation of it, the 'process of abstraction', a process unintentionally adopted, drawing him 'gradually away from the horse' (1986: 227).

This is a recurring theme in Lopez's writing. Seek to know place with all the senses, and beware too great an emphasis on the partial knowledge provided by interrogative processes. 'Put aside the bird book, the analytic frame of mind, any compulsion to identify and sit still', he writes, for 'the purpose of such attentiveness is to gain intimacy, to rid yourself of assumption' (1997: 25). In seeking to know a place, one must approach reverentially, with openness, without guile: 'the key, I think, is to become vulnerable to a place. If you open yourself up, you can build intimacy. Out of such intimacy may come a sense of belonging' (1997: 25). And intimacy is multi-dimensional, multi-sensual:

> where in this volume of space are you situated? The space behind you is as important as what you see before you. What lies beneath you is as relevant as what stands on the far horizon. Actively use your ears to imagine the acoustical hemisphere you occupy. How does birdsong ramify here? Through what kind of air is it moving? Concentrate on smells in the belief you *can* smell water and stone. Use your hands to get the heft and texture of a place — the tensile strength in a willow branch, the moisture in a pinch of soil, the different nap of leaves. Open a vertical line to the place by joining the color and form of the sky to what you see out across the ground. Look *away* from what you want to scrutinize in order to gain a sense of its scale and proportion ... Cultivate a sense of complexity, the sense that another landscape exists beyond the one you can subject to analysis (1997: 25).

The relevance of such a phenomenology of place to the wider concerns of ecological thought will be obvious. Empathy with place — conceived here primarily as *natural* place — enjoins a deep concern for the processes of life integral to, and defining and shaping the character of a given place. It conduces to environmentalism's stress upon living in accordance with ethical precepts: 'a specific and particular setting for human experience and endeavor is ... critical to the development of a sense of morality and human identity' (1997: 23). It also has a political edge that, as we will see, accords with the political and economic analysis of the environment movement:

> the real topic of nature writing, I think, is not nature but the evolving structure of communities from which nature has been removed, often as a consequence of modern economic development. It is writing concerned, further, with the biological and spiritual fate of those communities. It also assumes that the fate of humanity and nature are inseparable (1997: 23).

And it privileges the insights of indigenous peoples, peoples who have not developed the sophisticated philosophies of separation from nature that so characterise western thought:

> as a rule, indigenous people pay much closer attention to nuance in the physical world. They see more. And from a handful of evidence, thoroughly observed, they can deduce more. Second, their history in a place, a combination of tribal and personal history, is typically deep. This history creates a temporal dimension in what is otherwise only a spatial landscape. Third, indigenous people tend to occupy the same moral landscape as the land they sense. Their bonds with the earth are as much moral as biological (1997: 24–25).

In light of this, it is surprising that the *literature* of place has had so little impact upon mainstream ecological thought. The visceral meaning of much defence-of-wilderness direct action is hereby rendered explicable — as Chaloupka and Cawley note, 'each defense of each wilderness cites the virtues of that location' (1993: 13), insisting upon its specialness *as place*. But, among the poets and essayists of the nature/place tradition, only Gary Snyder also doubles as a widely read environment movement theorist. This is not, however, the only perspective from which phenomenological understandings of place have evolved.

THE COMMODIFICATION OF SPACE

The most prominent developments of phenomenological insights within environmentalism have occurred in studies of the dynamics of 'place' and 'space'. Not that all 'place and space' writing merits lodgement within the corpus of environmental thought. Much environmental thought proceeds in apparent ignorance of this theoretical tributary and, conversely, not all the contributions to the study of authentic place-making that neatly complement the values of the environment movement seem aware of the congruence. Prominent among the exceptions are David Seamon, a theorist of architecture, and Edward Relph, a geographer, each of whom overtly identifies with environmentalist values (for example, Seamon 1993: 17; Relph 1981: 161–64; 187–95; see also R.B. Hay 1988; 1992), and their work is taken as emblematic of this feeder stream below.

Beyond that, not all writing on the dynamics of place and space is housed within phenomenological investigation. A large proportion of such writings fall within Marxist geography, and focus more on the production and commodification of space via the processes of capitalism than they do upon place and perception. As much Marxist scholarship perceives itself to be in *epistemological* conflict with phenomenology, and as much Marxist activism perceives itself to be in *political* conflict with environmentalism, there is considerable hostility to phenomenological perspectives generally, and environmental critiques specifically, within much of this literature. Neil Smith's work (1990), aspects of which were considered earlier, exemplifies this position. Not all Marxist accounts evince hostility, however. Probably the best known work in this intellectual genre is David Harvey's *The Condition of Postmodernity* (1989). Harvey holds that 'there can be no politics of space independent of social relations. The latter give the former social content and meaning'. Moreover, 'the pulverization of space' is a means to 'facilitate the proliferation of capitalist social relations' (1989: 257; on the 'annihilation' of space, see also Urry 1995) — bad news for those seeking person–place relationships that are imbued with meaning.

Unlike most Marxist geographers, Harvey does not dismiss as unimportant the perceptual components of person–place relationships (see his 'grid of social practices'; 1989: 220–21), though from such a perspective a phenomenological approach to establishing the conditions for authentic place seems a futile and somewhat naive exercise. For their part, phenomenologists make little reference to Marxist analysis, though they sometimes develop critiques of the role of capital in the obliteration of special places that are not, in essence, much at odds with Marxist analysis (for example, Relph 1976: 114–17). I do not intend to pursue this bifurcation here, except to say this: whilst I consider the contribution of phenomenologists of place to environmental thought to be immensely valuable, it is a pity that environmentalism has not, to date, seen fit to explore the relevance of Marxist explanations for the commodification of space, and thus the alienation and obliteration of place.

PHENOMENOLOGY AND THE SEARCH FOR AUTHENTIC PLACE

Applied to questions of the human-in-environment relationship, most phenomenological investigation advocates nothing more complex than seeing particular places or environments from the inside out; from the empathising perspective of a particular place itself. From such a viewpoint a place is not merely 'the sum of the various psychological, social, economic, and political forces working on an environment at a particular point in time' (Seamon 1982:131). Phenomenologies of place proceed from an understanding similar to that of the Norwegian architect Christian Norberg-Schulz (1980). He argues that a place — and he has in mind a *human*-created place, though what he writes can also apply to places in which a human presence is less central — is dynamic, perhaps organic, and greater than its transitory individual components. Place, thus, has an *essential* quality: Norberg-Schulz (not originally) calls this its *genius loci* (literally, 'spirit of place'). To speak of a place having a *genius loci* is to assume a certain constancy through time; to see places as tenacious unities that self-perpetuate while people and historical events come and go. Thus, the essential nature of place changes only slowly, outlasting the people (and the other life components) to be found therein at any given moment. Such a view is in stark contrast to the conventional approach to 'environment', where a place is seen to be the sum of the various components (economic, political, cultural) identifiable within finite bounds at a given point of time. Norberg-Schulz's view is that a place is essentially 'what it is'; human intervention should take account of this and only seek to modify it in a way that works harmoniously with it, rather than confronting it with aggression and discord (Seamon 1982: 131).

Edward Relph provides a different approach. Norberg-Schulz seeks the essence of place inherently within itself — in qualities that are place-intrinsic. Relph's phenomenology, by contrast, stresses the experiential bonds that people establish with place. Traditionally, studies of person–place experiences have viewed place as a context within which basic questions of survival must be addressed: hence, as Seamon notes, there has been much emphasis on such emotions as territoriality and aggression. But 'a phenomenological perspective enlarges the emotional range of feelings that attach to place to include care, sentiment, concern, warmth, love, and sacredness' (Seamon 1982: 132). Place ceases to be a mere backdrop for survival; it is imbued with meanings that transform it from a theatre of fear and struggle to a haven; a positive context for living that evokes affection and a sense of belonging. Gaston Bachelard (1969) and Yi-Fu Tuan (1974a) have written of 'topophilia', described by Tuan (1974a: 4) as 'the affective bond between people and place', and by Relph as 'a homeward directed sentiment, one that is comfortable, detailed, diverse and ambiguous without confusion' (cited in Seamon 1982: 132). To 'topophilia' can be added 'topophobia' — 'ties with place that are distasteful in some way, or induce anxiety and depression' (Seamon 1982: 132).

Relph's is perhaps the most lucid attempt to reinterpret the person–environment relationship phenomenologically (in *Place and Placelessness* 1976; and *Rational Landscapes and Humanistic Geography* 1981). Finding the topophilia/topophobia dualism too polar, he posits instead a continuum with way-points between the extremes. For Relph, the acme of authentic place experience is 'insideness': 'to be inside a place is to belong to it and to identify with it, and the more profoundly inside you are the stronger is this identity with the place' (1976: 49). Along the continuum of insideness and its experienced opposite, outsideness, a single place assumes different meanings according to the experiences of a range of observers. Here is the major difference between Relph and Norberg-Schulz. For Relph it is not a question of place having an essential character: places will be interpreted differently by different people and, hence, will have an infinity of meanings. 'Existential insideness' is a situation 'in which a place is experienced without deliberate and selfconscious reflection yet is full of significances'; it is the experience most people feel 'when they are at home ... when they know the place and its people and are known and accepted there' (Relph 1976: 55). 'Existential outsideness', by contrast, involves 'an alienation from people and places, homelessness, a sense of the unreality of the world, and of not belonging' (1976: 51). Here, then, is a dialectical view of place-meaning, one that is in contrast to the Norberg-Schulz view of place as having an essential character independent of the human observer. For Relph, the dialectic between the human observer and place creates the place. A place will be variously interpreted, and hence have no objective and universally acknowledged meaning.

It would be wrong, though, to depict Relph's view of place-meaning as entirely the product of the unmediated individual psyche, for Relph is also concerned with the *social* process of place-identification. Place has a collective identity in addition to differing but interconnected private identities. Place-perceptions are constructed within social contexts, and it is upon shared perceptions of place that Relph lodges emphasis. The stability of the character of a place is thus related to continuity in communal experience, and also to the way the community experiences and reacts to changes to place (Relph 1976: 34). Shared perceptions of change often generate community articulation of ties to place that otherwise remain latent. It is through communal response to change and threat of change that much of the sense of attachment to place is articulated. Thus, authentic place evolves, is not static, has an organic quality — and along the way it becomes infused with meaning (on this, see Harries 1993).

Gary Snyder provides a more cosmic conception of place as 'a mosaic within larger mosaics' passing, in palimpsest, through time and space (1990: 27). Such a view posits, with Norberg-Schulz, an essentiality of place, within which its human components are a reinforcing part. Snyder is with Relph, though, on the question of ritual and custom — the constructions of stylised meaning devised and adopted by cohesive communities — as crucial for 'strengthening attachment to place by reaffirming not only the sanctity and unchanging significance of it, but also the enduring relationships between a people and their place' (Relph 1976: 32–33). This is a key function of ritual: to provide the individual with a sense of being part of a larger, meaning-infused context for living. Place-attachment lessens as rituals and myths decline in potency. 'In cultures such as our own', says Relph, 'where tradition counts for so little, places may be virtually without time', cut adrift from any historical context and, in reality, 'non-places' (1976: 33).

In both our communal and personal experience of living authentically in place, a sense of deep concern for that place will develop. Simone Weil, writing in the 1950s, argued that:

> to be rooted is perhaps the most important and least recognized need of the human soul ... A human being has roots by virtue of his real, active

> and natural participation in the life of a community which preserves in living shape certain particular treasures of the past and certain particular expectations for the future (1978: 41).

The need for the assured identity that roots provide is fundamental, then; it is the equivalent of such needs as liberty, the exercise of responsibility, and civil order (Relph 1976: 38). These intangible but crucial insights into the relationship between person and place are little understood. Perhaps most significantly, they are little understood in planning circles, where the view seems to prevail that a place is no more than its people, that these are in any case endlessly interchangeable, and that the rest is mere backdrop of relatively trivial importance — best expressed in terms of 'development potential' (Relph 1976: 81, 87–89). But, according to phenomenological analysis, there is a potent relationship between people and place, within which they dialectically shape a common identity. A 'deep relationship' with place is 'as necessary, and perhaps as unavoidable, as close relationships with people; without such relationships human existence is bereft of much of its significance' (Relph 1976: 41). Though the depths of this relationship may only become apparent at times of stress, many people so identify with their place that they organise within citizen action groups to defend it against externally imposed 'making over' in the name of development. The 'home' landscape contains messages and symbols that can serve as mobilising foci for a politics of place. In such a politics places are claimed as 'public', for they are known and created through common experiences and involvement in common symbols and meanings, and the public is entitled to claim a public right, superior to the use-rights of capital, to have the prime say in determining their future.

PLACE AND 'AUTHENTIC BEING': THE HEIDEGGERIAN LEGACY

Phenomenological theorists of place owe, and acknowledge a profound debt to Martin Heidegger. Relph, for example, has described this influence as all-embracing, its impact experienced both consciously and subliminally:

> I have tried to understand and absorb his thinking, and I hope that this re-emerges in a not-too-distorted form in my own writing. Sometimes, of course, this re-emergence is self-conscious; often, however, it seems to be unself-conscious, and only subsequently do I become aware of how much Heidegger's philosophy has coloured what I have done and written (1984: 219).

Heidegger's influence is foundational then (and other strands of environmental thought also draw on him). His essential concern was to critique modernity and, from that critique, to establish a basis for living authentically. He wanted the nature of 'being' to be recognised as the central question of philosophy, specifically a non-abstract conception of being as things in their particularity, their 'realness'. For Heidegger the question of being is not a metaphysical one; it is a question of how we 'dwell'. The abstracted preoccupations of post-Enlightenment modernity blind us, render us indifferent to the question of being. Our most urgent need is to overcome that indifference.

Heidegger's potential link to radical ecological philosophies will be apparent. To take an example, the arguments mounted in defence of tropical rainforests are usually utilitarian: tourist potential, or yet-to-be-discovered food sources or medicinal products. In the Heideggerian position, by contrast, with the emphasis placed upon 'authentic being', everything has a right to 'be' in the way in which it is proper for it to be. Much recourse can be had — and has been had — to such a position when attempting to find a philosophical grounding for ecocentrism (for example, Evernden 1985a: 60–72; C. Taylor 1992; Zimmerman 1983. There is a negative side to this, too, considered below.).

But it is within attempts to philosophically ground a defence of valued places against the obliterating dynamism of the capitalist market that Heidegger's impact has been most deeply felt in environmental thought. Heidegger's central concept of 'dwelling' is important here.

To dwell authentically is to dwell in place. It is to dwell within one's home. As Heidegger sees it, the essential character of modernity is homelessness; and we are doubly homeless, because not only are we estranged from home but we do not know that we are estranged from home. This is why we readily tolerate the obliteration of places we hold in affectionate regard. We feel pain and loss; but we are unable to find a reasoned justification for our pain or a reasoned argument against the right claimed by developers and governments to impose pain and loss upon us. We need to become aware of the responsibility that dwelling entails. In Sikorski's words, 'to dwell, in its most profound sense, is to preserve things in their peace, to spare them actively from anything that might disturb them, make them different from what they are', and this requires of dwellers that they be 'Guardians of Being' (1993: 32). The responsibility is one of 'sparing'. 'Sparing' is 'a tolerance for places in their own essence'; it is 'a willingness to leave places alone and not to change them casually and arbitrarily, and not to exploit them' (Relph 1976: 39). Put otherwise, sparing is 'the kindly regard for land, things, creatures, and people as they are and as they can become' (Seamon 1984b: 45). An essential aspect of human living is, then, to help maintain the world's processes of evolutionary change; the time-worn assumption that our role is to help ourselves to the world should now be jettisoned. Our task is to care for places, even 'through building or cultivation, without trying to subordinate them to human will ... It is only through this type of sparing and care-taking that "home" can be properly realized'. This is what it means to 'dwell', which is, for Heidegger, 'the essence of human existence' (Ralph 1976: 39).

A place about which one feels so deeply must become what Evernden (1985a: 63–65), Relph (1976: 38), Tuan (1974b: 241–45) and others — after Heidegger — call a 'field of care', and to care for a place involves more than holding it merely in affectionate regard; it also involves taking responsibility for that place. We have seen that Heidegger calls this responsibility 'sparing'; but it extends beyond a passive commitment to personally spare, to a duty to actively resist the vandalism others would inflict upon one's home. A 'field of care', in other words, entails a steward's duty of protection. To sit passively by and acquiesce in the destruction of one's home is to fail one's duty to take all steps possible to 'care' for one's dwelling.

The experience of 'home', then, is foundational — and it has largely been lost. It is:

> an overwhelming, inexchangeable something to which we were subordinate and from which our way of life was oriented and directed even if we had left our home years before. Home nowadays is a distorted and perverted phenomenon. It is identical to a house; it can be anywhere. It is subordinate to us, easily measurable and expressable in numbers of money-value. It can be exchanged like a pair of shoes (Vycinas 1961: 85).

Though Relph is less pessimistic than Heidegger — hence his relationships-to-place continuum (existential insideness to existential outsideness) — he endorses Heidegger's assessment of the need to *be*, at home, as a first-order human priority: 'home is the foundation of our identity as individuals and as members of a community, the dwelling place of being. Home ... is an irreplaceable centre of significance ... It is the point of departure from which we orient ourselves' (Relph 1976: 39–40; see also Tuan 1977: 149–60). Here is Heidegger's most significant contribution to environmental thought: his insistence upon the need to live authentically, to *be* at home, and to take responsibility for the defence of that home in all its aspects — human, natural, and the intangible particulars that constitute a place's essence.

NON-PHENOMENOLOGICAL THEORIES OF PLACE IN ENVIRONMENTAL THOUGHT

Through the 1980s place was linked to environmental thought via the phenomenological media we have just surveyed. In the 1990s, however, the notion of place was accorded a central role by theorists working within other traditions of ecological thought, and these contributions, though not usually incompatible with phenomenological assumptions, have proceeded without reference to such frameworks.

From an ecosocialist perspective, for instance, Michael Jacobs has argued that people require the knowing of places for identity: 'people do not simply look out over their local landscape and say "this belongs to me". They say, "I belong to this"' (1995a: 20). Like the phenomenologists, Jacobs argues that 'the attachment to place — not just natural places, but urban places too — is one of the most fundamental of human needs' (1995a: 21). He develops this case in order to argue the radical incompatibility between place attachment and neo-liberal economics. 'The neo-liberal vision sees the person primarily as a buyer of utility: a consumer, the individual as a stomach', he writes. 'But homeness is about identity, the individual as soul'. Similarly, 'the sense of place resists neo-liberalism because it implies diversity. All homes are different; that's how we know we're home', and so 'protecting the environment is about protecting identity: the things which make us who we are, in opposition to the standardising forces of the free market'. On these points Jacobs's case is similar to Relph's, though generated from different intellectual antecedents, and presented for different political and pedagogical purposes.

In constructing a case against the free-market principles of environmental economics, Mark Sagoff also champions 'place', noting that it 'brings together human, environmental, and natural history; it is particularly valuable in helping us to understand what we deplore about the human subversion of nature and what we fear about the destruction of environment' (1993: 6–7). Furthermore, 'the concept of place applies to landscapes that do more than satisfy the consumer preferences of individuals', thereby undermining the insistence of neo-classical economics (including its 'green' variant known as 'environmental economics') that all desires, needs and impulses can be rendered as number-value market preferences. Indeed, says Sagoff, 'much of the discussion about preserving resources might be better understood in terms of protecting places' (1993: 7; see also 1992).

Ecofeminists have also theorised place. The best-known feminist theory of place, Luce Irigaray's notion of body-as-place (1993), would seem, in its deliberate de-territorialisation of the concept 'place', to be incompatible with environmentalist formulations. More promising is Susan Griffin's ecofeminist take. 'One is dependent for coming into existence not only on a mother and father but on an intricate web of life', she writes. But there is nothing abstract about this web of life; it is immediate and specific: 'one is born from the ground, the tree, the bird in the tree, the body of water feeding the roots ... all that one sees'. It also includes social immediacies: 'others in the family, each of whom contributes daily to make one's life what it is, neighbors, villagers, the farmers, the baker ... are part of one's existence', and 'in this matrix one defines oneself finally ... by a layered complexity that includes the process of exchange ... by which one's life comes into being and continues' (1995: 91). Griffin draws together a belief in the importance of the local and particular with the ecofeminist stress upon material embodiment, a fertile conjunctivity that is under threat from the universalisation — and, hence, placelessness — of post-industrial knowledge. Much is endangered or already lost: 'the knowledge of place that is being all but erased in the technological consciousness is also a knowledge of the necessities and limitations of natural existence' (1995: 95). Griffin describes this endangered realm as 'a larger coherence to which we all belong', a coherence in which 'the rounds of birth and death from which life emanates, the rising and setting of the sun, the course of seasons, every need of the body, all partake of the infinity of natural cycles and so can enlarge consciousness to infinite domains' (1995: 96).

Some of the early ecophilosophical writings featured 'place' prominently (for example, Evernden 1978), but it remained undeveloped and fell from view. There are, however, some recent attempts to foreground 'place' within ecophilosophy. Norton and Hannon argue that 'place orientation is a feature of all people's experience of their environment'. They seek a central role for place in environmental ethics. A problem with particularising rather than abstracting ethics is that it legitimises the NIMBY ('not in my backyard') syndrome. To counter this, Norton and Hannon distinguish between 'NIMBY A: You may not do x in my backyard; therefore do x in someone else's backyard', and 'NIMBY B: You may not do x in my backyard; furthermore, if you cannot find some other community that democratically chooses to accept x, then x will cease' (1997: 244). NIMBY A, of course, is unacceptable, whereas NIMBY B is legitimate. Norton and Hannon see prospects for a new environmental ethic in such a formulation, one which involves 'an end to the *ex cathedra* pronouncements of the environmental expert', with the scientist instead enjoined 'to emphasise study of local ecosystems' (1997: 245; for other recent philosophical speculation on place, see Casey 1998; Malpas 1999).

There are two other strands of environmental thought within which the particularity of place is privileged. The first of these is bioregionalism, a variant of environmentalism owing much to the reformulated anarchisms that were in vogue in the 1980s. We look at bioregionalism in chapter 9. But the current 'rehabilitation' of place owes most to its privileging within postmodernist and deconstructionist modes of thought. There have been several applications of postmodernist perceptions of place to environmentalist thought. An influential contribution is made by Jim Cheney, who argues that the postmodernist paradigm gives much credence to bioregionalism, and within that, 'the idea of place as the context of our lives and the setting in which ethical deliberation takes place', as well as giving credence to 'the epistemological function of place in the construction of our understandings of self, community, and world' (1989: 117). We will return to Cheney's essay. A second wedding of postmodernist and environmental ethics is provided by Mick Smith, who uses an 'ethics of place', which is a 'discourse of relativity, proximity, dimensionality, distances' (1997: 339), to 'counter the current enclosure of the moral field within economistic and legal bureaucratic frameworks'. Postmodern thought seeks a dissolution of all rigidly authoritative frameworks, and an ethics of place, Smith argues, 'reconnects moral and physical spaces in such a way as to subvert our present ethical agendas' (1997: 340). It does this because it is grounded in different conceptions of relationships and in specific, non-abstract interpretative frameworks.

Though postmodernism's canonical writings cannot be counted within environmental thought, the importance of local place within these writings will be noted in chapter 10. Meanwhile, non-phenomenological defences of place remain unsystematised within the broader body of environmental thought, the theories sketched above (as well as others) laying claim to a territory of still uncertain status.

AUTHENTIC PLACE AND COMMUNAL VITALITY: AN ESSENTIAL LINK?

Much writing that elucidates the qualities of place seeks to re-establish the bonds of communal living that have been rendered tenuous under liberal capitalism.

This is sometimes more evident within activism than theory. Much green activism focuses upon the familiar 'home range', marrying notions of community invigoration with defence of communal environmental 'amenities': locally occurring wild species and their habitats, communal open space, clean air and water, and so on. Projects are promoted that are both environmentally benign and communally integrative, that endorse the relevance of the 'local' rather than assume its irrelevance. Elsewhere, I have written:

> it is the alienation from home and homeness that is the most telling consequence of global technology, global communications, global architecture, global religion, global bureaucratisation and global economy. None of this is to be confused with Marshall McLuhan's 'global village'. The global village never came. Villages are human agglomerations at a scale conducive to community — but global community never came. On the contrary, along with globalisation came the antithesis of community — the atomisation of daily life. As structures, technologies, forms and processes became remote and indifferent to unique place, so society was privatised out of existence. To recover 'home' is thus to recover 'community', by which is implied not simply meaningful human interaction, but the built fabric and natural processes that are essential components of one's 'significant environment'. To fight for home and community is thus to fight the debilitating and degrading alienation that, so many contemporary prophets have rightly informed us, is the modern condition. There can be few more urgent tasks (P.R. Hay 1994a: 11).

Green activist organisations that join a focus upon local action to a concern for community refurbishment occur throughout the western world. England's 'Common Ground' movement can serve as exemplar here. Taking its name from Richard Mabey's widely read plea (1980) for a more determined approach to countryside conservation, Common Ground seamlessly welds a concern for the human with the natural within place:

> excite people ... to savour the symbolisms we have given nature, and to revalue our emotional engagement with places and their meaning, so that we may go on to become actively involved in their care. We have chosen to focus attention, not singularly upon natural history, archaeology, architecture, social history, legend or literary traditions, but upon how each of these combine to form people's relationship with places (Clifford 1994: 16).

The stress upon place-uniqueness that is evident in the work of the phenomenological geographers also emerges within the activism of Common Ground. Here, too, is the Heideggerian stress upon 'home', conceived as a primary environment of indeterminate but wider range than mere 'house'; and, in keeping with the green activist's penchant to value maxims forged in struggle rather than principles derived from abstruse theorising, this is conceived without apparent recourse to Heideggerian scholarship (King and Clifford 1987: 2–4). Nor is this an individualised conception of 'home'. The projects undertaken by Common Ground — for example, the 'Parish Maps Projects', wherein people were 'encouraged to chart the wild life, landscape, buildings, history and cultural features which *they* value in their own surroundings' (King and Clifford 1987: i) — aim at community revitalisation by generating affection for common reference points within the local environment.

There is significance, too, in the very notion of 'common ground'. What is imputed here, against the connotation of 'space' as property or real estate, is a conception of space as primarily social in character — as belonging in essence, whatever is on the title deeds to individual properties, to the community. This is close to Relph's position, wherein it is the community that gives place meaning and identity (Relph 1976: 34). It is also close to the conception of the 'commons' vigorously championed by the Ecosystems Ltd team writing under the aegis of *The Ecologist* (1993). They maintain that the notion of the 'commons' was wrongly characterised by Garrett Hardin in his classic 1968 essay 'The Tragedy of the Commons' as an open-access regime, the only alternatives to which are privatising the commons out of existence or its forcible maintenance by illiberal but ecologically benign authoritarian government. For the Ecosystems Ltd team:

> the concept of the commons flies in the face of the modern wisdom that each spot on the globe consists merely of coordinates on a global grid laid out by state and market: a uniform field which determines everyone's and everything's rights and roles. 'Commons' implies the right of local people to define their own grid, their own forms of community respect for watercourses, meadows or paths; to be 'biased' against the 'rights' of

> outsiders to local 'resources' in ways usually unrecognized in modern laws; to treat their homes not simply as a location housing transferrable goods and chunks of population but as irreplaceable and even to be defended at all costs. (*The Ecologist* 1993: 12. Snyder also notes this flaw in Hardin's characterisation of the commons; 1990: 35–37.)

The Ecosystems Ltd team provides perhaps the most forthright defence of communal vitality as essential to creating and sustaining global ecological health. This is because 'the environment itself is local; nature diversifies to make niches, enmeshing each locale in its own intricate web': and so, 'enduring human adaptations must also ultimately be quite local' (Richard O'Connor, cited in *The Ecologist* 1993: 16). High utopianism? Not so. The Ecosystems Ltd analysis ranges across the globe, and focuses more on the defence and/or rehabilitation of what already works than upon the creation of a new network of economic and political organisation. It remains akin to the 'Common Ground' framework of working with 'what is', and in this is to be distinguished from the utopian thrust of environmental anarchism (considered in chapter 9).

Finally, it is worth noting that this concern for community, far from distancing its proponents from the nature-focused mainstream of the environment movement, can be held to dovetail with it. Eric Katz, arguing against the quasi-Gaian model of nature as organism, posits 'community' as a superior metaphor for natural relationships, for it allows for meaning and purpose to inhere within the individual components of those relationships (1992: 58; see also Rodman 1973: 583).

SOME PROBLEMS: HEIDEGGER AND THE NAZI TAINT; THE CLAIM FOR THE ESSENTIALITY OF 'NATURE'

An objection is persistently made to the inclusion of the themes of this chapter within the corpus of green theoretical concerns.

This consists of a rejection of any strand of thought seen to drink too deeply at the Heideggerian well, on the ground that Heidegger's never unequivocally repented engagement with German National Socialism renders him irredeemably tainted so far as any movement claiming a central place within the reconstitutive politics of a new millennium is concerned. The facts of Heidegger's practical and philosophical involvement with National Socialism are largely established. He manipulated Nazi sympathies to ruthlessly advance his academic career (Glaser 1978: 109). He supported Hitler in speeches in 1933–34, mixing 'his own philosophical vocabulary with the street language of National Socialism' (Zimmerman 1990b: 37). He never, in his lifetime, pronounced against the Holocaust. He produced an 'official story' after the 1939–45 war, in which he claimed a mere ten-month association with the Nazis in 1934, an association attributed to 'political naivete'. Subsequent scholarship has shown the 'official story' to be a fabrication, however; that Heidegger did not officially sever his ties with the party in 1934, and that he remained an active supporter of National Socialism, if a somewhat quirky one, through the 1930s and on through most of the war years (Zimmerman 1990b: 40–45).

Three questions are posed by Heidegger's involvement with National Socialism. What was there about National Socialism that induced Heidegger to throw in his lot with the Nazis? To what extent can it be claimed that his thought is quintessentially National Socialist, so that any subsequent Heideggerian philosophical influence necessarily entails a continuance of the doctrines and values of Nazism? And, as far as the green movement is concerned, does any of this matter?

The first of these questions is simply answered. Heidegger's *bête noire*, the target of his entire elaborate philosophical edifice, is the Enlightenment tradition of progressive modernity, particularly as it manifests through human-diminishing and nature-obliterating industrial technology (on this, see Zimmerman 1990b; 1994). The anti-technology theme within German Romanticism upon which Heidegger fed also influenced National Socialism, such that Nazism, like Heidegger, took aim at liberal capitalism and Marxist socialism — the two dominant embodiments of the Enlightenment progressive tradition. For Heidegger, Nazism offered 'an authentic "third way" between the twin evils of capitalist and communist industrialism', and he came to believe that it 'would renew and discipline the German spirit, thereby saving Germany from technological nihilism' (Zimmerman 1990b: 34).

The second question is more difficult. Some writers have argued that there is a seamless link between Heidegger's thought and his politics (for example, Farias 1989; Wolin 1990); some have claimed that Heidegger's thought is so incompatible with Nazism that his politics must be regarded as the dismaying folly of a political *ingénu* (for example, Lyotard 1990); others, such as Habermas (1989), Hindess (1992) and Zimmerman (1990b; 1994) take a middle path, *semi*-detaching the politics from the thought, condemning the former, and conceding value for his thought beyond the political context within which Heidegger placed it, whilst not denying a link between the two.

The latter position seems the only one capable of successful defence. Except when texts are foundational, constituting a body of received wisdom against which all subsequent contributions must be tested (as in the Bible, for Christianity; and *Das Kapital*, for Marxism), it is almost impossible to set contributions to Western thought *exclusively* within one or other philosophical or political tradition. These traditions are not tightly integrated, vertically; they are not stand-alone pillars, reaching their straight and self-generating ways through time. They interweave and cross-fertilise. There are Romantic elements, or at least elements that would reject the Enlightenment tradition of progressive modernism, within both liberalism and socialism; just as the conservative tradition no longer locates entirely outside the Enlightenment paradigm. I am, then, much taken with the carefully articulated position of Zimmerman — albeit this is a position that he has since moved away from — that 'Heidegger's texts ... can be read profitably without regard to their political implications', and that 'his thought cannot be reduced to the level of an ideological "reflex" of socio-political conditions' (1990b: 38). This is wise: great writings generate a plethora of interpretations and inform a wide range of subsequent political and philosophical ends. But, given the service to which he put his own thought, 'we must learn to read Heidegger with a deeper concern about how his thought may be appropriated and applied politically' (Zimmerman 1990b: 38). And this, too, is wise counsel.

And so to our third question: as far as the green project is concerned, does any of this matter? Yes it does. The more general question of the potential for the ecology movement to transmogrify into a form of neo-fascism will be considered later. Let me say here, though, that I do not think that general proposition to be as formidable a threat to the green movement's credibility as is the same question-mark specifically over Heidegger, whose place within the environmental corpus is real and not mere 'what if'. Heidegger has been an important influence upon green thought. Phenomenologies of place apart, he has been a strong presence within deep ecology (for example, Devall and Sessions 1985: 98–99), and it is largely through this connection with deep ecology that Heidegger has proven a political 'problem' for the green movement. As Zimmerman explains it:

> critics use the potential Heidegger-deep ecology link as evidence that the latter may lean toward ecofascism. Such critics employ the following logic: Heidegger supported National Socialism; his thought is at least partly compatible with deep ecology; therefore, deep ecology must be compatible with National Socialism (1994: 105).

That there is much compatibility between Heidegger's thought and deep ecology is indisputable. As Wayne Cristaudo observes:

> the attempt by deep ecologists to popularise the idea that all existing things are part of a common seam of life, that the fate of our species is inextricably linked to our ability to participate harmoniously within our planetary network, is identical with Heidegger's attempt to rethink beings within, and not as entities to be dislocated and torn apart from Being (1990: 302).

Yet, Zimmerman — the person most responsible for securing a prominent place for Heidegger at the deep ecologist's table (for example, 1979; 1983) — has since reconsidered his earlier assessment of Heidegger's deep ecological credentials, and his own commitment to deep ecology specifically and radical ecology generally. In addition, whilst he retains a belief in Heidegger's selective relevance to the green project, he is no longer convinced that Heidegger's view of the human–other nature relationship is sufficiently non-anthropocentric for him to merit an 'intellectual founder' status within deep ecology. This is despite the existence of such seemingly unequivocal observations within Heidegger's work as this rhetorical question: 'are we really on the right track toward the essence of man as long as we set him off as one living creature amongst others in contrast to plants, beasts and God?' (Heidegger 1978: 221).

Zimmerman's reassessment of his once-strong support for deep ecology and the anti-modernism of radical ecology is still instructive:

> I once believed Heidegger's thought would provide a way out of technological modernity's nihilistic disclosure of everything as raw material. Today, because I see that his total critique of modernity was in many ways consistent with the critique advanced by Nazism, I am more cautious about abandoning the political institutions of modernity, though I remain critical of its dark side (1994: 105).

It was argued earlier that there has been a tempering of support for radical ecology in the wake of recent assessments of philosophical ecocentrism as politically unpalatable or otherwise incapable of implementation. This trend has seen a diminution of Heidegger's stellar position within environmental thought, because it entails a more complex attitude to both the technological and political inheritance of the Enlightenment than Heidegger allows.

Though this reappraisal has largely taken place within the context of Heidegger's influence upon deep ecology, theorists of authentic place-making can take little comfort therefrom. The aspects of Heideggerian thought that have so appealed to deep ecology — the critique of advanced industrial technology as entailing the obliteration of authentic 'dwelling', and the message that we are to 'let the earth be as it is essentially, to let beings be' (Cristaudo 1990: 302) — are precisely those aspects that have appealed to phenomenologists of place. Moreover, deep ecology draws upon diverse antecedents: though it is faced with several recurrent criticisms, the uncertain status of Heidegger's legacy is not the most urgent of them. Phenomenologists of place, however, are crucially dependent upon their base in Heidegger: the wider status of his thought is, thus, a question of prime concern for this green tributary. I think it wise to be vigilant against the pathological turn which Heideggerian anti-modernism can take, and did in the case of Heidegger himself. But it needs to be remembered that Heidegger was mistaken in his assessment of the Nazi position on technology. National Socialism's anti-modernist feeder stream was ultimately a feeble one, readily — and necessarily — discarded. Far from being opposed to modernism's project of technological advance,

Nazism promoted — and glorified — an unprecedented explosion of technological development. As Janet Biehl notes, the Nazis extolled 'the return to simpler, healthier, and "more natural" lifeways ... even as they constructed a society that was industrially more modernized and rationalized than any German society had seen to that time' (1994: 133). This is what fascisms are ever bound to do, for they take their energy from the demonisation of enemies, against whom as formidable a combative technology as possible must needs be deployed. Thus, Harvey describes Nazism as 'reactionary modernism', noting that it 'simultaneously emphasized the power of myth ... while mobilizing all the accoutrements of social progress towards a project of sublime national achievement' (1989: 209). In any case, there *is*, as Giddens (1991; 1994), Beck (1992; 1995a; 1995b) and others have pointed out — without embracing a politics of nihilistic despair — a profoundly negative and dysfunctional side to the modernist project. The practice of damning via the 'logic of contamination' (Zimmerman 1994: 105) all those who criticise the pathologies of modernism's technological project is arguably as totalitarian in its use of smear to suppress dissent as any crudely totalitarian process of thought control, and to be rejected on that account.

A second objection to including phenomenologies of place within green thought can be made. As we have seen, most theorists of place, including those whose identification with environmentalism is overt (such as Relph and Seamon), assume a *human* environment. Though Norberg-Schulz's 'place essentialism' privileges neither natural nor cultural place, the place–person dialectic described by Relph would seem to assume that 'place' is primarily 'cultural place'. At least, following Heidegger, it is primarily concerned with the question of how *humans* should *be* in the world.

But against a focus on the quality of human life as the central green concern is the argument that the *foundational* green project is to guarantee, in perpetuity, the ongoing presence of 'nature' on earth. If this project is foundational, other considerations must be marginal within, or irrelevant to the primary project.

Robert Goodin is of such a mind. He posits a 'green theory of value' wherein sits a single principle upon which all else is built. This foundational tenet is 'an abiding concern that *natural* values be promoted, protected and preserved' (1992: 120). Goodin identifies several movement preoccupations that are needless encumbrances upon the definitive green principle of value. One of these is the concern to establish theoretical principles of right living, prominent among which Goodin identifies 'authenticity' (1992: 76–77). This 'needless encumbrance' is of concern here because 'authenticity' figures centrally within phenomenologically based theories of place. At one level Goodin finds little of which to complain: there is no problem if authenticity is understood as 'the naturalness of the processes involved'. But something more is usually intended when greens talk of 'authenticity':

> 'authenticity' admits of another interpretation, namely, the absence of pretence. And that rather suggests that it is not so much the naturalness of a thing's history of creation but something else — its simplicity, its lack of affectation or contrivance, its lack of artifice in that sense — that is valued (1992: 77).

Goodin does not specifically discuss the Heideggerian tradition or the more recent literature on place and perception. His main target is the 'alternative' or 'simple living' movement. But his strictures concerning authenticity also gather up the contributions to environmental thought considered in this chapter, for these, too, are primarily concerned with establishing principles for 'authentic' living in which 'authentic' is understood as 'lack of affectation or contrivance'. Given that the practical focus of most of the writings considered here is the human-created environment rather than the natural environment, they fall even more emphatically beyond the ambit of Goodin's foundational 'green theory of value'.

Nevertheless, he does provide a justification for preserving some human artefacts: 'they can qualify for protection on account of value derived from their being part of *nature*, somehow construed' (1992: 49). Though this seems a strange proposition on the face of it, it is arrived at in the following way: Goodin's 'green theory of value', with the protection of nature as its bedrock principle, is derived from the need to set one's life 'in the context of something larger than yourself'. This given, 'things that have value on account of their (purely human) history might well derive value from a source *akin* to, if not strictly identical to, that imparting value to naturally occurring objects in the non-human world', and a case can thus be made 'for the conservation of things in general on account of their history, whether human or natural' (1992: 50).

From a very different perspective, Anthony Giddens also develops an argument that would seem to challenge the inclusion of phenomenologies of place in a reconstitutive scheme of thought (though this is not a position that Giddens himself articulates). The notion of 'authentic place' presumes stability, low dynamism, ongoing tradition. Giddens, though, argues against seeking to:

> defend tradition in the traditional way. We might very well want to preserve old buildings, but we wouldn't want to, and mostly couldn't in any case, sustain the ways of life with which they were associated. Yet without those ways of life, the old buildings are scarcely 'larger than ourselves' — they are symbols of the past, relics or monuments (Giddens 1994: 212).

Giddens is critical of the repressive historical role of 'tradition', arguing persuasively that it is of tenuous relevance within the fluid and rootless world of technologically advanced capitalism. Arguments along these lines are not uncommon: David Lowenthal, for instance, maintains that 'the age-old appeal to tradition is generally obsolete because past and present now seem too dissimilar to make it a safe or valid guide' (1985: 370). Giddens insists that the key questions confronting humankind cannot be answered 'through tradition, understood in the traditional way' — but that we can 'draw on tradition to do so' (1994: 217). Whilst the defence of tradition 'in the traditional way' leads inexorably to fundamentalism and is to be rejected on that account, 'succouring traditions means preserving a continuity with the past which would otherwise be lost and doing so as a way of achieving a continuity with the future as well' (1994: 48).

The imputation that can be taken from these passages (though the latter qualification clouds the issue) is that a concern for the durability of 'authentic' place articulates a conservative's nostalgia for the social rigidities (and hence, inequities) embodied within relic landscapes, as well as a Tory suspicion of the progressive forces of industrial change. It thus ideologically mislocates the value 'authenticity'. But the argument fails, I think. It may be that the 'place' of Olde England *has* been defended in terms of its embodiment of social relations that are thought to correspond to the age of British greatness. But this is hardly an appeal to place 'authenticity', for authenticity lodges in the especialness and uniqueness inherent within each uncompromised place; whereas the conservative nostalgia for place as an embodiment of past social relations generalises place rather than particularising it. Place cannot be valued in the abstract; it can only be valued in the concrete (as it were) — in the celebration of individual uniqueness. Relph insists that to generalise place is to renounce the distinctive and especial (1981: 168–75). He is at one with Annie Dillard: 'landscape consists in the multiple, overlapping intricacies and forms that exist in space at a moment in time. Landscape is the texture of intricacy' (Dillard 1976: 126).

Now, having spent much of this chapter considering the charge that Heidegger-influenced ecological thought is likely to contain the seeds of fascism, it seems appropriate to turn to the broader question of whether the 'natural' politics of the environment movement is a politics of the right.

7

GREEN POLITICAL THOUGHT: THE AUTHORITARIAN AND CONSERVATIVE TRADITIONS

THE AUTHORITARIAN TRADITION IN ENVIRONMENTAL THOUGHT: THE 'TRAGEDY OF THE COMMONS'

When the modern environment movement came into being in the 1960s, it was heavily 'scientistic' in orientation. Its spokespersons were scientists, and its impetus came from people trained in the sciences, particularly the natural sciences. For most of these people this was an initial politicisation — few had backgrounds in conventional, let alone unconventional politics. The political pronouncements of the growing environment movement were, thus, characterised by an incongruous naivety. Exhortations to a tame politics of letter-writing and petition-collecting sat uneasily atop a devastating doomsday scientific prognosis. Some, though, quickly put such touching faith behind them — to embrace a politics that was powerfully illiberal and authoritarian.

In a sense the authoritarian tendency was a reaction against the failure of conventional political activity. There was, in the trusting simplicity of the early politics, a failure to appreciate that liberal democratic disputation has more to do with brokerage between competing interests than the search for objective truth. Politicised by the perceived urgency of the environmental crisis, the early proponents of the burgeoning environment movement assumed that they simply needed to demonstrate the validity of their analyses, after which governments would, almost reflexively, take the required action to rectify matters.

When this failed to happen, the doomsdayists concluded that democratic politics was a dismaying and unhelpful business. They turned in despair to an authoritarian politics as the only way to transcend the 'corrupt' politics of interest brokerage, which, in their view, would always lead to the subversion of the general good.

The result was some dire political prescription. Thus, in his influential *The Population Bomb* (1972), Paul Ehrlich argued that, as the population bomb was already primed and fused and as there was no means of defusing it, the affluent world should save itself, leaving those countries which had already passed the point of no return to face the consequences. He advocated that the United States use its economic and military power to rearrange underdeveloped nations according to their capacity to mend their destructive ways — in particular, their capacity to stabilise or reduce their populations — and reverse the slide to environmental catastrophe. 'I know this sounds very callous', he wrote, 'but remember the alternative. The callous acts have long since been committed by those who over the years have obstructed a birth rate solution. Now the time has come to pay the piper, and the same kind of obstructionists remain. If they succeed we all go down the drain' (Ehrlich 1972: 160; on Ehrlich's seminal contribution to environment debates, see J. Young 1991: 3–6).

Important contributions to the politics of Leviathan have also been made by Robert Heilbroner and William Ophuls. Arguing for an environmentally benign, centralised authoritarianism as the only solution to chronic environmental crisis, Heilbroner (1974) accepted as proven the case advanced in the 'doomsday-is-nigh' reports of the early 1970s — *Blueprint for Survival* (Goldsmith *et al.* 1972) and the Club of Rome's *Limits to Growth* (Meadows *et al.* 1972) — that the world faced grave crises of resource depletion and biophysical breakdown. He also accepted that human nature was, as Thomas Hobbes had described it in 1651, characterised by a brutish propensity to violence and bitter struggle and incapable, in the absence of external restraint, of maintaining ordered social relations. It is for this reason that, in the Hobbesian view, people contracted together to set up the structure — the state — necessary to the creation and maintenance of civil society. This given, Heilbroner argued, only an ecologically benign central authority would evince the political will and wield the incorruptible power necessary to reshape agricultural and industrial production, impose strict controls on population, and regulate production and distribution at levels that would entail a dramatic drop in standards of living. Democratic government, by contrast, is too weak-kneed: faced with the short-term political fallout occasioned by people being forced to make sacrifices today — even to ensure their very survival tomorrow — it goes to water.

The mood of the 1970s, set in place by predictions of imminent, ecologically wrought doom, was grim and desperate. The environment movement was Hobbesian, deeply despairing and anti-democratic. It judged human nature harshly. That environmentalism could be a positive force for refurbishing the quality of life on earth was scarcely considered. Liberatory possibilities were nowhere in sight. The vision was one of scarcity, frugality and fortitude to face the rigours to come, whilst the priority was survival; in the name of scarcity and survival the movement's spokespersons called for harsh and restrictive government action. Thus, the mood at this stage in the development of the environment movement, dubbed its 'survivalist' phase by Eckersley (1992: 11–15), was frightened and defensive — a collective state of mind much suited to authoritarian prescription.

These assumptions concerning the absolute urgency of the need for drastic action, and the dismaying block to such action imposed by the poor clay of human nature, determine the cut of Heilbroner's argument. But he does reach his conclusion with regret, expressing a preference, in the best of all possible worlds, for small-scale political systems along the lines of the Greek city-state (1974: 135). He even hopes that the powerful centralised state might only be necessary in the period of transition to ecological sustainability, and that possibilities for decentralisation might thereafter open up. Ophuls expresses a similar preference for social organisation that is small in scale; but, for him too, the choice is 'Leviathan or oblivion' (1973). Again, human nature is deemed to be selfish and concerned above all with private gain, and it follows that people will never voluntarily accept constraints on consumption and on the licentious misuse of habitual freedoms. In keeping with the Hobbesian character of the survivalist phase of the environment movement, Ophuls calls for a new social contract, in this case an 'ecological contract' (Ophuls 1977: 164–65). As Eckersley observes:

> unlike the social contract of John Locke (which was based on cornucopian assumptions), the ecological contract would be based on the Hobbesian premise of scarcity and would therefore require an all powerful Leviathan, not just a limited goverment. That is, if certain freedoms were not voluntarily surrendered by citizens, then restrictions would have to be imposed externally by a sovereign power (1992: 16).

Heilbroner and Ophuls arrived at their conclusions with reluctance; it is less clear that this is the case, however, with Garrett Hardin. His is the most explicit formulation of an illiberal, authoritarian ecological politics.

In 1968 he published his famous essay, 'The Tragedy of the Commons'. A biologist, he shares the views of the other doomsdayists that population growth must be curtailed. But population growth, he argues, is a problem for which there is no purely *scientific* solution. We must look to politics for answers.

He explained the 'tragedy of the commons' via the metaphor of a medieval herdsman who has free grazing access, along with all his neighbours, to common-property pasture. There is a point at which the introduction of just one additional cow will permanently degrade the pasture. As the herder sees it, if he decides to add in his own extra cow he will personally maximise the benefits, while the costs will be evenly distributed across all the other users of the commons. On the other hand, if he altruistically withholds his own cow he is likely to be personally disadvantaged, because, following a process of identical logic, all other herders are likely to add in their own extra cows. What is rational behaviour in these circumstances? It is to add the cow to the commons, even when the consequences of doing so are perfectly understood. Other herders decide similarly, because they have also acted according to the dictates of rational behaviour — and the overgrazed commons rapidly deteriorate. When faced with a choice between one's selfish interest and a more diffuse public interest, says Hardin, the rational course is the selfish one. The degradation of the commons is thus inevitable, and therein lies the tragedy.

To avoid inevitable tragedy, the free commons must be transformed into something other than a free commons. The problem is one of ownership. It is because environmental goods are common property resources — that is, they have no individual owner with a stake in their maintenance into the future — that tragedy occurs. No one owns the air, or the sea (or the fish in it), or, in most cases, fresh water. One solution is to strengthen the scope of individual control over resources: to enlarge the private domain at the expense of the public domain. This is a position with much appeal in a time such as our own, when the public sphere is everywhere in decline. Another solution, preferred by Hardin, is to drastically restrict the freedom of private individuals. Many of the rights we take for granted must be abolished.

We must also jettison the notion that all people have an equal entitlement to a share of the earth's bounty. The most extreme application of Hardin's commons thesis is the 'lifeboat ethic' (1974a). Here he argues that to try to assist the despairing and the destitute of the world is not only doomed to long-term failure, but will imperil the lot of everyone else. In this view, metaphoric lifeboats (Hardin is rich in metaphors) containing the citizens of the affluent west are surrounded by the poor and desperate of the third world, all of whom are clamouring to climb aboard. What is 'ethical' behaviour under such circumstances? Because the lifeboats have a finite carrying capacity, Hardin argues, not all who seek to clamber aboard can be allowed to do so. If the decision is taken to admit some, there is then the impossible question of what criteria to adopt for selecting the fortunate. In any case, to take even a chosen few on board is to destroy the lifeboat's spare safety capacity. The only appropriate — and 'ethical' — response is, thus, to maintain the safety margin and ignore the cries of the increasingly desperate.

And in the rich countries themselves? An authoritarian government in which the democratic scope is much reduced — via a contractual acceptance of a state of 'mutual coercion', brought into being by 'mutual agreement' of 'the majority of people affected' (1968: 1248). The right to parenthood, for example, must disappear as a simple entitlement, because the revised morality of the commons requires that reproduction be strictly regulated (Hardin 1973: 177–89). Ditto for the presumed right of access to recreational commons — especially wilderness (Hardin 1974b). Hardin argues that collective responsibility in such matters must be enforced from above, for it will not spontaneously emerge from below (1973: 128–32). To avoid irretrievable ecological breakdown, a powerful regime possessing the authority to enforce draconian restrictions upon the democratic scope must be installed.

Here is an ecopolitics entirely without community. It assumes a politics based wholly upon individual self-interest, free of behaviour-moderating community processes, and leading inexorably to the tragedy of the commons. The motivation to act in one's private self-interest must therefore be suppressed, and this requires a supremely powerful, ecologically informed authoritarianism. Though the libertarian counter-culture of the late 1960s was important in providing the impetus to environmental concerns that was necessary to hoist them up the political agenda, the politics emerging from this earlier phase of the environmental movement were of the political right. Small wonder that many on the left took one look at this and cried 'neo-fascist' and, later, 'ecofascist' — a label Hardin has had to wear more than most.

But the early 1970s was the high-tide of an authoritarian ecopolitics allied to doomsday science. Since then that form of environmentalism has been eclipsed by a politics which holds that the globally systemic nature of environmental crisis means that we cannot do a modern equivalent of retreating behind the walls of the medieval keep. Moreover, it would not be right to do so, for global problems are more of the affluent west's making than of the impoverished and exploited third world's. The typical position within western environmentalism today is less that of the 'lifeboat ethic' than that articulated by Porritt:

> the poor have little time or inclination to worry about global environmental trends, and yet in many ways they are more affected by the ecological crisis than the affluent who can just drive away from it. Many Third World people are forced by circumstance to destroy the very resources on which they depend ... those who are working for a better environment must simultaneously devote themselves to working for social justice. There is not only the moral imperative that compels us to seek ways of sharing the world's wealth more effectively; there is the ecological imperative to remind us that the protection of the Earth's natural systems is something we *all* depend on (1984: 98).

The earlier draconian authoritarianism has been nudged aside then, but it has not vanished. Ehrlich has acknowledged the transnational nature of environmental pathologies and has retreated from some of his direst remedies. But the influential founding editor of *The Ecologist*, Edward Goldsmith, guiding spirit behind the much-discussed *Blueprint for Survival*, and subscriber to a similar illiberal survivalism, has remained largely unreconstructed. In his contribution to a companion piece to the *Blueprint*, Goldsmith advocated 'a sort of national service for conservation' (1972: 286) and a reduction in the world's population 'by probably at least half' (1972: 286). In 1988, in his *The Great U-Turn*, he continued to put these positions, the former now a 'restoration corps' in which the unemployed would be compulsorily enrolled (1988c: 214). Among his other prescriptions were the use of subliminal advertising to create ecologically responsible patterns of behaviour (1988c: 212); and an end to labour-saving domestic technology (1988c: 210–13, 215–17). The point of this latter prescription is to shore up the nuclear family, which Goldsmith sees as the basic unit of a 'correct' form of social organisation, one sanctioned by nature (Goldsmith 1988b: 182).

Hardin, too, has remained committed to the essentials of his 1970s position. There is, however, one important difference between Goldsmith and Hardin: though both are proponents of rigorous social control, Hardin sees this as only achievable through the medium of strong central authority, whereas Goldsmith believes this is best achieved through the closed, culturally static, monastic commune, admittedly state-created, but relying thereafter upon self-regulation. This, Goldsmith argues, would force people to confront environmental problems at source, rather than passing them on for others to deal with (Goldsmith 1972: 286–88).

Despite their best efforts, the illiberal authoritarianism of Hardin and Goldsmith slipped from prominence. Their legacy has nevertheless lingered; for quite some time after the bloom had gone from authoritarian ecologism it continued to be seen by many ill-informed commentators as the dominant form of environmental politics. There are critics of environmentalism who prefer dealing in stereotypes to genuinely coming to grips with this most complex and fluid of social phenomena; even today it is common to find people who, unfamiliar with environmentalist writings of the 1980s and beyond, still assume the politics of the environment movement to be the politics of Garrett Hardin. But such a politics is rejected by the 'new' green movement with its contrary stress upon strong, participative democracy, on 'One World' inclusivism, and on 'greening' as a process that is positive rather than defensive, reconstitutive rather than besieged, and liberatory rather than constraining.

ENVIRONMENTALISM AND THE CONSERVATIVE POLITICAL TRADITION

Here we enter treacherous terrain: for 'conservative' has become a term of confused currency, this stemming from the fact that the word has long connoted different ideological positions in North America and Europe. I will deploy 'conservative' in the classical sense, as a system of belief having faith in the accumulated wisdom of tradition. A conservative will, thus, stress the social and political importance of venerable institutions — family, the law, religion, the nation — that have evolved as the repositories of that wisdom. Stability is valued because social dynamism, what the conservative might call chaos, is seen as a threat to these traditional sources of wisdom, which are only truly secure when society is calm and settled. Conservatives also tend to hold a low opinion of general human capacities, preferring to trust the judgement of outstanding individuals. They therefore place a high value on leadership, and a low value on popular participation.

There are obvious points of congruence between these values and the assumptions of authoritarian environmentalism. In its illiberalism, survivalist environmentalism shares the conservative's scepticism of the capacity of ordinary people to act beyond their small, personal sphere of competence without thoroughly muddying waters in which they should never have dabbled. And its preference for strong, benign leadership squares with conservatism's faith in leadership by a minority uniquely fitted for rule — fitted, in this case, by their uniquely superior ecological wisdom. Hardin certainly sees it this way. He labels his thought 'ecoconservatism' and places it squarely within the conservative tradition. 'Continuity', he writes, 'is at the heart of conservatism; ecology serves that heart' (Hardin 1985: 233).

Anna Bramwell is another who sees 'the moral and cultural ecological critique' as 'intensely conservative' (1989: 4). This conservatism resides in humankind's limited scope (within the green critique) for ongoing progress via technological change. A species modesty is in order, wherein humans accept a less ambitious conception of their place in the world. An especially conservative variant of this outlook is the neo-Hobbesian argument that, because humans have not the genetic programming to abide by the 'stasis and harmony' that is to be found in nature, humankind should institutionalise its limitations and force itself, almost perversely, to seek only minimal change: 'it is precisely man's lack of a fixed genetic inheritance that makes stable institutions essential as a substitute' (1989: 8).

There are also other ways in which conservative political thought and environmentalism might come together, and here we depart from the authoritarian environmentalism of the 1970s to consider possible sites of overlap that exist even today.

There is the special case of romanticism, already considered. The point was made that the political expression of romanticism has, in the past, usually lodged within the conservative tradition, and, whilst the modern environment movement is not romanticist in the main, pockets of romanticist environmentalism do exist.

Another congruent conservatism — a pragmatic, or 'gut' conservatism — is assumed in the argument that the environment movement is primarily concerned with the defence of middle-class privilege. This is a case that left-wing critics of environmentalism often make. The environment movement's political role, it is claimed, is to prevent working-class encroachment upon what Fred Hirsch (1976) has called the 'positional goods' of the middle class: those social desiderata that are valued because they are not available to most people, such as a living environment that is freeway- and factory-free, or recreational access to unspoiled coastline. Middle-class NIMBYism is a manifestation of it: 'don't build that freeway through my neighbourhood, but I'm happy to have it go through the adjoining precinct'.

An interesting form of conservative environmentalism is to be found in the notion of rural stewardship — a conservative landowner's ethic of generational responsibility to the land. Rose *et al.* put it thus: 'the current title-bearer to a particular piece of land ... is a custodian whose task is to hand on the land, preferably in better heart than he received it, to the next generation ... The land is therefore held "in trust" for the nation' (1976: 713; see also Newby 1985: 67). In Britain this position has cachet within the Conservative Party itself. Flynn and Lowe (1992: 17) quote Stanley Johnson, a Conservative Party member of the European parliament: 'the sense that we hold land on trust for posterity and that we should not therefore permit random destruction and degradation is very much part of the Conservative spirit'. Here, then, is an environmental equivalent of *noblesse oblige*: an ethic of responsibility that has been brought to popular attention through the satires of Tom Sharpe and the television comedy *To the Manor Born.* Rose and his collaborators quote a proponent of this position describing his creed as 'responsible Toryism'. They are, nevertheless, highly critical of it, seeing it as a justification for land oligopoly via a spurious claim of 'environmental responsibility' (1976: 718–19).

This notion of rural stewardship — and it is not just a landowner's faith, for it is also in evidence within rural and village communities rich in history and 'atmosphere' — is largely confined to societies, such as England, where considerable cultural value is placed upon those landscapes and upon strong generational ties to land. It is less in evidence in Australia, for example, where the *nouveaux riches* and the 'newly landed' (often the very opposite of rich) are more likely to support the doctrine that private property rights are unconstrained, and that ownership thus entitles one to do whatever one chooses with one's property.

Finally, conservative–green correspondences are to be randomly found at a quite general level. Thus, the conservative's emphasis on the social importance of tradition, continuity, stability, and organic (as opposed to human-blueprinted) change exactly replicates the values environmentalists defend within ecosystems. Conversely, the notions of 'balance' and 'stability' as central ecosystemic goods are echoed by conservatives as central social principles. It is no coincidence that 'conservation' and 'conservative' both derive from the verb 'conserve'.

The extent to which this strong congruence at the level of first principle establishes environmentalism as *essentially* a conservative political movement is another matter. Clearly, a *prima facie* case along these lines can be made; in some systems of environmentalist thought — Hardin's, for example — these general-level correspondences are taken seriously and promoted. Thus, an environment-promoting spokesman for the British Conservative Party's Bow Group speaks of 'basic Tory principles which are readily adapted to environmentalism: efficiency, order, patriotism, tradition, thrift, self-help, individual responsibility and international leadership"' (Tony Paterson, cited in Flynn and Lowe 1992: 32). In similar vein, John Gray, arguing that, 'far from having a natural home on the Left, concern for the integrity of the common environment, human as well as ecological, is most in harmony with the outlook of traditional conservatism of the British and European varieties' (1993: 124), warns against 'the danger of novelty; in particular, the sorts of innovation that go with large-scale social (and technological) experimentation' (1993: 137). Thus, 'Tory scepticism about progress' is identified as 'a significant point of conservative–green congruence' (1993: 124).

I agree. But this general level of green–conservative coincidence can be put to other purposes. It may be that the project of resisting relentless, purposeless change, of insisting on the importance of projecting the past into the future, has, under current circumstances, destroyed the notion that 'conservative' and 'radical' are conceptual opposites (Giddens argues this point at length; 1994: 22–50, 117–33). We have seen how the defence of authentic place is both 'conserving' and, in asserting the priority of community rights over capital, 'radical'. That discussion has relevance here. Blackwell and Seabrook, coming at the same issue from a slightly different perspective, argue that the modern experience of ceaseless change has set up a deep-seated yearning for a less frenetic existence:

> we begin to wonder if the reason why parties advocating radical change were so unsuccessful was because they were striking against the resistance of people who had changed, who had been compelled to change, too much. The experience of industrialisation had been of driven and relentless change, and continues to be so ... So why should we expect that exhortations to change will be welcomed by those who have known little else for at least two centuries? (1993: 3)

Historically, the radical project was one of resistance to the tyranny of inertia. But the world of industrial capital is not a world of inertia. It is a world characterised by a level of dynamism without precedent in human history, and in that dynamism's carelessness to deep-seated human need is to be found another tyranny. The radical project now is to oppose this tyranny. Blackwell and Seabrook argue:

> in this context the desire to conserve, to protect, to safeguard, to rescue, to resist becomes the heart of a radical project. A form of conservatism — to be most sharply distinguished from its multitude of imitations, its travesties and caricatures, and scarcely known to those who carry the banners of conservatism in the modern world — becomes indispensable to this work of resistance ... It may be a paradox that the only radical politics left to us should be based upon resistance, recuperation, remembering. But in a social system which requires the reverse of all these things, to oppose means to conserve (1993: 3–4; see also Giddens 1994: 10).

Giddens maintains that 'the political force which can lay greatest claim to inherit the mantle of left radicalism is the green movement' (1994: 10), for the green movement is best placed to articulate a re-jigged radical tradition, one needing to borrow certain tools from the conservative toolbox: 'the preservation and renewal of tradition, as well as of environmental resources, take on a particular urgency' (1994: 49). Blackwell and Seabrook, by contrast, claim only a qualified sympathy for the green paradigm; nevertheless, the passages cited above give a most apposite description of the political locus of the green movement. Some proponents of earlier left-radicalisms have been hard put to see how an anti-modernist movement can lay claim to a site within the radical space of the ideological terrain. The collapse of the conservative/radical dichotomy along the lines suggested by Giddens and Blackwell and Seabrook provides an answer. It also suggests that, in one important sense, the legacy of the current environment movement is at least in part conservative.

NATURE AND FASCISM

Prominent among the criticisms levied against environmentalism is the charge that it is a form of incipient fascism or neo-fascism. Fascism is perhaps best defined as a pathological conservatism: the elements of conservatism warped and twisted into a parody of themselves, by concentration on certain selected symbols (particularly the nation), which are mystified and deified.

For some, such as Richard Neuhaus, to conjoin nature and social theory is sufficient unto itself to warrant application of the label 'fascist' to such theory. Whilst such a crude application is insupportable, the claim that environmentalism shares dangerous affinities with fascism must be taken seriously, for there are elements which, in certain combinations, might conduce to a form of right-wing irrationalism.

These include the use of biological metaphors, a stress on the organic community and the individual's need to merge with it, the elevation of ritual, intuition and the mystical, and a corresponding distrust of the rational — all of which are to be found in ecological writings. They are also key elements of National Socialist ideology; Hitler used biological analogies, Bookchin reminds us, 'to underpin his viciously racist theories' (1986a: 10).

The ecological view that humankind is an integral part of nature can also be used to defend a social Darwinist position, with its justification of the intra-species aggression that is fostered by the totalitarian right. As Bramwell puts it:

> it is possible to assert that if man is part of the natural world, subject to the same laws as the animals, then he is, like them, entitled to compete to survive. Because he cannot hope to escape from his animal nature, he is justified in aggression. Thus is the social Darwinist argument associated by many with a politics based on nature (1989: 7).

In fact, Bramwell argues, the possible correspondences between ecological precepts and right-wing totalitarianism are even more explicit than this. Though distinguishing between National Socialism and fascism generally, 'a distinction not valid in all spheres but important here', for 'orthodox' fascism 'did not have or attract a green component' (1989: 20–21), she portrays a close congruence between today's environmentalism and the ideas and programs of some key Nazis. Her account of Richard Walther Darre, the Third Reich's minister for agriculture, describes a man who held now-fashionable ideas in favour of small, organic farming (1985; 1989: 201–05). How, Bramwell asks, did the Nazis come to be 'the first radical environmentalists in charge of a state' (1989: 11, 161–94)? To what degree did their idea of ecology conform with ours, and were the similarities more than 'embarrassing accidents' (1989: 11, 195–208)?

As Bramwell describes it, after World War I, in Germany and elsewhere, yearnings for a rural utopia hardened into a belief in the socially regenerative capacity of direct contact with the land. The life of the yeoman was exalted and the soil itself was accorded a quasi-mystical character. One outcome of this back-to-the-land urge was the appearance of new scout movements which Bramwell discusses under the heading 'Green Shirts' (1989: 105–12). The Kibbo Kift Kin, for instance, was founded in England by John Hargrave, a man with a bewildering mix of ideals. A Quaker, socialist and pacifist, he was also a pantheist with a leaning towards eastern religions, and to this he added a fervent belief in Anglo-Saxon nationalism. The Kin were expected to do more than simply expose slum children to the fresh air. Hargrave envisaged them as an elite which would, in time, lead to a counter-government, and to that end he recruited H.G. Wells, Julian Huxley, Havelock Ellis and the Indian poet-theosophist, Tagore.

Bramwell does not report whether he persuaded these luminaries to wear the Kin uniform — a Saxon cowl and jerkin beneath a Prussian army cloak — but she does observe that his writings were widely read in Germany in the 1920s (1989: 106).

Here is a queer mixture of left and right, and Bramwell shows how, in England, 'ecological' ideas were taken up and reinterpreted as much by the Tory right as by the utopian left (1989: 105, 161). Hargrave, for example, was much influenced by Rolf Gardiner, friend of D.H. Lawrence (himself an important figure in the English 'blood and soil' movement). Gardiner, though a socialist in his youth, was a Nordic racialist, a pagan, and a keen supporter of Nazi rural policies. He was also a pioneer of organic farming, running his Dorset estate on organic principles as early as the 1920s.

As might be expected, anomalies also abounded in Nazi Germany, where much the same mixture of utopianism, mysticism, folk-myth and scientific insight was set in place. 'Ecological' ideas appealed to the Nazis because they, too, believed that the laws of nature were immutable, and, Bramwell argues (1989: 205), with their economics of state-managed rather than market capitalism, they approved ecology's opposition to the laissez-faire market. The Nazis were persuaded by Darre's enthusiasm for organic farming; Himmler set up several experimental farms, including one at Dachau (1989: 204). Rudolf Hess was a convert, an enthusiastic supporter of bio-dynamic farming, as well as a homeopath and naturist. But, following his flight to Britain, all such ideas were tainted (1985: 178–79) and the SS ordered organic farmers to be harassed along with nudists and other 'minor deviants'. Darre, however, never lost his faith in the small farmer and organic methods, and after the war he tried to set up a Soil Association in the new Germany.

All this is in the past. But Germany has recently experienced an upsurge of neo-Nazism, and a right-wing ecology has been employed by several neo-Nazi groups as a mobilising device. Janet Biehl (1994) surveys these groups and their 'ecological' positions. In common is a depiction of environmental despoliation as the consequence of 'the arch-rationalist creeds of liberalism and Marxism' (1994: 133). 'Ecology' is used as a weapon in the ideological offensive against non-German immigration and the guest-worker program by linking nationalism 'mystically' to an 'ancestral' landscape — 'the message being that people should be where they are 'biologically embedded' (1994: 134). She writes of the neo-Nazi position:

> American cultural imperialism is genocidal of other cultures around the world, and its technological imperialism is destroying the global environment. The fascist quest for 'national identity' and ecological salvation seeks to counter 'Western civilization' — that is, the United States, as opposed to 'European civilization' — by advancing a notion of 'ethnopluralism' that seeks for all cultures to have sovereignty over themselves and their environment (1994: 133).

Biehl notes, though, that rather more passion is accorded the larger nationalist/cultural projects than their ecological subsidiaries.

Despite conservative and neo-fascist adoption of some ecological ideas, today the mainstream environment movement is firmly committed to the social justice concerns traditionally associated with the left, and most green-leaning people express puzzlement when confronted with the arguments of Bramwell and Biehl. Could the green movement become a form of rightist irrationalism? This question is considered below.

SOME PROBLEMS: CONSTRUCTING CASES FROM GREEN EPHEMERA

Much could be written in response to the matters discussed above, but we will consider just two questions in relation to the forms of environmentalism discussed here: the extent to which they set the tone for environmental thought generally, and the extent to which they could come to carry the environment movement's standard in the future, if they do not already do so.

Unable to restrain myself, I have already foreshadowed the answer to the first of these questions. The defensive doomsday phase of environmentalism — with its penchant for authoritarian prescription and its inability to recognise any potential for progressive theory-building in the responses to environmental crisis — passed from prominence in the late 1970s. It was replaced by a politics in some respects its precise opposite: a politics of small-scale participatory democracy rather than large-scale centralised control; a politics of compassion and support for the peoples of the third world rather than the lifeboat ethic; and a politics that sees humankind in terms of unfulfilled emancipatory and altruistic potential rather than a brutish, unrestrained human nature.

Such a sea-change was undoubtedly impelled by doomsday science's credibility deficit after the much-publicised debunking of the *Limits to Growth* Report's computer-generated scenarios for exhaustion of the earth's stock of non-renewable resources. Predictions of the imminent demise of environmental politics were even thicker on the ground at this time, the mid-1970s, than is usually the case; they proved just as ill-considered. Instead, green politics metamorphosed. 'Conservation' became 'environmentalism' (and shortly afterward, 'green'), and the doomsday scientists with their naive and illiberal politics were displaced at the movement's cutting edge by philosophers and social theorists concerned to demonstrate, not just how the world could be saved, but how life might be better lived.

In his 1798 *Essay on Population*, Thomas Malthus had argued that population growth increases exponentially and, therefore, always outstrips food production, which can only be increased at a constant rate. Famine among the rapidly breeding lower orders is thus inevitable, for it is the mechanism whereby carrying capacity is restored. Malthus's dismal vision earned him the opprobrium of the left, and the apparent acceptance by doomsday scientists and the early environment movement of Malthusian precepts and of his disdain for 'the masses' drew sharp criticism from both the left and from free-market radicals. The broad abandonment of Malthusianism was one of the first visible consequences of the sea-change of the late 1970s; though Malthusianism did not lose corresponding prominence within anti-environmentalist criticism (possibly because some high-profile groups, Earth First! in particular, continued to adhere to it). As Eckersley points out (specifically of ecocentrism, though the position she articulates would also apply to many non-ecocentric participants within the movement):

> an ecocentric perspective would necessarily reject the Malthusian response and support the case for a more equitable distribution of resources among the world's existing human population alongside the development of more 'appropriate' food sources and production methods. This is because an ecocentric perspective is concerned with human *and* nonhuman emancipation (Eckersley 1992: 159).

The rejection of Malthusianism can be traced to the waning of the influence of doomsday scientists, for whom the 'law of nature' formulation of Malthus greatly appealed. Bramwell characterises the 'moral and cultural ecological critique' as 'intensely conservative' whilst identifying the *radical* impulse within the ecological movement as scientifically inspired: 'the essential characteristic of ecology, while it does not fit happily into any one ideological category, is that it draws many of its conclusions from scientific ways of thinking, and is not conservative' (1989: 7). Here Bramwell gets the current environment movement very much askew. As we have seen, the ingredients for an 'intensely conservative' moral and cultural ecological critique certainly exist, and Bramwell may be right in claiming that there are specific historical instances in which these ingredients have potently coalesced. I think it is also correct that, thought through, the consequences of ecological *science* are profoundly radical (though, as we have seen, this assertion can certainly be disputed). But the scientists who determined the agenda and values of the early modern environment movement did not think through these consequences in this way. They developed a politics that was undemocratic, authoritarian, pessimistic, repressive, illiberal, static and closed. Whereas the subsequent phase of the movement — one in which the purveyors of the 'moral and cultural ecological critique' have determined the agenda and the values of the movement — has in most respects been the very opposite: strongly democratic, optimistic, dynamic, and, as a result, visionary, progressive and emancipatory in its politics.

Given the revolution in the core assumptions of environmental politics, there is little point considering critiques of the environment movement that assume the politics of the pioneering doomsday phase. Such analyses (for example, Beresford 1977; Wells 1978; Holsworth 1979) are now as obsolete as writings that seriously engage with phrenology or flat-earthism. Of course, this observation only applies to critiques that so characterise the ecopolitical core rather than outrigger segments. Given the diversity of position under the rubric of environmentalism, it is important to identify the values of the mainstream, and distinguish the voices that lay claim to a place in the movement, but which cannot in any sense be said to speak on its behalf. Thus, it is entirely legitimate to critique individuals and particular groups — such as Hardin, who proffers what is now a dissident voice, and Earth First! — as long as the critic does not claim to present an analysis of the wider movement.

A more difficult question is whether, given that the early doomsday analysis still has its flag-bearers, such views could one day regain the ascendancy. It is to this we now turn.

There is a sense in which the dominant emancipatory current within green thought is not completely secure.

Critics of Hardin's 'tragedy of the commons' formulation have argued that he is describing 'not a commons regime, in which authority over the use of forests, water and land rests with a community, but rather an open access regime, in which authority rests nowhere, in which there is no property at all' (*Ecologist* 1993: 13). We have already noted this argument; it is a position which is persistently put. It is argued, for instance, that Hardin overlooks the capacity of commons users, faced with degradation, to put in place remedial institutional arrangements. Societies do have 'the capacity to construct and enforce rules and norms that constrain the behavior of individuals' (Feeny *et al.* 1990: 13). This can be done, moreover, at the level of the commons-using community itself.

Despite the frequency with which this commons–community nexus is championed, there is nevertheless some consensus within the environment movement that Hardin's 'tragedy of the commons' thesis insightfully identifies an important 'social law' and explains much environmental degradation. It is the *political* prescription that Hardin appends to it for which there is little evident support.

Similarly, the retreat from doomsdayism is usually not based on a rejection of the *essence* of the Limits to Growth case. There is widespread acceptance that limits do exist, that there *are* points of no return, before reaching which far-reaching remedial action will need to have been taken; that the prospect of catastrophic breakdown of global biophysical systems as a consequence of human-induced overload is real. It is conceded, though, that the 1970s scientists overstated the imminence of catastrophe; the grim political prescriptions that were appended to their science are also rejected. But green activists work within a climate of quiet despair, a sense that there is not enough time in which to mobilise and win either local contests or the larger systemic one. Should evidence build to the point that catastrophe is surely imminent, and should the politics of interest brokerage continue to prove muddlesome and ineffectual, it is entirely possible that a resurgence of an ecopolitics of centralised authoritarianism could occur.

There are also sound reasons why it should *not* occur. The evidence suggests that getting the happy result of a quasi-dictatorship imbued with ecological wisdom is not easily achieved. Though not formally associated with the political right, the countries of the Soviet empire were certainly centralised authoritarianisms: and they have the worst environmental histories in the postwar world. Similarly, developing countries that have centralised political authority and surrendered democratic freedoms in order to mobilise for rapid industrialisation are also incurring massive environmental despoliation. Michael Redclift argues that the tendency for centralised authoritarian regimes to provide scant attention to ecological considerations is not coincidental, and that only an open society can supply a climate of values that will conduce to environmental protection: 'extending the role and influence of the state has only served to exacerbate environmental problems, removing their solution from the hands of those worst affected, and putting it into the hands of those for whom immediate economic growth must have priority' (1989: 182), whereas:

> green movements have arisen in response to strong, existential needs, for the utopian as well as for the pragmatic, for citizenship and 'the people' rather than the ubiquitous 'masses'. Achieving environmental objectives requires more democratic control and the espousal of different values (1989: 183).

The lesson seems to be, then, that the bureaucracies empowered by centralised authoritarianism contain operational assumptions that actually militate *against* environmental preservation. Intervening, changing, doing — rather than maintaining, preserving, sustaining. Thinking in the short term rather than the long term. Technocratically enthused rather than focused on the strange world of green life 'out there' beyond the city office. Building empires and mobilising resources rather than 'going gently'. The larger and more centralised the political system, the greater that society's technological capacity; and the greater that capacity, the more likely that those in charge will succumb to the temptation to fulfil it — especially as to do so will provide additional and better resources to use against the regime's opponents (external *and* internal). Orr and Hill, in an early but still unbettered essay on this topic (1978), develop these and similar points. They note that 'authoritarian groups are not as creative as their democratic counterparts, especially when confronted by complex problems' (1978: 463), whilst:

> without mega projects and war there is something of a void. How does the state justify its existence? Implicit in Heilbroner, Ophuls and Hardin is the questionable belief that an ecological authoritarian state could derive its legitimacy from its alleged capacity to preserve the environmental balance. But maintenance of complex and subtle ecological balance affords little opportunity for visible achievement and thus a minimal base for government legitimacy (1978: 464).

Theorists who argue along lines similar to those advanced here — that there is a direct and negative correlation between the power of the state and the attention given to environmental quality — often draw upon the classical study of Karl Wittfogel (1957), who examined the political importance of control of water in the construction of large Asian civilisations (see Walker 1985; Worster 1983). Wittfogel termed these 'hydraulic societies', the 'elaborate apparatus of water control' so rigid and powerful that these societies could properly be designated 'totalitarian' (Worster 1983: 169). Worster's 1983 study of the operations of a powerful water authority involved in dam projects on the Colorado River is strongly Wittfogelian in its conclusions. He also argues that a culture of bureaucratic control leads to the sponsorship of destructive technologies and the politically motivated manipulation of people and nature, echoing one of Orr and Hill's themes: 'large organizations, whether the state or the corporation, will further increase the reliance on "high" technology, itself a major source of ecological hazard, while they simultaneously lower social adaptability' (1978: 465). Whether it is the capacity provided by a large centralised authority to dominate people that leads to a commensurate domination of nature, or whether the causal relationship works in the other direction, the message is the same: large centralised authoritarianism seems, in its very bones, to conduce to ecologically irresponsible government.

This being the case, those advocating such a form of government as the last and best hope for a seriously ailing world are under an obligation to demonstrate how a dictatorial system can be brought to pass that will be, against the odds, ecologically responsible. Such strategy has not been forthcoming. Similarly, one of the main planks in the case against the messy politics of democracy is that the planet has so little time in which to get things right that we just cannot afford the luxury of drawn-out democratic decision-making. Those making this case need to explain where and how the time will be found to undermine, overthrow and replace a series of political systems that still, by and large, have widespread legitimacy. In the absence of such a demonstration, those arguing a 'there isn't time' case against the gradual processes of liberal democracy are likely to have their argument turned against them. They may well be right that there is not the time for democratic processes to take their slow course; on the other hand, there *certainly* is not time to wait around for the installation of a nowhere-in-sight, political alternative.

The authoritarianism of Garrett Hardin is often what critics have in mind when they use the label 'ecofascist'. But the stern rationalism in which Hardin and the doomsdayists place such store is the antithesis of the crusading *rejection* of rationalism that is central to fascist ideology. Is it possible that an ecologically responsible fascism could become the dominant form of environmentalism?

A reading of Bramwell's earlier books certainly conveys that impression. 'There are Greens today', she writes, 'who feel unease at some of their ideological forebears' (1989: 7). Some certainly do evince unease: though it is my impression that green awareness of the existence of these purported 'ideological forebears' is a singularly rare phenomenon (in contrast to the situation within anti-green think-tanks, where Bramwell seems to be required reading!). There is no need to 'feel unease', though, because the Nazis are not the 'ideological forebears' of the green movement. The point has been made that the current environment movement is discontinuous with pre-1960s proto-ecologisms (philosophical and activist), and that those who pioneered the current movement in the 1960s did so in historical ignorance, unaware of precedents upon which they might, in some cases, have usefully drawn. This given, there is no umbilical link, no *legacy*, connecting German Nazism and the current green movement. At best, there is a coincidence of some ideas: and it is worth asking how strong even this coincidence is.

A reading of Bramwell suggests a very strong coincidence. The Nazis, she claims, were 'the first radical environmentalists in charge of a state'. Bramwell's books reveal a capacity for meticulous research — let down by some strange analysis. For example, a reading of the 'ecological' basis of English fascism and German Nazism shows her case to be built around a 'blood and soil' folk mysticism, a belief in the cleansing, recreational value of contact with the outdoors, and, most importantly, an attraction (and that temporary) to organic farming and holistic medicine. But even at best these are peripheral matters within today's ecology movement, and each of them is, as well, capable of being taken up by people otherwise *hostile* to the greens. Instructive are two letters-to-the-editor written by Timothy McVeigh, the man convicted of the Oklahoma City bombing. McVeigh was a member of the burgeoning, armed-and-angry, populist, anti-government right. Taking the brunt of this backwoods anger are officials charged with implementing federal environment legislation, which has attracted the angry right's especial detestation. Many officials have been threatened and molested whilst carrying out their duties. Some have been run out of town. Yet McVeigh's letters reveal an opposition to the factory farming of animals, whilst the brothers Nicholson, arrested as 'material witnesses' in connection with the bombing, were reportedly living on an 'organic farm'.

Again, what this suggests is the importance of getting to the core values of the movement, and to avoid the temptation, into which

Bramwell falls, of categorising it in terms of its epiphenomena. Hence my earlier claim that it is not fascism generally, but the case of Heidegger particularly, that a green movement wishing to distance itself from the totalitarian right must combat most urgently — for Heidegger's place within green thought is *not* epiphenomenal. The difficulty of nailing down a more general argument linking Nazism (or any form of fascism) with today's environmentalism is thrown into sharp relief if we consider the so-called 'four pillars' widely adopted throughout the green movement as the bases upon which activist organisations should rest: ecological interconnectedness, non-violence, social justice and grassroots democracy. On only one of these, 'ecological interconnectedness', do fascisms sometimes register, and the extent to which even this is so is highly debatable. It is difficult to reconcile a belief in ecological interconnectedness with a willingness to engage in massive civil and state violence, for nothing is more destructive of ecological relationships than the devastation wrought by modern warfare.

And here we come to the heart of the problem. Fascism is exclusivist: it sets people against each other. Some superior people are to inherit the earth (or chosen parts thereof), at the expense of other, inferior people. The green movement's globalist inclusivism is inimical to such intra-species characterising, making a present-day conjunction of ecology and fascism highly unlikely.

But a qualification is in order. Unlikely — but not impossible. There are those in the environment movement who prefer a centralised authoritarian state, and, as we have seen, many of the anti-rationalist beliefs that are manifest within fascism are to be found among the ingredients within the green kitbag. To bring these together is to bring about ecofascism.

This is precisely Janet Biehl's fear. Her purpose in surveying the resurgent Nazi sects is to demonstrate that:

> an ecology that is mystical, in turn, may become a justification for a nationalism that is mystical. In the New Age milieu of today, with its affinities for ecology, the ultra-right may well find the mystical component it needs to make a truly updated, modernized authoritarian nationalism (1994: 134).

Biehl's essay falls within the Institute for Social Ecology's campaign against the anti-modernist, spiritualist wing of the green movement. She recounts with relish the religious ideas of Henning Eichberg, the founder of one neo-Nazi sect, the 'National Revolutionaries'. Eichberg advocates 'a new religion that mixes together neopagan Germanic, Celtic and Indian religions with old volkish-nationalistic ideas. It is to be based on "the sensuality-physicality of dance and ritual, ceremony and taboo, meditation, prayer, and ecstasy"' (1994: 136). But a position such as Biehl's ignores the specific and uniquely Germanic conditions that have made a destructive conjunction of nature mysticism and violent nationalism possible. Bramwell draws attention to this. So does Simon Schama, who describes how forests have served the full gamut of ideological symbols within western nationalisms: in the United States 'the woods' are God's immanence on earth; in Poland they are the spirit of liberatory nationalism; in England 'the greenwood' was, for the 'freeborn Englishman', the 'bulwark of liberty', the 'Heart of Oak'. Only in Germany did the Black Forest come to symbolise racial purity and 'militant nationalism' (1995: 118–19).

We can conclude that a credible path from the gently pacifist, lightly interventionist, *pre*-nationalist mysticism of some sections of the western ecology movement to the aggressive, warring, high-tech nationalism of neo-fascism cannot be plotted. So, we can discount Biehl's conclusion that 'when "respect for Nature" comes to mean "green reverence", it can mutate ecological politics into a religion that "Green Adolfs" can effectively use for authoritarian ends' (1994: 170).

If an authoritarian-right green movement does evolve, it is more likely to emerge from a resurgence of despair in the face of biophysical breakdown than from a within-the-movement evolution from new age ecological spirituality. Certainly, an ecological disaster of global proportions could render redundant much of the foregoing reasoning, perhaps especially as it applies to prospects for a non-fascist, ecologically informed authoritarianism. But there is no guarantee that either a fascist or non-fascist authoritarian political order arising from a breakdown in global order would be ecologically responsible; indeed, given fascism's penchant for scapegoating in times of crisis, ecological wisdom seems an unlikely outcome. As Bramwell has pointed out, it was only the National Socialists of Germany and the fascists in England who, among the European fascists in the 1930s, showed any interest in 'ecological' ideas. We have seen that Germany's neo-Nazis continue to hold such an interest. Within other resurgent neo-fascisms the most pronounced interest in ecological concerns seems to be found within Jean-Marie Le Pen's National Front in France (though Bramwell also cites Italy and Belgium; see her discussion on the 'ecological' component within the European *nouvelle droite*; 1989: 231–33). Whether, and in what form neo-fascist political forces take on an ecological focus thus seems entirely random.

It also seems all but impossible for an 'ecologism' of this type to evolve from today's environment movement. Any such development would be separate from, and strongly resisted by the lineal heirs to what is currently recognised as *the* green movement.

Turning, finally, to conservatism, I am inclined to be less forthright, noting that there is an important congruence between the conservative's stress on social order and his/her counsel against running the unpredictable risks of political disruption, and the ecologist's stress on ecological balance and his/her counsel against running the unpredictable risks of ecosystem disruption. I am, too, much in sympathy with the Blackwell and Seabrook and Giddens arguments that to resist change — to resist the breakneck, devil-take-the-hindmost character of change within the increasingly global capitalist market — is to engage politically, not on the side of the elite and the privileged, but on the side of the exploited and victimised (not excluding the imperialised and increasingly impoverished domain of non-human life). History has, thus, imposed upon us a contradiction in terms — the radicalising of conservatism.

Not everyone is taken with this argument. Seeking to establish the emancipatory credentials of the green movement, Eckersley concludes that 'the conservative political tradition may be ruled out', despite having some 'resonances with emancipatory ecopolitical thought', because 'conservatism's endorsement of the established order, hierarchical authority and paternalism and its resistance to cultural innovation and social and political experimentation put it at considerable odds with the egalitarian and innovative orientation of emancipatory green thought' (1992: 30). There are certainly aspects of Gray's green conservatism that will not find ready acceptance in the green mainstream. Gray emphasises compacts between the generations, but he has little to say about human responsibilities to other lifeforms. He endorses the steady-state economy (1993: 140–42), but is critical of green 'animus' against 'the distinctive technological and social forms of modern life'. He expresses support for nuclear power, which is 'one of the least invasive of modern technologies' (1993: 126) — though the notion of a nuclear-powered, *steady-state* economy certainly asks much of human credibility! Gray also criticises green rejections of 'the environmental benefits of market institutions' (1993: 126), whilst himself seeming to reject the very market dynamism — some would say the *intrinsic* market dynamism — that inexorably promotes the 'untried novelty', 'large-scale social, and technological experimentation', and the fragmentation and obliteration of 'the common life' that he so deplores. It is, thus, likely that the 'emancipatory', 'egalitarian' and 'innovative' current seen by Eckersley as central to the emerging green tradition will continue to preclude significant conservative inroads into green allegiances.

8

ENVIRONMENTAL LIBERALISMS: GREEN THOUGHT MEETS THE DISMAL SCIENCE

ENVIRONMENTALISM AND THE LIBERAL RIGHTS TRADITION

The 1990s witnessed an upsurge of interest within ecological thought in the liberal tradition, an interest that stands in contrast to the 1970s and 1980s, when environmentalist thinkers saw little ground for compatibility between their own core values and those of the liberal and neo-liberal traditions. But even then there were alternative voices.

John Rodman, for example, explored the possibility of deriving an ecological ethics from the thought of the nineteenth century English liberal, T.H. Green, as early as 1973. Rodman argued that the principles of diversity and toleration, in conjunction with the liberal tradition of establishing the community of rights and duties at the broadest range for which justification can be found, can, by extension, be so drawn as to include the sphere of non-human nature. For Rodman:

> there are vital elements in Green's thinking: the notion that rights and obligations are functions of community and that property rights are subordinate to the general welfare; the conviction that a community has a responsibility to make the rights of its weaker members operative; the view of the outer boundaries of the ethical community as expanding in time; the vision of reality as a dialectically differentiating and unifying system of relations that includes humans and nonhumans as members; and even the postulate of an 'eternal consciousness' transcending anthropocentric and historical perspectivity and yet at least partially available to man: these suggest possible paths to a new world view (1973: 584).

Not all ecologically based ethicists are comfortable with ethical extensionism, and Rodman himself became a prominent early critic. What we have here, though, is an argument that a basis for an ecological ethics *can* be found within the liberal tradition. Rodman's project, in fact, was nothing less than the ecological rehabilitation of liberalism:

> certain elements will be selected for extinction - e.g., mere humanism, the contract metaphor, and the economic ethic of acquisition, production and reproduction - while others will be selected for survival - e.g., the notion of an on-going process of liberation by way of an ever-expanding sphere of rights, and the basic Liberal principle of toleration. To the original justification for toleration can be added an ecological support: diversified ecosystems are more stable, more vital, less susceptible to disease, disaster, and death, than are monocultures (1973: 585).

Since 1973, however, liberalism's economic stream has triumphed comprehensively over the elements within it that Rodman wished to see elevated. Concerning an economistic liberalism, Rodman noted: 'it seems unlikely ... that a moral/political philosophy that remains within the economic Liberal paradigm of salvation through the insatiable transformation of nature into property will suffice today' (1973: 580). Yet, the near-total hegemony that has been attained by 'the economic Liberal paradigm' is such that some observers (most prominently Fukuyama 1992) have been moved to proclaim the end of political and economic evolution, deeming liberal capitalism and liberal democracy to constitute the acme of economic and political advance. Speak of liberalism and the version that most readily comes to mind is the free-market economism in which Rodman placed no more store than did most of the green thinkers who followed him. It is for this reason that the environment movement of the 1980s saw 'liberalism' as the arch-opponent of its own life-privileging philosophy.

But there are divergent strands within liberalism. The utilitarian school of English liberalism, through the medium of John Stuart Mill's *Principles of Political Economy*, stands as progenitor of a compassionate, humanist liberalism that informed the creation of the welfare state (and, hence, the regulated economy). Following a usage traceable to Isaiah Berlin (1958), Wissenburg refers to this as 'positive' or 'social' liberalism (1998: 76; see also P.R. Hay 1990), which is essentially 'freedom to pursue one's chosen path', and is distinguished from 'negative' liberalism, which is essentially 'freedom from external constraint'. Mill has ecological credentials as well. These, too, were advanced by Rodman (1977a), and, though Mill now receives only passing attention, in the 1990s some of these unfashionable liberal traditions began to attract attention from ecological thinkers, albeit it remained the case that 'for much environmental thought it is the very values and practices implicit within liberal justice theory which now constitute the key environmental danger' (Vincent 1998: 443).

Much of the discussion of the relationship between liberal and ecological values takes place within a critique of liberal democracy. But there are many liberalisms. The animal rights philosophy of Tom Regan, for example, properly belongs within 'deontological liberalism' (the rightness of an action is independent of its outcomes). The animal liberationism of Peter Singer also belongs to a liberal tradition, but a rival one — utilitarianism (the rightness of an action is to be judged by its consequences) (diZerega 1996a: 702). Here we are concerned only with defences of a green liberalism that engage with principles in the abstract, setting aside consideration of 'actually existing' liberal democracy until chapter 10. We can do this because the institutions of liberal democracy as we know them in the west are not the only possible political arrangements that can be derived from liberal principles. 'As it has evolved', writes diZerega, 'the liberal order tends to respect neither the natural processes necessary for the individual flourishing of nature, nor individual plants and animals' (1996b: 174; see also 1995). diZerega laments this state of affairs; but his project is to demonstrate the relevance of liberal precepts to ecological imperatives. He also makes positive reference to 'liberal democracy'. He can do this because, as with all the writers considered in this section, he is less concerned with *the* liberal democracy as he is with *a* liberal democracy.

Distinguishing between principles and their familiar applications can be made with even more confidence in the case of the capitalist market. We have seen that Rodman, in 1973, contrasted his liberalism with the 'narrow' liberalism of free-market economism, and this distinction is retained by the theorists considered below. This even holds for Wissenburg, who is the most 'orthodox' of our ecologically minded liberal theorists:

> political liberalism is not predestined to a laissez-faire attitude towards the market, nor does it hold that life is all about making profits or the satisfaction of material self-interest; it sees trade and commerce as one way of life, one means to the realization of a plan of life, among others. The critique of economic liberalism must therefore be judged on its own merits, and cannot reflect on political liberalism (1998: 213).

Wissenburg notes that the sustainability critique has presented liberalism with its stiffest challenge yet, but the ensuing legitimacy crisis is to be welcomed, because it gives liberalism the incentive to work through some grey areas. In Wissenburg's view liberalism can meet all the requirements of sustainable existence by drawing upon John Rawls's 'savings principle', reworked by Wissenburg into his 'restraint principle'. To construct his 'restraint principle' Wissenburg takes Rawls's idea that 'society is a scheme of cooperation among different generations existing next to one another', and the familiar case for intergenerational equity (the 'obligation to treat the next generation no worse than you want the previous one to have treated you'; 1998: 88). According to this restraint principle, 'every physical object should be dealt with as if it is insubstitutable, at least as much as humanly possible' (1998: 88). To this, he adds an 'inverse restraint principle': 'if no good may be destroyed (unless necessary) then no good may be wasted (unless necessary): in other words, no ungood may be produced (unless necessary)' (1998: 166). 'Against our expectations' an apparently surprised Wissenburg concludes, 'this leaves us ... with a far better protection of nature or natural capital against exploitation than could have been expected' (1998: 88). 'Discounting the future' is a familiar tool in the economist's preferencing repertoire, and one often used in somewhat cavalier fashion. But not under the Wissenburg restraint principle. This principle 'can only be overruled by the quest for survival, i.e. if there is no other way to preserve a present life worth living other than by harming a future life' (1998: 136).

Wissenburg is by no means the only theorist to adapt aspects of Rawls's theory of justice to environmentalist ends. Others to do so include Brent Singer (1988), Peter Wenz (1988), Wouter Achterberg (1993) and Roger Taylor (1993); though Wissenburg is critical of the use made of Rawls by some of these theorists (1993:17). We will consider just two of these: Achterberg and Taylor.

Like Wissenburg, Achterberg sees the challenge to the legitimacy of liberalism induced by ecological crisis as an opportunity for refurbishment — though the refurbishment he envisages goes somewhat further than Wissenburg's. It entails 'a far-reaching correction of the market mechanism — more far-reaching than we have been used to and than is considered desirable these days' (1993: 90). But change should only occur if it accords with aggregated individual preferences and, as we can expect 'strong opposition against the limitation of market processes', there is a need to establish 'a shared public basis on which to ground the *legitimacy* of restriction and corrections' (1993: 91).

Achterberg finds this in the liberal principle of neutrality: there are normative principles about which such common agreement exists that their claims are effectively removed from political contention (1993: 92). But this principle cannot be pushed too far. This is because — and here Achterberg turns to Rawls — democratic societies are characterised by 'an enduring pluralism that cannot be suppressed by the government except by forceful means' (1993: 94). Given, then, that clashing views of the meaning of life are endemic, Rawls argues that 'an agreement about the fundamental organization of society is only possible as an "overlapping consensus"', a conception of justice that can 'survive a shift in the balance of power' between partisans of rival visions of 'the good' (Achterberg 1993: 94).

Achterberg proposes to add 'new elements' to Rawls's minimalist conception of overlapping consensus (1993: 95). These pertain to the requirements of sustainability. Achterberg concedes that this will involve some clashes of rights and liberties, but he argues (from Rawls) that qualification upon some liberties 'are acceptable for the sake of other liberties which would come off badly otherwise'. What is needed is a more elaborated system of rights and liberties, constructed, moreover, 'with an eye to those of future generations' (1993: 96). Market rights are among those that will require modification, but (still staying with Rawls), 'the right to own means of production does not belong in the list of basic rights'. Achterberg, thus, foresees 'a regulation of the right to own means of production and natural resources', and this is likely to also require 'regulation and the introduction of more planning elements' (1993: 96).

Roger Taylor, the third of the Rawlsians considered here, is rather more ambivalent about the liberal tradition than either Wissenburg or Achterberg. Whereas they saw the challenge to the legitimacy of liberal values posed by environmental crisis and environmental thought as an opportunity for renewal, Taylor candidly observes that 'liberalism may not be the best theoretical approach to protecting the environment; it may indeed prove inadequate to the task' (1993: 279). Nevertheless, he argues that the extent of environmental obligation arising from liberal theory, 'while perhaps unsatisfactory to ecocentrists, may nevertheless be surprisingly extensive'. Though liberal approaches must, axiomatically, eschew ecocentrism and seek environmental protection 'purely as an instrumental means to the end of achieving justice for human beings', general ecological protection could be secured 'through substantive and sufficiently strong obligations to future generations' (1993: 266).

Taylor examines several important contributions to recent liberal thought, and finds them variously useful. Rawls, for instance, provides formidable reasons for protecting natural systems in the name of the rights of future generations (we have seen that Wissenburg also holds this view). On the other hand, Taylor rejects libertarian Robert Nozick's defence of a right to pollute, finding it inconsistent even with Nozick's own core principles (1993: 277). Considering the libertarian position more generally, however, Taylor concludes that 'although libertarianism places no intrinsic value on the environment, its absolute opposition to coercion would appear to condemn virtually any form of environmental degradation, since such degradation almost always affects non-consenting individuals involuntarily' (1993: 279). In sum, then, 'since liberalism requires that all members of society be treated equally in some respect ... it gives birth to obligations to protect the environment in the interests of those who are to be treated equally' (1993: 279).

Taylor regards an ecocentric liberalism as a logical impossibility and this is also Wissenburg's view. It is not possible, Wissenburg thinks, to further extend our recognition of 'obligations to reference groups other than future generations of humans'. Even were it possible, the incorporation within liberal institutions of 'a more eco- or biocentric, more distinctively holistic, less materialistic, less technocratic ethic' would not be desirable, because 'liberal democracy would then cross the line between democracy and authoritarianism' (Wissenburg 1998: 89). But there are those who disagree; they discover points at which radical environmental thought can make common ground with liberalism, to the mutual advantage of both. For such people the project is less to demonstrate that liberal theory contains all that is necessary for sustainable existence — in which case environmental thought is, by and large, unnecessary — as it is to discover points of intersection and opportunities for hybridisation.

Robyn Eckersley asks whether liberal precepts can be made to serve ecological needs and finds grounds for optimism in an enlarged concept of autonomy and the liberal rights discourse. The key liberal principle that rejects the rightness of any act that 'infringes the rights of humans to choose their own destiny' can be developed in 'an ecocentric direction' via a broader concept of autonomy, here reconstituted as 'the freedom of human and non-human beings to unfold in their own ways and live according to their own "species life"' (1996a: 179–80). Whilst only humans may be 'moral agents', 'if the general moral premise of respect for the autonomy of *all* beings is accepted', then humans 'must collectively acknowledge that human choices need to be constrained by a recognition and consideration of the interests of non-human beings' (1996a: 180). Of course, there will be problems of competing rights claims, involving some 'adjustment, restriction or redefinition of human rights' and 'the challenge for green theory would be to develop principles that would enable the reconciliation of human and non-human autonomy' (1996a: 180; Eckersley points to Naess's principle of 'vital needs' as 'a useful starting point' here).

There are significant problems with this sort of formulation, and Eckersley acknowledges and works through them. The rights discourse is an individualistic one, and it is difficult to see how it can apply beyond individuals to the relational entities recognised by ecology. Rejecting the 'extreme holism' of Leopold's land ethic, Eckersley turns to autopoietic intrinsic value theory as affirming 'the moral considerability of both individuals and ecological entities'. In terms of the rights discourse, 'the rights of individual organisms would need to be framed in the context of the requirements of larger autopoietic entities, such as ecosystems', and in terms of operationalising such rights, we come back to 'the basic normative requirement that the range of choices open to humans must be circumscribed and conditioned by broader requirements of ecological sustainability on the grounds that these requirements provide the conditions that maximise the autonomy of all life-forms' (1996a: 189).

Ultimately, though, these problems do limit what is achievable via recourse to the rights discourse. There are unresolvable boundary problems involved: 'while it may be a relatively simple matter to identify individual organisms, populations and species, it is no easy matter to determine the boundaries of ecosystems or other collective entities *with the degree of precision that would be required for the purposes of rights ascription*' (1996a: 190). This does not demand 'the total abandonment of the rights discourse. Rather, it requires further elaboration of the circumstance in which rights ascription might be appropriate in a world of nested communities with soft and overlapping social and ecological boundaries' (1996a: 190–91). Eckersley, thus, locates grounds on which rights may be ascribed to individual animals and communities of animals. Beyond this, though, 'the individualistic premises of the rights discourse', faced with 'essentially open-ended and highly contingent biological and ecological processes' prove somewhat intractable (1996a: 193).

Eckersley separately considers a case for 'human environmental rights' (1996b), and here she finds cause for greater confidence. If liberal society is founded on the principle of the 'inherent dignity and autonomy of each and every individual' (1996b: 222), it could be argued that a precondition for such a life is 'the human requirement for a physically situated sense of self' — what Eckersley, following Benton (1993b), terms 'ecological embodiment'. A case can thereby be claimed, on such a basis, for rights to environmental health, to a grounding in place, and to all other conditions conducive to one's unconstrained bodily development. It amounts to a right to unimpaired habitat. It constitutes something of a challenge to individualistically grounded rights, too, for it frames individual rights within a more primary context and makes common cause with an emergent social reworking of liberal rights discourse:

> although the social and ecological critiques of liberal autonomy (and associated liberal rights) differ in scope and application, they are structurally similar. That is, both proceed from the premise that the well-being of individuals is indissolubly linked with the well-being of the broader social and/or ecological communities of which they are part. Individuals do not simply enter into social and ecological relationships; rather they are constituted by these relations ... both social and environmental rights must therefore be seen as part and parcel of citizenship rights (1996b: 226).

It seems likely that non-economist traditions of liberal thought will increasingly suggest synergistic possibilities to environmentalist thinkers. On the other hand, there is no sign yet that these minority traditions within liberalism are making significant headway against forms of liberalism which hold that human interactions should be governed by the preference-aggregating mechanisms of the capitalist market. Time, then, to examine the interplay between current economic thought and the values and imperatives of the environment movement.

ENVIRONMENT, ECONOMY: FIRST PRINCIPLES IN CONFLICT?

Attempts to reconcile the economics of the capitalist market with environmental imperatives are many and increasing. These attempts at reconciliation almost invariably leave the assumptions of market economics intact: it is the environmental paradigm that is required to do the adjusting. Perhaps this is because economists, more than any other category of social scientist (possibly more than any category of scientist, period) are wont to insist that economics proceeds from no contestable assumptions, that it is value-free, that it is in some fundamental sense 'natural'. So 'natural', in fact, that if biophysical laws are seen to be out of true with economic laws, it is the presumably less fundamental biophysical laws that are required to 'bend'. But such attempts at bringing ecology into line with economics might simply prove impossible. It may be that environmental thought contains an *inherent* rejection of the logic of the capitalist market.

Such a view might be reasoned this way. The dominant market paradigm proceeds from three core assumptions. The first of these is the model of the so-called 'rational economic man', a model used to account for individual economic behaviour. According to this model, the capitalist market's dynamism comes from the multitude of individual costings of potential courses of action undertaken by each participant, or potential participant, in market transactions. The costs of each course of action are weighed against the expected benefits, and the result is a decision rationally reached in the individual's self-interest. The market is the sum total of an infinite series of such rational calculations made by individuals as consumers.

The second assumption follows from this. If economic activity is driven by the pattern of rationally determined decisions about what individuals wish to consume, it follows that production will organise itself to meet these wishes. This is the classical formulation of the law of supply and demand. Consumers demand certain goods; producers, mere cyphers of articulated demand, supply them. Hence 'consumer sovereignty': supply (production) is determined by — subordinate to — articulated consumer demand (as registered in the market by trends in consumption).

Each of these assumptions of the standard classical market paradigm is used to explain individual market behaviour. A different level of explanation is needed to explain how producers acquire the necessary capital to respond to consumer demand. The logic runs thus. Of the profits generated from the supply of goods, a certain amount is spent by the producer acting as consumer, and a certain amount is returned to the enterprise in the form of new equipment, plant, premises, or research into superior ways of conducting business. Some of the profit, though, is set aside as accumulation, and the impetus to set aside a component of profit in this way has been termed 'the logic of capital accumulation', or 'the accumulation imperative' (J. O'Connor 1987). This accumulated capital is eventually reinvested in new productive enterprises: that is to say, for enterprise *growth*. It follows that *the motor driving the capitalist market is economic growth*; that growth is not a mere consequence of market activity, but is instead a primary factor, a market *essence*. This insight recurs constantly within environmental thought. For example, from perspectives within ecological anarchism, Takis Fotopolous notes that 'the growth objective and the implied principles of economic organisation (efficiency, competitiveness) derive from the logic and dynamics of the system itself' (1994: 14), whilst Murray Bookchin decribes the 'competitive marketplace spirit' as '"grow or die" — a maxim that identifies limitless growth with "progress" and the "mastery of Nature" with "civilization"' (1986b: 49). And John Dryzek, observing that 'ecological crisis' is now 'the most likely agent' for upheaval on the scale necessary to threaten the very existence of capitalism, says of the role of growth within capitalism: 'capitalism is a "grow or die" system. Capitalist economies that do not grow automatically experience rising unemployment, reduced investment — and therefore economic contraction and all its associated political pathologies' (1996a: 147; see also 1992: 19–20; McLaughlin 1993: 41).

Less dramatically, it can be argued that without growth there can be no logic of capital accumulation. This is because there is then no expansion of demand categories or increase in the level of demand within existing demand categories and, thus, nothing to require future investment. The motive that impels the system is missing.

Those who defend the paradigm based upon these assumptions — overwhelmingly the majority of people in the western world — argue its virtues in the following terms.

First, it is held to maximise human *freedom*. Freedom is primarily an economic category. All other freedoms are dependent upon market freedom, which provides people with a capacity to pursue material conditions conducive to the sort of life they wish to lead unencumbered by restrictions imposed by external parties presuming to decide what is good for them. Only the sovereign individual is entitled, the argument runs, to pronounce upon what is good for himself or herself.

Second, market transactions are *just* transactions. The market delivers justice by rewarding ability and effort, in the form of greater access to market goods, precisely and unerringly.

And third, the market distributes resources *rationally*. It does so because it is the sum of the multitude of rationally constructed individual decisions. It is also the best mechanism for determining the use of *scarce* resources, because these will receive a market value that appropriately reflects that scarcity. As this will deter profligacy, the market is the best conserver of the rare and the scarce.

The case presented here is admittedly simplistic (though it is no more a cardboard cut-out than are the stereotypes invariably presented in the hostile analyses of environmental thought mounted by defenders of dominant market economics). But, in juxtaposition with what follows, it points up the contrast between the core principles of environmental thought and the core principles of the dominant economic paradigm.

Over the years the logic of classical market economics has been subjected to sustained attack, yet it is probably in a more dominant position now than at any other time in its history, mainly owing to the eclipse of the socialist intellectual tradition in the 1980s. The major opposition to the logic of classical (and neo-classical) economics now comes, not from the declining socialist tradition, but from the burgeoning field of environmental thought.

Some there are on the edge of the environment movement who accept the premises of classical economics and try to work within them. Thus, one management tool that is often advocated as an appropriate mechanism for preserving environmental goods — cost-benefit analysis (for example, Kneese 1980; Markandya *et al.* 1990; Pearce 1983; Pearce *et al.* 1989; Pearce *et al.* 1990) — is a tool based in the logic of 'rational economic man'. It assumes that a money value can be assigned to a wild area by (say) weighing the money-measured ecological importance of preservation against the greater, but short-term economic gains accruing from intrusive activities (such as mining).

But most ecological analysis rejects the premises and the logic of classical economics. First, the 'rational economic man' model is rejected as an ideological construct that does not accord with real consumer behaviour. Irrationality, rather than rationality, it is held, is every bit as characteristic of consumer choice. Advertising gives the game away. Were economic rationality governing consumer choice, advertising would be much more focused upon the dissemination of factual data. Instead, it concentrates upon image, upon spurious constructions of 'style', and on subliminal rather than rational appeal. And it does so because advertisers know that these are the factors that really govern consumer choice.

Second, the laws of supply and demand have irretrievably broken down. Consumer sovereignty, it is held, is a myth. Again, by recourse to the manipulative power of advertising, producers may now decide to produce what is most technologically and economically convenient and rely upon advertising to create the necessary demand.

And third, there is resistance to the notion of environmental 'goods' being 'reduced' to cyphers of the market. Some things, the argument goes, are beyond value (Ehrenfeld 1988). To determine the fate of a species in accordance with a notionally assigned money value besmirches the meaning and status of life itself; it is less a process of assigning value than it is a process of *de*valuation. It also does not work. Wild areas are likely to be already doomed if they are not valued highly until they become scarce, because too many ecological thresholds will have been passed *before* the market assigns a high value. The market cannot adequately value ecological goods, then, because it cannot recognise ecological imperatives. (One school of thought, 'environmental economics', argues a contrary logic: that it is precisely the *absence* of a tangible valuation that leads to ecological degradation.) A variation on this argument holds that the market cannot respect the interests of other species, or even of humans as yet unborn, because it is a register of *human* preferences in the here and now. Other species and future generations of humans have no way of registering their interests in the market, for neither have any buying power, and that is the prerequisite for involvement in the capitalist market. Thus, only humans, and only living humans, are represented by market decisions.

The argument that capitalism is impossible without growth, and that growth occurs through the transformation and consumption of an ever-increasing amount of ecological stock, seems to lead to the position that capitalism as an economic system can never be ecologically benign. This view predominates within ecoMarxist thought. For John Bellamy Foster, 'the absolute general law of environmental degradation under capitalism' is 'the second contradiction of capitalism'; furthermore, this 'second contradiction ... increasingly constitutes the most obvious threat not only to capitalism's existence but to the life of the planet as a whole' (1992: 78; see also J. O'Connor 1991b; 1994; M. O'Connor 1994a: 54). For Jean-Paul Deleage, capitalism's assumption that nature's bounty is endless ignores unavoidable entropic limits to waste assimilation (1994: 37–45). And for Martin O'Connor, global ecological crises constitute both a crisis of supply and a crisis of legitimation for capitalism. Though Marx himself 'contributed to the optimistic view that these problems were resolvable technologically if only the social relations of production were appropriately reformed' (1994b: 3), capitalism is in crisis because of its tendency to 'impair or destroy its own social and environmental conditions and to retrieve itself from self-induced crisis through mechanisms and measures that cumulatively tend to worsen the damage' (1994a: 54).

Though the gulf in first principles was apparent from the start, early green analyses assumed economics to be redeemable and were not uncompromisingly hostile to capitalism. Some of the first developments in environmental thought focused upon questions of economics and it is to these we now turn.

'CLASSICAL' ENVIRONMENTALIST RESPONSES: THE CRITIQUE OF ECONOMIC GROWTH

The founding contributions of Adam Smith (1723–90), David Ricardo (1772–1823) and others are now referred to as 'classical economics'. The environmentalist critique of liberal economics has existed for only three decades, but it, too, has its 'classical' school. Though the green critique of conventional economics has since broadened, its classical focus fell upon the linked factors of *growth* and *scale.*

It has been suggested that growth is fundamental to capitalist market economics. Yet, neither the founders of classical economics nor the founders of the environmentalist critique believed in the essentiality of growth within capitalism.

In the seminal work of classical economics, *The Wealth of Nations* (1776), Adam Smith argued that economic growth would deliver improvements in the human condition, but that it would eventually become impossible to sustain (O'Riordan 1981: 39–40). Profit and incentive would fall away as production tapered off, as it inevitably must. And for Smith a competitive free-market need not be an unregulated one, whilst his concept of improvement embraced 'better social and psychological conditions as well as monetary wealth' (O'Riordan 1981: 40). Ricardo also saw endless growth as impossible (O'Riordan 1981: 45), whilst Thomas Malthus (1766–1834), having none of Smith's large-heartedness and little sympathy for the plight of the burgeoning industrial poor, built his thought around a growth-negating law: 'population, when unchecked, increases in a geometric ratio. Subsistence only increases in an arithmetic ratio' (1969: 7). The impossibility of food supply keeping pace with population growth thus precluded an endlessly expanding economy.

Whilst there are echoes of Malthus in the doomsday environmentalism of the 1970s, the environment movement's 'classical' economic critique is less influenced by Malthus, though here, too, the possibility of endless growth is rejected. An outline of 'classical' green economics follows.

What we would now call a 'green' economic critique was in place before a comparable body of political or philosophical thought came into existence (as the 'eco' in 'ecology' and 'economics' derives from the same Greek root — *oikos*, 'household' — this is perhaps not to be wondered at). In the 1960s and 1970s Kenneth Boulding, E.J. Mishan, Herman Daly and E.F. Schumacher developed a critique of the growth paradigm of mainstream economics that much discomfited defenders of the conventional paradigm (indeed, one economic theorist, Wilfred Beckerman, has devoted much of his working life to the rebuttal of no-growth economics; for example, 1974; 1975; 1995a). The account that follows concentrates on Daly and Schumacher.

Herman Daly notes that debate over growth had dissipated by the middle of the nineteenth century (1973b). The desirability of, and infinite capacity for economic growth was thereafter taken for granted, and we developed the mind-set that 'there is no such thing as enough, that cannot conceive of too much of a good thing' (1973b: 149–50). The consequence is that economists 'have built elaborate theories around the dangerously circular reasoning that growth is the solution for problems caused by growth' (O'Riordan 1981: 85).

Daly maintained that society must cease to look upon economic growth as uncomplicatedly desirable. We need instead to aim for economic equilibrium, an economics of zero growth referred to by Daly as 'steady-state economics'. The attainment of this state of 'dynamic equilibrium' will require 'major changes in values, as well as radical, but not revolutionary institutional reforms'. A change in values is absolutely crucial: 'unless the underlying growth paradigm and its supporting values are altered, all the technical prowess and manipulative cleverness in the world will not solve our problems and, in fact, will make them worse' (Daly 1977: 2).

Such problems as pollution and urban decay — and their concomitant social consequences such as increasing alienation and unemployment — are, it is contended, beyond technological resolution. They are pathologies of growth and will not be solved by more doses of technologically fuelled growth. Mishan and Schumacher have stressed the seductive unattainability proffered by visions of endless growth. Expectations of growth, Mishan (1969) argued, set up a spiral of unrealisable private expectations on the one hand, and an erosion of the satisfactions of community and collective amenity on the other (because the pursuit of private satisfaction transforms public goods into private ones, thereby materially and psychologically impoverishing the public domain). Even when absolute needs are met we continue to consume, pursuing transient wants in the expectation that one more purchase will deliver the grail of contentment.

Opposition to growth can be construed as an attempt by those who have already attained affluence to deny the have-nots access to it (Coombs 1990: 52–53). But, whilst the environment movement's early doomsayers were not much given to sympathy for the world's poor, the classical environmental economists were inclined to greater compassion. This is particularly true of Schumacher (for example, 1974: 136–59; 172–84), an admirer of Gandhi, and with much experience of East Asia — experience that was crucial to the formation of his thought. The classical environmental economists were also more inclined to positive envisioning than doom-laden scientific prognosis. It is not good enough for the rich to use their market power to avoid environmental disamenity; they must be prepared to sacrifice unsustainable rates of consumption in the wider interest. This applies to rich nations and it applies to rich individuals within nations.

However distributional questions are resolved, the reality is (according to Daly 1977: 6; 1996: 104, and others) that a world of 4 billion people cannot sustain the consumption rates of the growth-dominated economies of the west. Nor can it sustain the prospect of an endlessly expanding world population, under *any* conditions of ecological achievement. Daly argues that the earth's resource stock — the 'natural capital ... that yields the flow of natural resources' (1994: 154) — is being treated as interest on capital when it should be seen as the capital itself. It is capital, not interest, that is being frittered away. This insight is central to the environment movement's economic critique: 'it is an obvious fact', write Ekins *et al.*, 'that this "capital" is the precondition not only of production but of life itself' (1992: 50). A steady-state economy, by contrast, 'respects impossibilities, and does not foolishly squander resources in a vain attempt to overcome them' (Daly 1977: 6).

From the first law of thermodynamics 'we know that matter-energy can neither be created nor destroyed'. For humans the contextualising environment is, thus, 'a source for its inputs of matter-energy and a sink for its outputs. Everything has to come from somewhere and go somewhere. "Somewhere" is the natural environment' (Daly 1977: 15; see also 1996: 65). Though economic actions proceed on the assumption of infinite resources and boundless waste-assimilating capacities, natural systems are actually closed loops. There is, moreover, a net global gain of *entropy* ('the energy unavailable for work') in all human transaction. Thus, the universe has a fixed store of energy (the first law of thermodynamics), but the component available for useful work is decreasing (the second law)' (Merchant 1992: 38). Whilst we rely upon finite fossil fuels for energy (on this, see Daly 1977: 129–46), production today is reducing the capacity for production tomorrow. 'While the rest of the biosphere lives off solar income', writes Merchant, 'human beings, since the transition to an inorganic economy, have been living off non-renewable geological capital. This means that humans are no longer in equilibrium with the rest of nature, but are depleting and polluting it, overloading the natural cycles' (1992: 37).

In the steady state this would not be permitted. Throughput must be contained within the coping capacity of the host ecosystem. The economy is best seen, Daly argues, as a throughput that begins with 'depletion of nature's sources of useful low entropy and ends with the pollution of nature's sinks with high-entropy wastes' (1977: 16). To keep the capital intact it is necessary that throughput be regulated and minimised.

This does not mean stasis. Knowledge, culture, the distribution of income and the allocation of resources may undergo qualitative improvement (Daly 1977: 16–17). But the ideal steady-state economy is one in which the stock of artefacts and people remains relatively constant. Artefacts — Daly also refers to 'physical capital' — 'wear out and must be replaced', whilst people (human capital) inevitably die. This outflow 'is offset by an inflow of births and production' and this, Daly says, must be kept constant rather than be allowed to decline or grow (1977: 15). In his later writings he distinguishes between 'economic growth' and 'economic development', the latter being 'development without growth beyond environmental carrying capacity, where development means qualitative improvement and growth means increase' (1996: 9).

In summation: a steady-state economy is characterised by approximately 'constant stocks of people and artefacts', maintained in equilibrium at low rates of 'maintenance throughput' — that is, 'by the lowest feasible flows of matter and energy' (Daly 1977: 17) from the natural capital to the last stage of consumption, the ejection of pollutant wastes. Such an economy seeks a responsible treatment of resources — one which allows for the needs of future generations *and* for the habitat needs of the planet's non-human denizens.

This, then, is the theoretical basis of steady-state economics. Noting its beginnings as 'a branch of moral philosophy', Daly argues that economics must return to a base in ethics and biophysics in order to counter the misguided assumptions of growth economics (1977: 3). We need a political economy that recognises ecological imperatives and puts economic discourse back within reach of ordinary people. In the key area of technology, for instance, dominant economics liberates technology from human control. Technology has, thus, been rendered autonomous, and it is people who make the adjustments and compromises. Steady-state economics, by contrast, would restore autonomy to people, and technology would become the servant again. Schumacher (1974) also argues that economic and political structures should not exceed what is humanly comprehensible. When the level of individual comprehension is exceeded, control over economic forces is lost, and an economics that is impervious to control poses problems when its dictates are environmentally degrading.

The notion of the steady-state here intersects the other major theme in classical ecological economics: scale. An emphasis on reducing and decentralising operational scale is evident in both theory and practice. In practice it leads to calls for local/regional production, using local/regional resources, and employing local people; for economic self-sufficiency is seen as the answer to the environmental, social, economic and political exploitation that enters when local autonomy is destroyed. Daly focuses upon growth, Schumacher on scale; but their respective projects closely correspond.

Schumacher, like Daly, argues that industrial civilisation is inherently unsustainable, treating what is natural capital as interest (1974: 12–13; 1980: 13–18). In developing a critique of 'natural capital', Schumacher and Daly laid the foundations for the potent though increasingly contestable notion of 'sustainable development', which took centre stage in the 1980s (and remains there).

Also like Daly, Schumacher insists upon the need to relegate economics to a position of subservience to ethics. Henderson's summation is hard to beat: 'higher levels of ethics and greater sensitivity to all living things would transcend the narrow, market view and that most dreadful reduction of human beings to a commodity called "labour" could be replaced by the concept of "Good Work"' (1988: 177). In his essay, 'Buddhist Economics', Schumacher draws on the Buddhist notion of 'Right Livelihood', wherein economic life would re-focus upon the needs and growth capacities of people involved in fulfilling work:

> the Buddhist point of view takes the function of work to be at least threefold: to give man a chance to utilise and develop his faculties; to enable him to overcome his egocentredness by joining with other people in a common task; and to bring forth the goods and services needed for a becoming existence ... To organise work in such a manner that it becomes meaningless, boring, stultifying, or nerve-racking for the worker ... would indicate a greater concern with goods than with people, an evil lack of compassion and soul-destroying degree of attachment to the most primitive side of this worldly existence (1974: 45).

The purpose of work, then, is to enable humans to develop and grow, to transcend narrow individualism through the cultivation of a moral sensibility, a sense of neighbourliness, and 'to be creatively engaged, using and developing the gifts that we have been blessed with' (1980: 116).

For the provision of 'Good Work' within a 'Buddhist Economics', production must become value-driven, not only in the way it conceptualises 'labour', but also in terms of its processes and end products. Products must be of such a nature that workers can take legitimate pride in their production. Work processes must complement and sustain biophysical processes. Centralised, intrusive and compartmentalised production processes cannot achieve this. Small or 'human' scale production using intermediate, decentralised technologies can. Of small-scale economics Schumacher has written:

> experience shows that whenever you can achieve smallness, simplicity, capital cheapness and non-violence ... new possibilities are created for people, singly and collectively, to help themselves, and that the patterns that result from such technologies are more humane, more ecological, less dependent upon fossil fuels, and closer to real human needs than the patterns (or life-styles) created by technologies that go for giantism, complexity, capital intensity and violence (1980: 57).

Organisations have become too large to deliver extrinsic or intrinsic satisfactions and efficiencies. Now, Schumacher argues, we have the technological capacity to de-scale economic institutions and processes (1980: 21).

Economics and technology are inextricable linked. Largeness of scale is only technically feasible if large, resource-intensive, centralised technologies are employed; conversely, human-scale economics requires a commensurately de-scaled technology. 'Intermediate technology' (or 'technology with a human face'; 1974: 122–33) is broadly accessible, not damaging to the natural world, and empowering and stimulating for workers. Schumacher envisages a development from what already is: 'making best use of the stock, and not wasting too much time in being heroic and reinventing the wheel' (1980: 128). And work and production will take place within a context of concern for the ecological connections that link all forms of life.

Finally, as well as being unable to provide for human needs *within* production processes, and too intrusive to deliver ecological sustainability, the large institutions of the market cannot deliver a broader social responsibility: 'there is no probing into the depths of things, into the natural or social facts that lie behind them. In a sense, the market is the institutionalisation of individualism and non-responsibility. Neither buyer nor seller is responsible for anything but himself' (1974: 36).

There are dimensions to Schumacher's thought that go beyond the economic, and these will be noted in chapter 9. Before moving on, though, we should note that the classical critique of growth has largely survived the 1980s–90s diversification of environmental thought. Though perhaps not holding the same pre-eminence, and though views on the matter have become more diverse and complex, Schumacher is still read and Daly continues to write (and be read). Others are also building on their achievements. Kassiola (1990), for example, has linked the Mishan-Daly case against growth to a larger transformative political project in a plea for a strongly participative 'transindustrial' (not to be read as anti- or post-industrial) society.

MEASURES OF HUMAN WELLBEING: BEYOND GNP

Daly's focus is on the dominant paradigm's faith in perpetual growth as the motor that drives economics along. This, he argues, is strongly reinforced by the devices used to assess levels of economic wellbeing. The measures of economic *health* (GNP and GDP) are actually devices for gauging *growth*, and each reading is required to register an increase over that which has preceded it. If it does not, gloom reigns, doom is prophesied, governments are required to resign, stockbrokers throw themselves from windows, and ordinary people lose jobs.

Gross National Product (GNP) is a slightly broader concept than Gross Domestic Product (GDP), but we can treat them as interchangeable here, because the faults consistently attributed to the one are also held to apply to the other. GNP is merely a measure of the sum total of money flows, and many economists defend it on these grounds, pleading that it is unfair to attack GNP for its failings as an index of wellbeing when this is not what it is intended to be in the first place. Perhaps. But, appropriate or not, GNP is accepted as *the* measure of overall economic wellbeing — it is, indeed, promoted as such by governments — and on this account it plays a potent ideological role within western market economies. Environmental thought takes issue with GNP at a number of points, but the case comes down to one powerful argument: that GNP adds but never subtracts; that it counts as positives what should be accorded debit-ledger status. This can include both 'defensive expenditure' — that used to prevent undesirable externalities in advance (the cost of maintaining police and military forces, as well as pollution control equipment and welfare and health services, for instance) — and remedial expenditure (the costs of cleaning up after Chernobyl and the *Exxon Valdez* spill, for example). 'GNP does not deduct waste, deterioration, the replacement of capital, and the exhaustion of physical resources', writes O'Riordan (1981: 97); whilst Manfred Max-Neef, the Chilean theorist of 'barefoot economics', notes: 'any process that generates a monetary flux is acceptable. It is totally irrelevant whether it is productive, unproductive or destructive' (cited in Schwarz and Schwarz 1987: 39). Jose Lutzenberger is even more forthright:

> GNP as a *measure of progress* is the most stupid, the most absurd and the most pernicious index that could ever have been thought of by a discipline that calls itself scientific ... Today in my country, Brazil, we are flooding thousands of square kilometres of pristine rainforest to make electricity for three mills that export aluminium. In our national accounts the foreign earnings from exporting the aluminium are added to the GNP, but nowhere do we deduct for the permanent loss of the ore or the demolition of the mountains (1996: 20).

Not to mention, he might have added, the rainforest.

Improvements in physical and mental health, the availability and security of meaningful work and of recreational and cultural opportunities, and increased life-expectancies as stress and pollution levels decline are all likely consequences (its protagonists claim) of the steady-state economy, and a more objective quality-of-life index would take these into account. This does not occur, says Schumacher, because 'GNP, being a purely quantitative concept, bypasses the real question: how to enhance the quality of life' (1980: 126).

It will not surprise to find that alternatives to GNP abound within environmental thought. Daly and Cobb (1989) propose an 'Index of Sustainable Economic Welfare'. This is a device for determining gross welfare, in which personal consumption is adjusted for distributional factors, thence calculated downward or up as a number of variables are added in: variables such as chronic environmental deterioration, decline in stocks of non-renewable resources, negative or remedial social and environmental expenditure (both private and societal), and net growth in financial and social capital and in communal resources. Although extremely influential in its time, this index was overly complex. Even its authors found it of limited operability (1989: 415–16). But this has not deterred others from attempting to devise a more satisfactory alternative to GNP. John Young suggests a measure of 'Net Human Benefit' (NHB): 'economics which look healthy in terms of GNP might look much less flourishing in terms of NHB if the violence, anomie, boredom and the diseases of affluence were set down on the ledger to show that not all growth is good' (1991: 115). Clive Hamilton argues that 'contributions to our wellbeing only count if they are transferred to the market sector; and the "side effects" of the market that diminish our well-being are sent back to the social and environmental sectors where they no longer count' (1998: 26). His proposal is for a 'Genuine Progress Indicator', which 'provides a balance sheet for the nation which includes both the costs and benefits of economic growth' (1998: 27). Christian Leipert, noting that 'the price to be paid in environmental protection and social costs' is then added into GNP 'and this higher level is hailed by economists, politicians and the business community as an achievement' (1989: 132), labels these remediation-of-externalities costs 'defensive expenditures'. He proposes an 'Adjusted National Product', which 'can be calculated from the GNP by subtracting from the latter the defensive expenditures of all sectors' (1989: 139).

These proposed substitutes (and several more could be adduced) identify a similar flaw in GNP and they are, despite their different labels, remarkably similar. What is particularly salient is that, whereas GNP for most western economies has climbed exponentially since 1970, calculations using these alternative measures do not. This indicates that, once environmental and social wellbeing become components of the calculus, there has been no growth in 'real' welfare, or even negative outcomes.

ALL HAIL 'SUSTAINABLE DEVELOPMENT'

The sustainability discourse is the lineal heir to the 1970s debate over growth.

O'Riordan traces 'sustainable development' to a series of conferences in Africa in the mid 1960s (1993: 48) and Redclift states that the concept was first run up the flagpole 'at the time of the Cocoyoc declaration on environment and development in the 1970s' (1987: 32), though its introduction to the discourse is traced by most to its appearance in the World Conservation Strategy of 1980. In 1987 the World Commission on Environment and Development, in the 'Brundtland Report', declared 'sustainable development' to be the priority objective of all economic policy, and it did so to immediate and near-universal approbation. 'Sustainable development' was endorsed by the G7 group of industrialised nations at the 1988 Toronto Summit (which gathering, Jacobs reminds us, included Margaret Thatcher and Ronald Reagan; 1991: 59), and it was written into the Framework Convention on Climate Change (1993) and the United Nations Convention on

Biological Diversity (1993), as well as Agenda 21 of the United Nations Conference on Environment and Development, the 'Rio Earth Summit'. Industry and governments welcomed it; so too, though only initially and somewhat more cautiously, did the environment movement.

The Brundtland Report defines 'sustainable development' as 'development that meets the needs of the present without compromising the ability of future generations to meet their own needs' (WCED 1987: 43). But the definition's larger context is crucial to understanding its controversial status within environmental thought. On the key question of whether sustainability is compatible with economic growth, Brundtland is emphatically affirmative. There must be growth, at least in some parts of the world, because, if there is not, the existing inequitable division of global wealth will be forever frozen. Growth, then, but with a strong ethic of global resource distribution contained within it; the yardstick of sustainability being employed was whether or not any given economic policy serves to reduce or maintain the existing stock of environmental resources for future generations. Environmental thought had long posited the interests of other life and of humans yet to be born as the two crucial sites of interest omitted from conventional economic welfare accounting. Though ignoring the former, Brundtland established near-unchallengeable legitimacy for the latter.

The contestable nature of the concept of 'sustainable development' can be seen by tracking the concept's 'official' path, post-Brundtland. Even in Brundtland 'sustainable development' was linked to the agenda of trade liberalisation. Following interpretation and mediation in various national policy responses around the world, it emerged in Agenda 21 of the Earth Summit as a strengthened commitment to a globally scoped open-access and free-trade regime: 'an open multilateral trading system, supported by the adoption of sound environmental policies, would have a positive impact on the environment and contribute to sustainable development' (UNCED 1992), whilst Principle 12 of Agenda 21 warns against the use of national environmental protection policies as 'a disguised restriction on international trade' (UNCED 1992).

It was apparent to most people in the environment movement that the goal of 'sustainable development' had been thoroughly subverted. The very term, 'sustainable development' was thought by many to be an oxymoron. For others, the formulation was a cynical exercise to accord a fake legitimacy to full-on, business-as-usual, global environmental rapine (for a particularly forthright expression of this view, see Doyle 1998). People inclined to this view maintain that the lack of definitional precision has been deliberately engineered to leave the concept open to endless manipulation. David Pearce *et al.* are not of such a mind, yet they also note that 'there is some truth in the criticism that it has come to mean whatever suits the particular advocacy of the individual concerned' (1989: 1). Michael Jacobs notes 'concern among some environmentalists that the lack of clarity of the definitions allows *anything* to be claimed as "sustainable" or as "promoting sustainable development"' (1999: 24). Jacobs well describes the range of misgivings:

> for many years Greens have struggled to show how industrial expansion causes environmental damage ... Now, just when the extent of damage threatens to give them a conclusive case, the term sustainable development appears like a magic wand to wave away such conflicts in a single unifying goal. We can have our cake and eat it, it seems to say (1991: 59).

It is on account of doubts such as these that in some countries environment movement spokespersons have insisted on routinely affixing the qualifier 'ecological' to 'sustainable development'. Insistence upon 'ecological sustainable development' as the non-negotiable, baseline term seeks to deny legitimacy to interpretations of 'sustainable development' as 'economic sustainable development', the latter being a reference to the economy's capacity for sustained expansion, or to an economic enterprise's robustness (or lack thereof) within the market, but having not one whit to do with environmental protection in either case. Environment-dismissing misuses of the concept are not uncommon, giving us the paradox described by Jacobs of business interests that 'claim they are in favour of sustainable development when actually they are perpetrators of *un*sustainability' (1999: 24).

In fact, though it is true that meanings of 'sustainable development' abound (Torgerson counted forty; 1994: 303), these can be distilled to two cleanly opposed interpretations. Jacobs (1999) terms these 'weak' and 'strong' (so do Pearce *et al.* 1993; Turner adds 'very weak' and 'very strong'; 1993: 9–15), Orr (1992) calls them 'technological' and 'ecological', and Diesendorf (1997) dubs them 'economist' and 'ecologist' concepts of sustainability. The 'weak' definition is the Brundtland one; the comparable exemplar of the 'strong' definition is that from WCU/UNEP/WWFN's *Caring for the Earth* (1991) — 'improving the quality of life while living within the carrying capacity of supporting ecosystems'. The latter is specifically ecological in its terminology; the Brundtland definition is not. In fact, the Brundtland definition dismisses by strong implication the notion of biophysical limits to growth, claiming that the only restraining factors are 'limitations imposed by the present state of technology and social organisation on environmental resources' (WCED 1987: 8). Daly, on the side of the 'strong' definition, rejects the Brundtland claim outright, arguing that the primary constraint upon development is imposed by the strains and crises rapidly accumulating within natural capital: 'the fish catch is not limited by fishing boats, but by remaining populations of fish in the sea' (1995: 50; see also Common and Perrings 1992). Jacobs, too, specifies this as the key axis of disagreement. He identifies four 'faultlines' along which 'weak' and 'strong' interpretations of 'sustainable development' divide, and the first of these concerns 'degree of environmental protection' (the others concern the degree of equity sought, levels of participation in economic policy, and the policy scope assumed by 'sustainable development'). The 'strong' interpretation 'adopts the more stringent idea of "environmental limits"', the 'weak' version does not:

> it is accepted that the environment is important and should where possible be protected, but the idea that economic activity should be confined within pre-determined 'environmental limits' is rejected. In this interpretation, environment-economy integration means balancing or trading off the benefits of economic growth against those of environmental protection (1999: 31).

The nub of the question is the status within sustainable development of 'natural capital'. Almost everyone writing within a green suite of values insists that, here, the bottom line must be drawn. Diesendorf, for example, argues for the centrality of 'maintaining biodiversity' and 'ecological integrity' (1997: 72–74); for Bill Hare, a basic principle of sustainable development is that 'conservation of biodiversity and the protection of ecological integrity should be a fundamental constraint on all economic activity' (1990: viii). Jacobs argues that sustainability is actually a simple concept based upon the familiar need to maintain income whilst not running down capital stock, in which 'the natural environment performs the function of a capital stock for the human economy, providing essential resources and services, including the assimilation of wastes'. The trouble is, 'economic activity is presently running down this stock', generating wealth but 'reducing the environment's capacity to continue to provide resources and services' (Jacobs/Real World Coalition 1996: 17). And Paul Ekins (1992: 412), noting that 'there is literally no experience of an environmentally sustainable economy anywhere in the world where such sustainability refers to a non-depleting stock of environmental capital', insists that the legitimacy crisis over 'sustainable development' will continue until this is rectified.

But the 'weak' or 'economist' concept of sustainable development does not concede this bottom line. 'For economists', writes Sharon Beder, 'sustainable development means incorporating the environment into the economic system' (1993: 9). She is right: it is that stark. At stake in this dispute over interpretation are rival claims over what is ontologically primary: the natural environment or the human economy. Is the environment a sub-set of the economy, or is it a prior and sustaining context within which, and dependent upon which, all human systems, including economic systems, reside? For proponents of 'weak' sustainability the former is clearly assumed. 'No aspect or "level" of the environment is regarded as inviolable', writes Jacobs, 'at least until countervailing economic benefits have been assessed'. Moreover, 'broadly speaking, this is the interpretation used by governments and business interests' (1999: 31).

Sustainable development, on the weak/economist interpretation, is a matter of manipulating the *total* capital stock (human labour and human artefacts, as well as natural resources), to get a maximised welfare return. Natural capital is not inviolate. In accord with the *substitutability* principle (proposed in the early 1980s by prominent environmental economist, David Pearce), it is quite in order to reduce natural capital; what is important is that there be no overall reduction in 'welfare', but the share of natural capital can be reduced as long as it is compensated by a commensurate increase in human capital. As Pearce *et al.* note, 'the requirement to keep the total of capital constant is consistent with "running down" natural capital'; it is consistent, for instance, 'with removing the Amazon forest so long as the proceeds from this activity are reinvested to build up some other form of capital' (1991: 1–2).

But Pearce and his team also note that 'for some environmental assets, which we term "critical capital", there is no question of an acceptable trade-off'. Here, 'the bias is towards conserving natural capital' (1991: 2). In fact, much of Pearce's writing insists upon the absolute need to preserve stocks of natural capital. 'Sustainability', he writes, 'requires at least a constant stock of natural capital, construed as the set of all environmental assets' (1992a: 69), for 'there are many types of environmental asset for which there are no substitutes' (Pearce *et al.* 1989: 37). And he has written:

> the definition advanced here suggests that sustainable development is based on constant or augmented natural capital stock. This approach appears to have derivable linkages to the social objectives of equity within and between generations, economic efficiency and resilience. It appears also to be consistent with rights in nature (Pearce 1992a: 76; see also Turner *et al.* 1993: 293–98).

But ambivalence on the matter is evident in some of Pearce's writing. This is because he argues that it is the economic value of natural capital, not physical stock itself, that must be kept constant; depending on how this is written up, it sometimes reads as closer to the substitutability position, and sometimes as closer to the 'maintain physical stock' position. Holland notes that Pearce seems to want to combine his 'economic value' position with the 'purely physical criterion' in order to ensure that physical stocks do not fall below a 'critical minimum' (1994: 171).

The points of contest between 'weak' and 'strong' were so sharpened by the substitutability principle that it sparked a celebrated debate in the journal *Environmental Values* in 1994–95. Taking Pearce's substitutability principle as the cornerstone of 'weak sustainability',

Beckerman (1994) attacked 'strong sustainability' as 'morally repugnant' because it countenances the withholding of resources that could be used to alleviate poverty and degradation in the here and now. He was answered by Daly (1995) and Jacobs (1995b), before himself having the last word by way of rejoinder (1995b).

Beckerman and Daly set up exaggerated types of the positions they wished to attack. For Beckerman 'strong sustainability' is 'a requirement to preserve intact the environment as we find it today in all its forms' (1994: 194), a definition denying the dynamism inherent in nature, and dismissed by Daly as 'absurdly strong sustainability' (1995: 49). But Daly may have constructed his own man of straw. He assumes that the substitutability principle implies 'perfect substitutability'; but, as Dobson notes in an insightful discussion of the Beckerman-Daly-Jacobs debate, 'nowhere, except in neo-classical economic theory, is such a belief to be found' (1996a: 410). In fact, it is not even the position adopted by the neo-classical Beckerman: 'the assumption ... of infinite substitutability between manmade and natural capital is not made by me or any economist that I know of' (1995b: 174). But Beckerman can be relied upon to make 'generous assumptions' regarding substitutability (Dobson 1996a: 410–11), and as Dobson, Jacobs and Daly all point out, in this conception 'the distinction between the human and non-human worlds collapses, and from this collapse derives the possibility of ignoring any claims the non-human world might have' (Dobson 1996a: 412).

There, for the moment, matters rest. With the apparently successful capture of 'sustainable development' by a neo-liberal worldview disinclined to make the slightest concession to ecological imperatives, the question needs to be asked: why do some in the environment movement continue to plumb the concept in search of intrinsic worth and political utility?

Michael Jacobs is one who is aware that the commitment to sustainable development can be entirely hollow, often cynically manipulated, often merely meaningless. Nevertheless, he argues, we should not throw the concept out just because it is contestable. Environmental behaviour *does* require 'constraints on economic activity' (1991: 59), and 'sustainable development is evidently not the path of development which has been followed by the global economy, or by most individual nations, over the past fifty years; even less over the last twenty' (1999: 27). And there *are* essences within 'sustainable development' that cannot be argued away by the sophistries of 'weak' sustainability. For instance, inherent in the concept is the notion that 'humankind has to live within the limits of the earth's capacity to support it: to find ways of improving human welfare that are not ultimately self-defeating through environmental degradation' (1995a: 5).

Three notions essential to 'sustainable development', says Jacobs, can be turned to good account. First, the concept entrenches environmental concerns in economic policy because it insists on the integral nature of what has previously been kept apart. Whilst 'traditional economic orthodoxy' ignored environmental imperatives, this is no longer possible (1991: 60). Second, the Brundtland formulation incorporates 'an inescapable commitment to equity', especially between nations and between generations: 'if something is sustainable it is able to last or continue' (1991: 60). Third, because 'development' is a much richer, more multi-faceted term than 'growth', 'sustainable development' places such intangibles as community health standards and access to high quality environmental amenity at the forefront rather than at the margins of economic policy (1991: 60–61). 'The core meaning of sustainable development is thus neither empty nor insignificant' (1991: 61), and this even applies to the Agenda 21 formulation — the final straw for 'sustainable development' as far as many people in the environment movement are concerned. For Jacobs, though, 'the important point is that, in signing Agenda 21, the majority of countries have rhetorically accepted that sustainable development does represent a new trajectory for development' (1999: 27). Contest its cynical interpretations then, but defend, argues Jacobs, the core concept.

Others argue similarly. Noting that 'it is not unusual for important concepts to be contested politically', John Dryzek also rejects the view that 'sustainable development' is merely 'an empty vessel that can be filled with whatever one likes'. Sustainable development discourse does have boundaries, and, 'because it requires coordinated efforts to achieve goals', it cannot be comfortably subsumed within the prescriptions of neo-liberalism (1997: 125). Thus, the notion of sustainable development can be effective in opposing governments in thrall to market ideology (1997: 134). For Paul Selman, the practice of sustainability requires a public-spiritedness that can rejuvenate civic culture and its defining characteristics, individual and collective civic competence and active participation (1996: 144–52). Julie Davidson maintains that 'sustainable development as ethical ideal challenges the view of the liberal state as neutral umpire', and 'demands a conception of human flourishing that recognises the intrinsic value of other nature, whose own flourishing is constitutive of a good human life' (1998: 246). And William Lafferty argues that 'sustainable development' is an idea appropriate to a time in which 'the politics of environment-and-development have created new models and opportunities for an emergent global civil society'. The processes set up in response to the valorisation of 'sustainable development' may be cynical and obfuscatory, or even honestly inadequate, but they have 'imbued the new agenda with a moral urgency and an ethical legitimacy which concentrated effort and political mobilisation should be able to exploit' (1996: 204).

Another who defends the *environmental* usefulness of 'sustainable development' is David Pearce. Though Pearce pioneered the case for natural capital–human capital substitutability, and though his typologies of sustainable development allow the possibility of 'perfect substitutability', he evinces clear concern for the fate of natural capital, including the integrity of wild lands. He acknowledges that some 'ecological assets' are crucial to human survival, and, that some, such as wild places, are essential for less tangible aspects of human well-being (Pearce *et al.* 1993: 16; see also Turner *et al.* 1993: 290–98). Alan Holland, in an unpublished essay, even locates 'an unacknowledged commitment to the intrinsic value of nature in Pearce' (cited in Dobson 1996a: 417). And, like Jacobs, Pearce stresses assumptions lodged in the non-negotiable core of 'sustainable development'. Sustainability involves 'a substantially increased emphasis on the value of natural, built and cultural environments'; it 'extends the time horizon' for economic planning and welfare assessment, as 'future generations must not inherit less environmental capital than the current generation inherited' (Pearce *et al.* 1989: 3); and it emphasises 'providing for the needs of the least advantaged in society (intragenerational equity), and on a fair deal for future generations (intergenerational equity)' (1989: 2). Other useful ideas would seem to inhere within the concept — for example, the idea of 'false sustainability', in which conditions of sustainability are met within one country 'at the cost of non-sustainability in another country' (Pearce 1992b: 406). And — this, too, is a sentiment hard to reconcile with support for 'weak' sustainable development — Pearce *et al.* identify as integral to the idea of sustainable development the belief that 'the economy is *not separate from the environment in which we live*' (1989: 4).

PROTECTING ENVIRONMENTAL AMENITY THROUGH MARKET-BASED MECHANISMS

In his 1994 attack on 'strong' sustainability, Beckerman also dispensed with 'weak' sustainability, arguing that the principle of substituting human for natural capital whilst maintaining overall levels of welfare was such a central notion of neo-classical economics that the two were thereby rendered indistinguishable.

Perhaps, then, we need not concern ourselves with environmental degradation or with constructing policy to protect the environment. It is frequently argued so. The price system will deliver optimal resource allocation and is in no need of any misguided assistance from the political system (Anderson and Leal 1991: 167–71). And if a resource begins to diminish — though this will be a much less common occurrence — market scarcity will force up its price, thus providing the necessary incentive for applying our vast technological ingenuity to the discovery or invention of alternatives. This has tended not to happen because markets have not been free to demonstrate their extraordinary capabilities. The penetration of the market into spheres currently denied them is, thus, an important part of the neo-classicist solution. It is a paradigm described by one of its proponents in this way: 'environmental problems are the natural consequence of an absence of property rights. Lacking individual protection, a resource is at risk. Because the problem is a lack of property rights, the solution is to expand the market system to include all environmental resources now at risk' (F.L. Smith 1991: 599). This commentator envisions 'a world in which every tree and animal would have an owner/defender' (1991: 598), because 'ocean reefs in the South Pacific, Andean mountaintops, whales, elephants in Africa, the shoreline of Lake Baikal — all deserve protection as the private property of some group or individual' (1991: 605). Why can a system (or non-system) of maximised private property rights achieve what is beyond the capacity of the regulative state? Smith gives these reasons:

> higher resource prices — when the resource is privately owned — pose little threat to the sustained use of that resource. Increased value may encourage 'poaching', but it also encourages greater antipoaching efforts, provides owners the financial resources to finance such protective efforts, encourages consumers to conserve, and leads others to find new supplies. Finally, markets encourage entrepreneurs outside the immediate economic sector to develop better 'fencing' and 'resource protection' technologies and management systems. Private ownership links a resource into the rich system of interrelationships that has made markets so effective at advancing human welfare and thus should be viewed as a primary means of environmental protection (1991: 599).

At its extreme, faith in the market can take the form of a defence of capitalism as 'ecological' or 'ordained by nature', a view that removes it from the realm of artefact and renders it organic (for example, Rothchild 1992; critiqued by Birkeland 1992; for a discussion of the social Darwinist position that the individualism, greed and competition of capitalism replicates nature and hence is 'natural', see Hodgson 1993: 29–31). But mine is a book about the thought of the environment movement, and such contestations as Smith's and Rothchild's do not emanate from the environment movement. I will pass on, though the reader may wish to refer to the free-market environmentalism of Anderson and Leal (1991) and Moran *et al.* (1991); and, though few accord it sufficient credibility to warrant serious analysis, there are some reasoned green ripostes to free-market environmentalism (for example, Eckersley 1993a; 1993b; Rosewarne 1993). Eckersley's sketch of free-market environmentalism is succinct and quotable:

> the ingenuity (some might say cunning) of free market environmentalism is that it turns the green economic analysis of environmental degradation on its head. Environmental externalities such as resource depletion, pollution and species extinction are seen as arising not from the operation of 'market forces' or self-interested behaviour for short-term gain (the

> green view) but rather from an absence of well-defined, universal, exclusive, transferable and enforceable private property rights in respect of common environmental assets (1993a: 239–40).

Free-market environmentalism must be distinguished from 'environmental economics', a school of thought which also lodges within the paradigm of neo-classical economics, but which advocates market mechanisms as instruments to secure desired public policy outcomes, a somewhat different conception to that of the market as an inexorable force which renders foolish the very enterprise of public policy. Redclift calls it a 'revisionist' neo-classicism: 'environmental economics has sought to extend neo-classical theory, by encompassing the environment, and attaching monetary values to losses in natural capital' (1993a: 11).

Unlike the 'pure' neo-classicism of free-market environmentalism, the 'revisionist' version — environmental economics — does concede that markets fail. That, in fact, is the point. 'Unfettered markets fail to allocate resources efficiently' because 'many environmental products, goods and services do not get represented in the price mechanism'. They are treated as free goods and, as they are 'zero-priced', are inevitably over-utilised. There is no incentive to protection because the value of such a good to humans and the sustaining environment does 'not show up anywhere in a balance sheet of profit or loss, or of costs and benefits' (Pearce *et al.* 1989: 5). In addition, 'economic goods and services "use up" some of the environment', so production costs 'tend to be a mixture of priced "inputs" (labour, capital, technology) and unpriced inputs (environmental services)', with the consequence that 'the market price for goods and services does not reflect the true value of the totality of the resources being used to produce them' (1989: 154; this insight is not original to Pearce; it was first made by A.C. Pigou in 1920). Turner *et al.* argue similarly. When 'prices accurately reflect the true value of resources, the free market encourages conservation'. However:

> market failure situations arise where a firm produces units of output which create private profits but also impose large external costs upon society. Only when these external costs are brought into consideration (for example, by forcing the firm to pay for the external costs it causes, in line with the Polluter Pays Principle) will firms act so as to prevent market failure occurring and move from a market optimal to a socially optimal level of output. An additional complication is that many environmental goods do not act like goods in the market-place; they are 'public'-type rather than private goods. This 'publicness' is one of the reasons why markets do not evolve naturally in environmental goods and services (1993: 77–78).

Environmental economics seeks to redress this deficiency *within* the framework of assumptions essential to neo-classical economics.

'The task that environmental economics has set for itself', writes Hamilton, 'is to develop analytical and practical means of internalising environmental costs that fall outside the orbit of markets' (1997: 46–47). The user should pay, then, and one way this might be done is via the creation of 'real' markets for environmental services 'so that private decision makers determine prices through supply and demand' (Hamilton 1997: 42). This is an emphasis within environmental economics that points away from environmentalist imperatives and back towards the vision of the market as prior and supreme. Consider a wetland that is breeding territory for many species of endangered waterfowl but is coveted by a transnational oil company. The problem is simply resolved. All the Outer Swampville Birdwatchers' Group needs to do is outbid Planetary Oil Corp in the market-place and the birds will be saved. What could be fairer?

Yet, this is ultimately a minor emphasis within environmental economics. Its main project is to assign a 'true' value to environmental goods and services in the interests of sustainability. The person who has contributed most to this project is undoubtedly David Pearce, particularly via his co-authored *Blueprint for a Green Economy* (1989). We have seen that Pearce's personal adherence to environmentalist goals is beyond dispute. He concedes, for example, the significance of the ethical dimension to questions of resource utilisation (Pearce *et al.* 1990: 3), and in this he has rather more in common with the advocates of ecological economics (considered below). Environmental economics nevertheless remains somewhat marginalised both within environmental thought and academic economics, notwithstanding the inroads it has made within real-world public policy.

Pearce's key claim is that economic modelling has become such an advanced procedure that it has placed an array of evaluation techniques at the disposal of economists, such that the full costs of 'free' environmental services and of environmental 'externalities' can be accurately accounted for. The need for such accounting is crucially important: 'if development is to be sustainable it must encompass a full appreciation of the value of the natural and built environment in terms of the direct and indirect contributions that environments make to people's wellbeing' (Pearce *et al.* 1989: 33). Such valuation is achieved primarily via a process known as *contingent valuation*: assigning a notional monetary value to goods not directly recognised by the market. This involves extending consumer preference valuation through the allocation of a 'willingness to pay' value to goods with no directly assessable market value (on the calculation of 'willingness to pay', see Pearce 1983: 9–10; 1993: 64–68; Turner *et al.* 1993: 94–96), and of 'shadow prices' to goods whose true value has not been allocated owing to market distortions. These can then be factored with or against those preferences which do have directly allocated, undistorted market value in a much-enhanced application of cost-benefit analysis (hereafter CBA; on the relationship between contingent valuation and CBA, see Grove-White 1997; Jacobs 1997). Though controversial, contingent valuation-enhanced CBA has become the 'orthodoxy' in many western policy domains (Foster 1997: 7; Grove-White 1997).

Whilst he understands that the project of 'reducing' environmental components to money values will, in Tim Hayward's words, 'offend conservationists and radical ecological critics', Pearce is actually 'seeking to put money values on certain aspects of environmental quality to underline the fact that environmental services are *not* free' (T. Hayward 1995: 99). Of the various devices that may be employed for this purpose, Pearce prefers the mechanism of CBA: a utilitarian process ('any course of action is judged acceptable if it confers a net advantage'; Pearce *et al.* 1990: 57) which can claim to be the *only* mechanism which 'explicitly makes the effort to compare like with like using a single measure of benefits and costs' (T. Hayward 1995: 99). As Hamilton puts it, 'the purpose of the method is to convert all impacts into a single unit of measurement (dollars) so that decisions can be made simply by comparing one number with another' (1997: 61). Whether we are entitled to doubt the motives of many advocates of free-market environmentalism, the case Pearce mounts on behalf of CBA is transparently intended to meet environmentalist aspirations; thus, '"cost-benefit thinking" would greatly enhance the chances of conservation competing with "development" on equal terms' (Pearce *et al.* 1991: 3). A case based in the hard currency of money values is much more likely to make political headway than is a case based in less measurable — and, therefore, less *familar* — theoretical or ethical abstraction. CBA (more specifically, contingent valuation-enhanced CBA), additionally, is able to recognise variations in *strength* of preference, a further advantage of adopting a money unit of valuation:

> the attraction of placing money values on these preferences is that they measure degree of concern ... What we seek is some expression of how much people are willing to pay to preserve or improve the environment. Such measures automatically express not just the fact of a preference for the environment, but also the intensity of that preference (Pearce *et al.* 1989: 55).

It can also count the 'fuzzy' values of non-use, what Pearce *et al.* refer to as 'existence' values (1989: 61–63; 75–80), defined as 'concern for, sympathy with and respect for the rights or welfare of non-human beings ... a value unrelated to use' (1989: 61). And it can even take cognisance of such important — though apparently intangible — ecological values as irreversibility (see Pearce's discussion of irreversibility in his case study of the Gordon-below Franklin Dam in what is now Tasmania's World Heritage Area wetland wilderness; 1983: 98–102). Though there is considerable debate over the validity of non-market preference allocations, this line of criticism leaves Pearce unfazed:

'whether there is an actual market in the asset or not is not of great relevance. We can still find out what people would pay if only there were a market' (Pearce *et al.* 1990: 8).

CBA also enables a modelling of the preferences of people yet to come, though there is much debate over appropriate rates of 'discount'. Discounting is explained thus: 'future benefits and costs are *discounted* — i.e. given a lower weight relative to a similar benefit or cost in the present' (Pearce *et al.* 1990: 3). By this mechanism 'we can compare the value of economic resources and services at different points in time' (1990: 24). But why discount at all? Because people, being impatient, exhibit a 'time preference' for an identical good here and now rather than postponing gratification to the future (and individual preferences, in the neo-classical paradigm, are sovereign). And because investment now yields enhanced value upon a resource in the future — what economists call the 'marginal productivity of capital' (Pearce 1983: 37–40; Pearce *et al.* 1989: 132–34; Pearce *et al.* 1990: 24–25; Turner *et al.* 1993: 99). Pearce *et al.* cautiously observe that in the past discount rates may have been set too high: that is, the interests of the future have been accorded insufficient weight (1990: 35). But high discount rates bring mixed environmental results. Higher rates protect natural areas because they make investment in such profits-for-the-future ventures as large dams in wilderness areas less attractive. On the other hand, high discount rates encourage more rapid development of non-renewable resources despite the consequences of scarcity and depletion (1989: 135). In the 1989 *Blueprint* Pearce *et al.* noted the environmentalist attack on discounting; after consideration (1989: 136–47; see also Turner *et al.* 1993: 102–06), and the part-concession of some points, they concluded that, rather than adjust discount rates, 'it is better to define the rights of future generations and use these to circumscribe the over-all cost-benefit rule, leaving the choice of discount rate to fairly conventional current-generation oriented considerations' (1989: 150). The cost-benefit calculation should, thus, take place within an ethically bounded context, as established by the criteria for sustainability.

But environmental economics is not simply about valuation and CBA is not the only device its practitioners advocate. 'Environmental economics' also signifies an orientation to environmental policy. We have noted that environmental economics works within a revised neo-classical paradigm, but begins from the premise that markets fail. The answer to market failure is not to turn away from the market. It is rather to 'modify market signals' by 'centrally deciding the value of environmental services and ensuring that those values are incorporated into the prices of goods and services' (Turner *et al.* 1993: 144). Turner and his co-authors refer to this type of regulatory regime as a 'market-based incentives approach'. They favourably contrast it with the more common 'direct ("command-and-control") regulatory approach' and predict that their preferred approach will be increasingly favoured by authority, because it has manifest cost advantages over direct regulation (1993: 144; see also Stavins 1989; Zarsky 1990–91). The principle for 'centrally deciding the value of environmental services' is the Polluter Pays Principle' (PPP). They describe its application thus:

> the basic tenet of PPP is that the *price* of a good or service should fully reflect its *total cost of production*, including the cost of all the resources used. Thus the use of air, water or land for the emission, discharge or storage of wastes is as much a use of resources as are other labour and material inputs. The lack of proper prices for, and the open access characteristic of many environmental resources means that there is a severe risk that overexploitation leading to eventual complete destruction will occur. The PPP seeks to rectify this market failure by making polluters internalize the costs of use or degradation of environmental resources (Turner *et al.* 1993: 145).

It will be clear how the valuation and market incentives aspects of environmental economics fit together. It will also be clear that what is proposed is still a regulated system. But the regulation takes the form of establishing a framework within which enterprises work, providing strategic flexibility within that framework, and placing more emphasis upon incentives than sanctions. And, though environmental economics writings are not necessarily at one on this point, it may be that traditional 'command and control' mechanisms will not be entirely supplanted. Pearce and his collaborators are at pains to point out that 'the instruments for securing development include the traditional "standard setting" regulatory approach *and* a panoply of measures which make use of the marketplace' (1991: 2).

Turner *et al.* provide a useful survey of incentive mechanisms favoured by environmental economists. These include charges on pollutant emissions and for recovery of treatment, disposal and administration costs; charges on products 'that are harmful to the environment'; marketable permits; 'deposit-refund systems'; and 'performance or assurance bonds' that are forfeitable if practice fails to meet agreed criteria for environmental performance (1993: 160–63). 'Marketable permits' are described thus: 'these are environmental quotas, allowances or ceilings on pollution levels. Initial allocation of the permits is related to some ambient environmental target, but thereafter permits may be traded subject to a set of prescribed rules' (1993: 161). Turner and his team are attracted to this mechanism's potential. There is no loss of environmental quality because the number of permits is set by the relevant authority to ensure that end, but the polluter is accorded flexibility in meeting the standard and compliance is achieved less expensively because no cumbersome bureaucratic infrastructure is required (1993: 188; see also Pearce *et al.* 1989: 165–66; Stavins 1989: 59–60; Zarsky 1990–91: 25).

Turner's team also compares pollution taxes with regulation non-compliance charges, noting that in practice charges tend to be set too low to achieve the environmental outcomes that are ostensibly intended, functioning instead as mere revenue-raising devices (1993: 164). Called 'Pigovian taxes' after Pigou, who saw them as the solution to imperfect market valuation of environmental services, Turner *et al.* argue that these have clear advantages over 'penalty attached to non-compliance' regulatory mechanisms. Pollution taxes are harder to evade (because they are built directly into the larger tax framework), and they deliver incentives not present in the direct regulation system as, under a tax system, a polluter is rewarded for achieving reductions in pollution with tax relief (1993: 170). As a device with even broader potential for securing environmental outcomes, much faith is being placed in taxation. O'Riordan and Voisey argue that 'eco-taxation will emerge by stealth, creeping through the finance ministries' (1997: 176; see also Ekins 1996; Pearce *et al.* 1989: 162–65; Robertson 1997; 1998; O'Riordan 1996; Tindale 1997; von Weizsacker and Jesinghaus 1992).

The path has not been easy for environmental economics. Not all economists agree that CBA can accommodate the extension Pearce and others envisage without sacrificing the high degree of precision which has made it a formidable evaluative instrument. Environmentalists have been even more critical, criticising the application of CBA to ecological 'unquantifiables' and challenging the argument that market-based incentives constitute a valid alternative to a direct regulatory approach. We will survey these criticisms in the final part of the chapter. We will also be looking, later in the chapter, at a rival school of environmentalist economic thought, ecological economics, which Hamilton presents as specifically constructed to counter the claims of environmental economics (1997: 61–62).

Before leaving market-based environmentalism, we should note that there are also forms of direct involvement in the capitalist market that command pockets of environment movement support. Some envisage change via the collective impact of 'green' buying power. 'Behaviour is in transition. The very success of the consumer society is breeding new opportunities', write Elkington and Burke. Thus, 'we are now witnessing the arrival of the "green consumer"' (1987: 248; see also Coddington 1993). Similar developments have also been described within finance markets, in the form of so-called 'ethical investment' (for example, Crawford 1991; Partridge 1987). Can one invest in sustainable economic practice and thereby escape from the linked tyrannies of environmental and social exploitation? Some within the environment movement say so. But environmentalist support for ethical investment as a private action does not extrapolate to support for an end to regulatory approaches to environmental protection. At best it evinces a view, by no means overwhelmingly supported, that there is a place for 'in the system' economic involvement within strategies to secure environmental goods.

There is also a body of literature that argues for the capacity of capitalist enterprise itself to meet the needs of planetary stewardship. The circle is complete when, to the greening of consumer behaviour and the greening of financial markets, the greening of business practice is added.

This literature covers quite a range of positions. The more common position is that of Elkington and Burke, for whom 'socialism, as an economic theory, is dead. The argument is now about what kind of capitalism we want'. And we are fortunate, because 'what we are seeing is the emergence of a new age capitalism, appropriate to a new millennium, in which the boundary between corporate and human values is beginning to dissolve' (1987: 250; see also Freeman, Pierce and Dodd 2000; Willums 1998). For Elkington and Burke, the system adapts; all will be fine. More radical is Paul Hawken, who also believes in capital's capacity to adapt, though he believes it has farther to travel than Elkington and Burke believe it has; he also believes in the free market, though he is critical of neo-classical economics, maintaining that the affectionately regarded rules that once governed the operation of the local market have minimal relevance as explicators of the dynamics of the global market (1995: 79–90). Richard Welford is perhaps the most thoroughgoing of all. Comfortable with the most radical variants of environmental thought, Welford argues for a reconstituted business ethics that assimilates many environmentalist ends: protection of ecosystems, implying 'minimal use of non-renewable resources and minimal emission of pollutants' and avoidance of species loss (1995: 199); an end to animal testing and a recognition that animals have rights; an acknowledgement of the rights of minorities and indigenous peoples, and a commitment to principles of global equity; an acknowledgement of optima in organisational scale and that decentralised decision-making and regional development are goals that should be promoted by business; and a commitment to sustainability in its strong sense, as well as to ecological auditing of a business's performance and the adoption of green marketing stategies involving 'a shift away from cynical forms of product promotion, sales hype and stereotypical advertising' (1995: 204; see also Coddington 1993; Hartley 1993; Schmidheiny 1992; Welford and Gouldson 1993).

Though this genre is the most overtly pro-capital within 'green' literature, it is, in some ways, closer to mainstream green concerns than much of the writing within environmental economics. Certainly, there is little in it to provide comfort to those whose neo-classical economics has remained unmoved by the challenge of the environmentalist critique (Beckerman, for example). Thus, Hawken argues for the complete elimination of waste from industrial production, for a shift 'from an economy based on carbon to one based on hydrogen and sunshine' (1995: 209), and for a 'restorative economy' in which 'growth and profitability will be increasingly derived from the abatement of environmental degradation, the furthering of ecological restoration, and the mimicking of natural systems of production and consumption' (1995: 210) — all ideas likely to induce apoplexy in an unreconstructed, neo-classical Promethean, believing, as Prometheans do, in humankind's capacity for boundless technologically driven progress.

Views promoting green capitalism are not entirely idiosyncratic. The idea that the business of capital has an adaptive capacity that can lead to a resolution of ecological crisis has been potently theorised in the 1990s, and thither we now turn.

ECOLOGICAL MODERNISATION

'Ecological modernisation' has, since its introduction in the 1980s writings of the Germans Joseph Huber and Martin Janicke, presented a formidable challenge to environment movement calls for structural social and economic change. Martin Hajer defines it as: 'the discourse that recognizes the structural character of the environmental problematique but nonetheless assumes that existing political, economic, and social institutions can internalize care for the environment' (1995: 25). It contrasts with ecosocialist views of natural limits as one of the rocks upon which capitalism is destined to founder. Indeed, in one of the most influential writings on ecological modernisation, Albert Weale explicitly considers and finds unsatisfactory the ecoMarxist view that environmental demands threaten the state's symbiotic relationship with capital, precipitating a legitimation crisis (1992: 47–51; 89). In Weale's view the state's capacity to lead opinion can avert a legitimation crisis. Insofar as such a crisis stems from conflict between environmental imperatives and spiralling private desires, the state can help restructure definitions of the good life by 'ecologising the economy' so that people come to see an increase in environmental amenity at the expense of private consumption as nevertheless an increase in overall standard of living.

The concept was formulated in response to observed developments in Germany and The Netherlands. Weale (1992), Hajer (1995) and Dryzek (1997) note the replacement there, in the early 1980s, of 'the central dichotomy of 1970s environmental policy' — the assumption of a zero-sum relationship between environmental protection and corporate sector costs (Boland 1994: 135–36) — with a regime in which 'industry itself cooperates enthusiastically in the design and implementation of policy' (Dryzek 1997: 142). Government and industry, then, seek co-operative solutions to environmental problems, on the understanding that 'high environmental standards create new markets in environmental technologies', and 'these in turn offer significant potential for new specialist industries and sectors, with the possibility of substantial employment growth' (Gouldson and Murphy 1997: 85; see also Boland 1994: 136; Dryzek 1997: 142). Dryzek notes that, as a real-world model, it is not to be found in political systems with a tradition of adversarial decision-making (such as Britain, the United States, Canada, Australia and New Zealand; see also Weale's unflattering comparison of Britain with Germany; 1992: 79–88). He also notes that these countries have dramatically poorer records on environmental protection than countries where the ecological modernisation paradigm is descriptive, not merely prescriptive.

At the level of public policy ecological modernisation is, one might say, almost beyond politics. It assumes a degree of corporate state comfort within public policy processes. But in the political arenas where the environment movement operates, the implications of this are profound. Assumed dichotomies dissipate and new ones are created. In Weale's words, 'a cleavage begins to open up, not between business and environmentalists, but between progressive, environmentally aware business on the one hand and short-term profit takers on the other' (1992: 31). It also divides environmentalists against themselves: on the one hand, there are those who reject ecological modernisation as just another dose of snake oil; on the other, there are those who, seeing in this an end to the need for confrontation, thereby re-position themselves within the corporatist fold. Hajer identifies four reasons why ecological modernisation is attractive to governments and this is one of them. It provides a heady defence of 'what is' against the unacceptable demands for fundamental change emanating from the radical wing of the environment movement:

> ecological modernization straightforwardly rejects the anti-modernist sentiments that were often found in the critical discourse of social movements. It is a policy strategy that is based on a fundamental belief in progress and the problem-solving capacity of modern techniques and skills of social engineering. Contrary to the radical environmental movement that put the issue on the agenda in the 1970s, environmental degradation is no longer conceptualized as an anomaly of modernity. There is a renewed belief in the possibility of mastery and control, drawing on modernist policy instruments such as expert systems and science (1995: 33).

Another of Hajer's four reasons for ecological modernisation's popularity with authority is also contained within this passage. In furnishing the political means to combat radical environmentalism, it defends potently the modernist project against the anti-modernist critique: 'ecological modernization does not call for any structural change, but is, in this respect, basically a modernist and technocratic approach to the environment that suggests that there is a techno-institutional fix for the present problems' (1995: 32).

Hajer's other two reasons are also worth noting. Ecological

modernisation 'makes the "ecological deficiency" of industrial society into the driving force for a new round of industrial innovation' (1996: 249), and it thus 'puts the meaning of the ecological crisis upside-down: what first appeared a threat to the system now becomes a vehicle for its very innovation' (1995: 32). Ecological modernisation, therefore, 'assigns a central role to the invention, innovation and diffusion of new technologies and techniques' (Gouldson and Murphy 1997: 75). Finally, ecological modernisation constitutes an alternative, welcomed by governments with relief, to the crudely punitive regulatory regimes that have failed to deliver environmental protection or improvement since the 1970s. It aims 'to shift industry beyond reactive "end-of-pipe" approaches towards anticipatory and precautionary solutions' (Christoff 1996a: 477; see also Hajer 1995: 31).

Let the 'regulatory' not be overlooked. This is a theory of environmental management that takes the capitalist market as given, but it is not a surrender of the state to the market. It is, rather, an acceptance of the need to institutionalise the precautionary principle, and a recognition that stewardship of the resource base is necessary to ongoing capital accumulation and, with it, the legitimacy of the state. As Dryzek observes:

> conscious and coordinated intervention is needed to bring the required changes about. It is no good relying on any supposed 'invisible hand' operating in market systems to promote good environmental outcomes. Yet this intervention does not take place in adversarial fashion, in terms of government imposing design criteria and other policy measures on industry (1997: 142).

This view is most fully developed by Weale. We have seen that he constructs ecological modernisation in opposition to ecosocialism; but he is even more concerned to differentiate it from the anti-statist individualism of neo-liberalism. Focusing particularly on public/rational choice theory (1992: 214–18), he argues that 'ecological modernisation is mercantilism with a green twist; libertarian conservatism is its antithesis' (1992: 88). Weale does think that market-based mechanisms (such as tradeable emission permits) can 'supplement the traditional forms of administrative regulation' (1992: 167), but the thrust of his argument is that the former cannot replace the latter (1992: 164–67; see also Boland 1994: 139. Mol, by contrast, sees a reduced role for the state; 1996: 314). In the corporatist marriage of state and capital envisaged within ecological modernisation, then, Weale places more faith in the adaptive state than in the instruments of the market.

PRECAUTION

Ecological modernisation institutionalises the precautionary principle. This has significance when it comes to positioning ecological modernisation within the complexity of environment–economy thought, because the validity accorded the precautionary principle is a key indicator of where one stands on one of its major divisions. 'Weak' sustainability qualifies the precautionary principle; 'strong' sustainability embraces it. A range of attitudes towards precaution is found within environmental economics; though, even where it is accorded credence, it tends to be seen as a vexatious concept. Less ambiguity exists within ecological economics, where its meaning and validity are largely assumed.

Much of the controversy stems from the variety of meanings attributed to the principle and from difficulties operationalising it. Its dimensions shift according to whether it is under discussion by economists (Pearce 1994), legal theorists (Cameron and Abouchar 1991; Gullet 1997) or ethicists (Attfield 1994). Gullet sees the precautionary principle as 'essentially a new legal response to the scientific uncertainties surrounding the capacity of the environment to cope with the increasing demands placed upon it', and, though 'the principle has definitional and implementational shortcomings', it nevertheless has 'the capacity to inform environmental practices systematically as the basis of a regulatory regime' (1997: 52). Pearce, noting that 'until recently economists have had little truck with the precautionary principle' (1994: 132), argues that it does have 'implications for the advocacy of market based economic instruments for controlling environmental problems' (1994: 149). The implication for cost-benefit analysis, for example, is that benefits have to considerably outweigh costs (rather than be merely 'greater than unity') to justify any project which damages the environment (1994: 144). Furthermore, where environmental problems 'are regarded as close to thresholds, and hence may have characteristics of irreversibility, then economic instruments will be inefficient ways of keeping just below the thresholds' (1994: 149).

O'Riordan and Cameron see 'six basic notions' in precaution:

- 'a willingness to take action in advance of scientific proof of evidence of the need for the proposed action';
- 'recognition that margins of tolerance should not even be approached, let alone breached';
- 'a bias to conventional cost benefit analysis to include a weighting function of ignorance';
- 'duty of care, or onus of proof on those who propose change';
- 'promoting the cause of intrinsic natural rights' by inclusion of 'the need to allow natural processes to function in such a manner as to maintain the essential support for all life on earth';
- and 'those who have created a large ecological burden already should be more "precautious" than those whose ecological footprints have to date been lighter' (1994a: 1–18).

These elements have not met with uniform acceptance — or even recognition — either in law or within policy fora. Particular difficulties stem from an ecocentric interpretation of the principle rather than viewing it as a future-for-humans measure (O'Riordan 1992: 13; Attfield 1994). Even O'Riordan and Cameron's fourth 'basic concept' (they list it fourth, but it could be regarded as the principle's essence) — 'duty of care, or onus of proof on those who propose change' — presents problems: 'formal duties of environmental care, coupled to an extension of strict liability for any damage, no matter how unanticipated, could throttle invention, imagination and growth' (1994a: 17; though there is also a capacity to stimulate imagination and creativity).

The mix of these elements determines the principle's standing. For 'very weak sustainability', which is 'based on the presumption that losses of environmental resources can be made up by innovation, ingenuity, imagination and adaptation', the only place for precaution is 'as a spur to innovation and managerial adaptation'. 'Weak sustainability', by contrast, introduces 'firmer measures of the value of safeguarding ecological and biogeochemical processes that are irrecoverable if lost'. Sustainability's strongest guise 'favours a more fundamentalist mode of ecological solidarity with the earth' and 'aims to build precaution into an approach to living that is altogether more in empathy with the natural world' (O'Riordan and Cameron 1994a: 19–20; see also 1994b).

Whilst disciplinary specialists perceive problems with operationalising the precautionary principle, and others wrestle with contesting definitions and interpretations of it, for most people in the environment movement it is much more straightforward. They would agree with Dryzek's simple definition — 'scientific uncertainty is not a good reason for delaying action against pollution' (1997: 68) — save only that they would extend 'pollution' to cover all forms of environmental degradation.

For such people the principle merely states what is blindingly obvious to anyone who is not an unmitigated knave. It is not beset with imponderables. 'The precautionary principle was strongly resisted — indeed, barely comprehended — by the national governments of Britain and the United States in the 1980s', writes Dryzek: 'in the US, the Reagan and Bush administrations used scientific uncertainty as an excuse for taking little or no action on the acid rain issue ... In Britain, the absence of conclusive science became the standard governmental excuse for inaction on every major regional and global pollution issue' (1997: 139). For most people of an environmentalist cast, these policy timidities are not the result of intractable operational problems or definitional obscurities; they are evidence of the gross planetary irresponsibility of our political and industrial leaders.

And they will be contemptuous of those Prometheans 'who regard the precautionary principle as tantamount to lunacy, as guaranteeing only that the wealth which is the real key to environmental health will be further dissipated by excessive and costly regulations' (Dryzek 1997: 152). The Promethean considers himself hard-headed and his opponents given to soft and fuzzy thinking. But, the environmentalist might say, to argue that an action should be proved potentially or actually harmful before it is stopped involves an almost breathtaking faith in technological ingenuity to get us out of any fix we get ourselves into. Only the most obsessed religious fundamentalist is likely to emulate it; in that case the consequences are not likely to be anywhere near so dire. This given, the Promethean's 'hard-headed' self-assessment starts to look extraordinarily self-deluding.

ECOLOGICAL ECONOMICS/ALTERNATIVE ECONOMICS

We come to the form of economic thought with which the environment movement, in its mainstream, is most comfortable. This is the tradition that has built upon the pioneering work of Daly and Schumacher (Daly, indeed, remains a key contributor to it). There are quite distinct academic and semi-popular or 'practitioner' versions of this thought. I have termed them (by no means originally) *ecological economics* and *alternative economics.*

Several dissident economists have, with Daly, developed an economics that is sourced to ecological imperatives (see, for example, Common and Perrings 1992; Costanza 1989; Costanza and Daly 1987; Costanza, Daly and Bartholomew 1991; Daly 1991; Gowdy and Olsen 1995; Khavari 1993; Martinez-Alier and Schlupmann 1987; Norgaard 1985; 1987; 1988; 1992; 1994; Perrings 1987; Stanfield 1982; Swift 1993).

'Ecological economics', Rosewarne notes, 'represents a significant shift insofar as nature is endowed with an ontology not previously acknowledged by environmental economists, one that has required a recasting of how economic process ought to be theorised' (1995: 109–10). Such theory is problem-focused rather than concerned with abstract modelling, and, in contrast to conventional neo-classicism, ecological economics shifts the focus from micro to macro and relevant time frames from the very short term to deep time (Costanza, Daly and Bartholomew 1991: 5). Ecological economics complements the relational and synergistic realities of ecology. It is, therefore, a holistic rather than a reductionist endeavour and gives due weight to process, change and flux rather than stasis (Costanza 1989; Costanza and Daly 1987; Costanza, Daly and Bartholomew 1991; Stanfield 1982). Such an economics also incorporates an ethical and visionary dimension — necessary because grounding economic thought within a broader and prior context requires strictures of 'ought' to govern contextual relationships.

Ecological economics rejects the valorised individual as the sovereign economic unit. Natural and social systems are processual and holistic, and an economics that is not constructed in accordance with this insight will be at odds with, and ultimately destructive of relational entities. This is precisely the case with conventional reductionist economics. It sees the essence of the human–natural systems relationship as one of exploitation of ecological systems to optimise material wealth. Natural processes are assumed to be infinitely flexible and manipulable; where they are not, technological ingenuity can supply substitutes for lost ecological goods and services (Costanza, Daly and Bartholomew 1991: 5). Ecological economics, by contrast, stresses endemic uncertainties in our knowledge of ecosystems and our interactions with them; it is 'prudently sceptical' (Costanza, Daly and Bartholomew 1991: 5). We neither know how ecosystems function, nor their degree of resilience, nor the precise nature of the dependence of social and economic systems upon them. In the absence of certainty, ecological economics presumes the need for human systems to adapt to ecological systems, because the slow rate of ecosystem adaptation renders the latter more susceptible to catastrophe as a consequence of even apparently slight perturbations.

There is an urgent need for reliable explanatory models of human–natural systems interrelationships. Knowledge is particularly needed, not only of the matter–energy interchange essential to both systems but also of their change–tolerance capacities. 'Once the macroeconomy is seen as an open subsystem, rather than an isolated system', Daly writes, 'the issue of its relation to its parent system (the environment) cannot be avoided' (1991: 34). The 'major task of environmental macroeconomics' is to 'design an economic institution' that will 'keep the weight, the absolute scale, of the economy from sinking our biospheric ark' (1991: 35). For instance, the goal of a global economic system larger than the present one by a factor of five or ten, as institutionalised in the Brundtland Report in the *name* of sustainability, is, says Daly, arrived at in ignorance of the fact that the economy already claims a quarter of the planet's 'net primary production of photosynthesis' (1993: 268–69). In his continued insistence upon the reality of entropic constraints and his concern for the unsustainability of the matter–energy flux within fossil-fuel industrialism, Daly does most to keep the doomsday paradigm at the forefront of environmental thought.

Despite its disavowal of the abstraction of neo-classicism, it might be thought that ecological economics is fixated upon establishing a *framework* for economic analysis. But most ecological economists do proceed beyond that. Costanza, Daly and Bartholomew, for instance, move from 'goals' to a research agenda and then to a series of policy recommendations. Their 'research questions' are grouped into the subthemes of sustainability (1991: 7–9), 'valuation of ecosystem services and natural capital' (1991: 9–11), ecological accounting (1991:

11–13), local, regional and global scale modelling (1991: 13–14), and 'innovative instruments for environmental management' (1991: 14–15). Policy recommendations include 'fees, taxes and subsidies ... to change the prices of activities that interfere with sustainability' (1991: 17). Whilst market incentives may be used 'where appropriate in allocation decisions', when it comes to questions of scale, 'individual freedom of choice must yield to collective decision making by the relevant community' (1991: 17). The bottom line is 'the maintenance of the total natural capital stock at, or above the current level', because, 'given our uncertainty and the dire consequences of guessing wrong, it is best to at least provisionally assume that we are at or below the range of sustainable stock levels and allow no further decline in natural capital' (1991: 16). Thus, mainstream and ecological economics differ strikingly over consumption. Whereas conventional economics seeks to guarantee ever-rising consumption, ecological economics demonstrates its unsustainable consequences for natural capital.

Though ecological economics extends ideas already present in Daly's early work, there are interesting variations on the familiar themes. For example, Martinez-Alier and Schlupmann (1987) develop a Marxist ecological economics, whilst Richard Norgaard writes of 'coevolutionary' ecological economics. Martinez-Alier's ecological economics, like Daly's, owes much to Georgescu-Rogen, and his themes do not depart substantially from those of other ecological economists, though he focuses more upon third world 'development', and he places greater emphasis on distributional and equity questions.

Norgaard conjoins economic and ecological paradigms in a perspective that is not dialectical in the sense of a new fusion displacing its constituent paradigms, but which draws simultaneously upon incongruous models in a 'co-evolutionary' dynamism. Norgaard is critical of the implicit faith economists and decision-makers hold in the assumptions of mechanistic science and modernist rationalism. The modern scientific outlook sees the world as an agglomeration of atomised parts which are integrated mechanistically and hence amenable to formal modelling, and which disclose knowledge that is universal and usable (1992: 77). Similarly, neo-classical economics views the economic system as atomistic ('land, labour and capital ... are like individual atoms. They are combined but not changed during the production of goods and services'; 1992: 78) and mechanistic ('economic systems can operate in equilibrium at any position along a continuum and move back and forth between positions. If more labour becomes available, the system adjusts so that more labour intensive goods are produced'; 1992: 78). Thus, 'the dominant vision of social organization stems from the Newtonian model of systems consisting of mechanistically related atomistic parts' (1987: 107; see also 1994: 3–7; 61–74; 204–09).

Against the mechanism of neo-classical economics, Norgaard's 'co-evolutionary perspective' draws upon the following insight:

> during the last few decades biologists have put even more emphasis on how ecological relations between species affect evolution, how species co-evolve. The shift from thinking of each species evolving in response to a changing physical environment to thinking of species co-evolving together has led to a new understanding of evolutionary dynamics (1987: 114).

Applied to the economy–nature relationship, the central insight is that 'the intertwining of knowledge, values, social organization, technology and resource systems is more or less symmetrical; no system dominates another' (1988: 616). And these sub-systems change in a co-evolutionary way: 'not only is each subsystem related to all the others, but each is changing and affecting the evolution of the others' (1988: 617). Random changes in any sub-system will be tested for evolutionary 'fit' against the other sub-systems, as well as the one in which the change occurs, and so, 'with each subsystem putting selective pressure on each of the other subsystems, they co-evolve in a manner whereby each reflects the others' (1988: 617). Until the age of hydrocarbons 'development was a process of social system and ecosystem co-evolution that favoured human welfare', but 'the era of hydrocarbons drove a wedge between the co-evolution of social and ecological systems' and it did so by freeing societies 'from immediate environmental constraints but not from ultimate environmental constraints' (1988: 617). And, because change ramifies across the sub-systems:

> value systems have been collapsing. Knowledge has been reduced to western understanding. Social organization and technologies have become increasingly the same around the world. The cultural implosion and environmental transformation have been closely interconnected. The switch to hydrocarbons allowed cultures to stop coevolving with their unique environments and adapt the values, knowledge, technologies and organization of the west (1992: 82).

The result is unsustainability.

The other form of green economics that merits consideration under 'ecological economics' is the thought of green movement activists. These are not, in the main, dissident academic economists (there are exceptions, and some are also refugees from the world of 'orthodox' business practice), and their economics is often presented within broader schemes of utopian envisioning. Characteristic devices include co-operative enterprises such as green banks (Benello *et al.* 1989: 163–67; Partridge 1987: 149–50), community credit co-operatives with non-professional management and ethics-framed charters (Partridge 1987: 142–43), land trusts (Benello *et al.* 1989 19–62), and local currencies or exchange systems (Jackson 1992; Linton *et al.* 1989: 196–203; Partridge 1987: 157–158; Schwarz and Schwarz 1987: 46–47).

Local currency schemes can serve as paradigmatic here. Self-regulating economic networks which allow their members 'to issue and manage their own money supply within a bounded system' have succeeded in several parts of the world. Such schemes retain local money within the community, make full use of local goods, services and skills, allow for greater local economic control, and preclude speculation, because 'green dollars only have local value' (Schwarz and Schwarz 1987: 46). They are touted as important mechanisms for community rehabilitation, and they have an even deeper symbolic importance; effectively realised, they constitute a functioning economy within the formal economy, and so give embodiment to the maxim of 'building the new society within the shell of the old'.

This is the economics of radical environmentalism in its movement guise. Its themes constitute a development, in popular or semi-popular form, of ideas explored in the pioneering contributions of Daly and Schumacher. They include:

- the need to divest economics of its disciplinary hubris and render it subservient to cultural and ecological process — in particular to place it within a context of ethical and/or spiritual need (Daly and Cobb 1989; Hamilton 1994; Lutzenberger 1985; McBurney 1990; Norberg-Hodge 1997);
- a damning critique of the economics of industrial capitalism (Hamilton 1994; McBurney 1990; Trainer 1985; 1989; 1995a; 1995b), including globalisation (McBurney 1998; Norberg-Hodge 1997; Robertson 1998; Sachs 1989) and the 'pathology' of growth (Chiras 1992; McBurney 1990; Robertson 1983; Trainer 1995a; 1995b; 1998);
- a call for an economics based upon simplicity and the satisfaction of core need (Elgin 1981; Galtung 1980a; 1980b; Max-Neef 1985; 1992a; 1992b; Trainer 1985; 1995a);
- a case for decentralisation, localism, community revitalisation and self-reliance (Dauncey 1989; Douthwaite 1996; Galtung 1979; 1980b; 1989; Max-Neef 1985; 1989; 1992a; Sachs 1989);
- a priority on environmental health (Robertson 1989a; 1998);
- the need for an economy based upon low matter–energy throughput (McBurney 1990; Robertson 1983);
- restructured concepts and practices of work (Hamilton 1994; Robertson 1985; 1989b).

There is hostility to the capitalist market, though little agreement over whether it is reformable or beyond redemption. Those taking the former view advocate such mechanisms as ecotaxation; here, the alternative economics agenda overlaps that of environmental economics, though it usually extends application of the Pigovian pollution tax. The less-radical wing shades into green consumerism. Schwarz and Schwarz, for example, observe approvingly that 'investment in ecologically sound ventures is an aspect of new economics that has got well beyond the drawing board' (1987: 47; see also Chiras 1992: 77–101), though this reads incongruously in a case for a society so revised along green lines as to constitute a revolutionary departure from what is.

The economic thought of radical green practice had its profile boosted by The Other Economic Summit (TOES), which was set up in 1984 and 1985 as an international forum for alternative economic thought to shadow the high-profile London and Bonn economic summits. Paul Ekins's *The Living Economy* (1989) showcases many of the contributions made at the 'alternative summits'. The legacy of Daly and Schumacher is clearly evident in the summit papers — Daly was himself a 1984 presenter — and the themes listed above are predominant among the matters explored. We will consider here the contributions of just two prominent contributors to TOES: James Robertson and Johan Galtung.

The nature of work, and the implications of changing conceptions of work for economic paradigms, was the subject of Robertson's contribution (1989b) to the 1984 TOES. Robertson is strongly critical of conventional economics. In an early work he called for a paradigm shift from an economics of 'super-industrialism' or 'Hyper-Expansion (HE)' (1983: 20–21) to an economics that is 'Sane, Humane, Ecological (SHE)'. In this view industrial society is 'a dangerous masculine fantasy — exploitative, elitist and unsympathetic', whilst 'the line of social progress shaped by the technological and economic imperatives of the industrial age has reached its limit' (1983: 23; see also 1989b). What is needed is a 'new economic direction' that shifts priorities to 'human need, social justice and ecological sustainability' (1983: 36), and in which 'the economic system will become an integral part of the larger ecological system, i.e., a closed loop of material cycles powered by the sun' (1983: 37). Economies will be decentralised, technology will be made humankind's servant rather than its master, much emphasis will be placed on self-reliance and mutuality, and greater income equality will be fostered (1983: 30–45). In his later work he develops these ideas, adding a critique of globalisation (1998: 12–13; 61–67) whilst shifting in part to a less radical concern with in-the-system economic reform, such as ecotax initiatives (1998: 42–43), a 'citizen's income' provided to all as a right of citizenship (1998: 44–45), and the greening of financial institutions (1998: 51–59). But it is for his development — in the tradition of Schumacher — of an ecological theory of work, 'ownwork', that he is best known:

> *ownwork* means activity which is purposeful and important, and which people organise and control for themselves. It may be either paid or

> unpaid. It is done by people as individuals or as household members; it is done by groups of people working together; and it is done by people who live in a particular locality, working locally to meet local needs ... The new work ethic will teach that to immerse oneself in today's organisational world is to sink into a world of abstractions and turn one's back on real life; and that real life means real experience, and real work means finding ways of acting directly to meet needs — one's own, other peoples' and, increasingly, the survival needs of the natural world which supports us (Robertson 1985: 69).

A noted theorist of development economics, Galtung was a presenter at the 1985 TOES. He argued (as he has done elsewhere; for example, 1979; 1980a) that the market's rationality is attained at the cost of moral constraints. 'No man or woman', writes Galtung, 'would behave within the family as they do in the market' (1989: 99). His solution is an economics which conjoins global interdependence to local self-reliance. There are four basic principles of local self-reliance. First, 'some mechanism in addition to the market must be found for the satisfaction of basic human needs' (1989: 100). Second, we must produce 'what is needed, relying on ourselves, on our own production factors', in opposition to the dominant tendency in centre–periphery relationships, where the centre treats the periphery as a store of resources for plundering and a sink for wastes and surpluses (1989: 100). Challenges must, therefore, be internalised, not resolved by displacement to somewhere else. Third, as not all needs can be met from local resources there must be trade, but governed by strict principles of equality: primary products for primary products, manufactured goods for manufactured goods, tertiary services for tertiary services (1989: 102). Fourth, the theory of self-reliance enjoins actors to transcend 'the nakedness of economic relations'. It is strongly normative, promoting self-respect and 'a feeling of compassion and a will to resist threats and the actual exercise of violence, direct or structural, from the outside' (1989: 102). The dimension of global interdependence involves application of the principles of self-reliance beyond the local level to nation-states and on to the larger transnational region (1989: 103–05).

Other contributions to 'alternative economics' could be presented, but most are variations on familiar concerns and ideas. One particularly influential contribution to 'alternative economics', one which draws together academic 'ecological economics' with activist 'alternative economics', requires acknowledgement. This is the 1989 book of Herman Daly and process theologian, John Cobb, *For the Common Good.*

Daly and Cobb challenge the hegemonic claims of economics as a discipline. They dismiss the exclusivist use of 'rationality' employed within economics: 'economists typically identify intelligent pursuit of private gain with rationality, thus implying that other modes of behavior are not rational. These modes include other-regarding behavior and actions directed to the public good' (1989: 5). Moreover, as a deductive discipline, economics is obsessed with abstraction. It 'abstracts a discrete subject matter from the totality, treating it as if its connections with the remainder of the world were not important' (1989: 121; similar challenges to the unworldliness of economic abstraction are to be found in Lawson 1992; McBurney 1990). Daly and Cobb, by contrast, call for a return to 'more purely empirical-historical studies of the course of events in the process of economic change' (1989: 123), because working at the realm of abstraction too easily enables one to avoid confronting real-world consequences — except via another abstraction, 'externalities', which also screens the reality of social and ecological dislocation from the person employing the term.

In arguing for the rationality of actions taken in the public interest, and for embedding a discipline that has become too big for its boots within the larger context it systematically denies, Daly and Cobb shift the meaning of *homo economicus* from the 'pure', sovereign individual to 'person-in-community' (1989: 7). An economics for community must resist the notion that social relations are an aspect of the economic system rather than economics itself deriving from those social relations. But Daly and Cobb do not seek an end to the market as the mechanism for allocating goods, wishing only to highlight 'the revisions required in neoclassical theory and in actual capitalist practice in order that the destruction of community be ended' (1989: 18). In fact, 'we are convinced of the general soundness of the account of markets ... we believe many public purposes could be better served by the application of market principles than the patchwork of government regulation now so prevalent' (1989: 19).

The authors of *For the Common Good* describe themselves as 'Christian theists', and the final chapter of their book is entitled 'The Christian Vision' (1989: 376–400). We have seen that a call to lodge economic thought and practice within a spiritual and/or ethical mantle is characteristic of alternative economics formulations, and it is also occasionally acknowledged within the more conventionally academic literature of ecological economics. 'Ecological economics recognises that the natural environment is important to humans not only because it provides services that ensure survival and provide enjoyment', writes Clive Hamilton (whose 1994 book is entitled *The Mystic Economist*), 'but because there is some deeper psychological connection between humans and the natural world' (1997: 56). Such a concern for the spiritual and ethical basis of life has always been a feature of Cobb's work, though it is more prominent in Daly's later writings than in his 1970s work. In 1993 he wrote, with Townsend:

> it is important to be very clear on the paramount importance of the moral issue. We could opt to destroy the spaceship in an orgy of procreation and consumption. The *only* arguments against doing this are reli-

> gious and ethical: the obligation of stewardship for God's creation, the extension of brotherhood to future generations, and of some lesser degree of brotherhood to the non-human world (Daly and Townsend 1993: 155).

What remains most tenacious in Daly's writing is the lesson he took from Georgescu-Rogen back in the 1970s: 'the idea that low-entropy matter-energy is the ultimate resource' (Daly and Cobb 1989: 11), and that, conversely, the material basis of industrial society is fundamentally at odds with the requirements of sustainability. In 1993, attacking 'sustainable growth' as absurd, he wrote:

> in its physical dimensions the economy is an open subsystem of the earth ecosystem, which is finite, nongrowing and materially closed. As the economic subsystem grows it incorporates an even greater proportion of the total ecosystem into itself and must reach a limit at 100 percent, if not before. Therefore its growth is not sustainable. The term 'sustainable growth' when applied to the economy is a bad oxymoron — self-contradictory as prose, and unevocative as poetry (1993: 267).

Though it does not fit smoothly under either the 'ecological economics' or 'alternative economics' labels, this seems the most appropriate place to note an ecofeminist political economy developed by Ariel Salleh (1997) — especially as Salleh herself deploys the term 'ecological economics' in relation to her work. She provides a *reproductive* critique of *productive* economics, arguing that capitalist patriarchy is not only constructed upon the bounty of nature, which is accorded only implicit value until it is appropriated. It is also constructed upon women, who are likewise expected to 'furnish the conditions of production', and whose 'main task area is "social reproduction", which term uncritically lumps together a variety of economic functions' (1997: 80). Women's potency 'is greatly diminished before the celebration of men's technological production', and women's bodies are '"resourced" for free by capital to provide ever-new generations of exploitable labour' (1997: 93). But women, she writes, are also 'the productive sex in addition to their generativity' — a reference to the fact that by far the majority of the work performed upon the planet is performed by women — so 'why are they universally assigned a position outside production?' Her answer is that, against all the evidence, 'production' is identified with masculinity; it is 'claimed by men as their own area of competence' (1997: 81).

It will be clear that Salleh uses 'reproductive' in a broader sense than human population generation. 'Reproductive activity' is, in Salleh's usage, a mode of acting in the world — nurturing, loving, life-sustaining, 'meta-industrial', deepening of our relationship with nature — and it is not 'accounted' in the orthodox economy. Relationships based upon 'reproductive activity' are to be found in the non-alienated, non-capitalised experiences of women in the third world — and in some aspects of women's labour surviving even within the industrialised west. They are a potent force for change. But the globalisation of patriarchal capitalism massively threatens this:

> while economic 'growth' appears to bring material benefits to some men and women in the North, in another sense it can be said that almost all women inhabit the South. The annexation of women's work is reinforced with industrialisation and consumerism, whether by computers, labour-saving gadgets, or new reproductive technologies. Meanwhile, in 'developing' regions, expropriation of farmlands for commodity markets, technocratic Green Revolutions, and now gene patenting undercut the very means of women's labour for subsistence. Continued capital accumulation, the expanding hegemony of transnational operations, and the rise of 'phantom-states' like international drug cartels all add up to deepen nature's and women's subjection (1997: 87–88).

Though 'women's work makes accumulation possible for all kinds of men, and the economic surplus women generate is crucial to the operation of capitalist patriarchal relations' (1997: 90), it is the case that, 'from a global capitalist patriarchal vantage point, the Third World debt crisis generated by World Bank and IMF policies has provided an excellent opportunity to restructure class and gender relations across continents, integrating new proletariats into the global economy' (1997: 91). The results are devastating: 'indigenous women's expertise developed over thousands of years — knowledge of seed stocks, the water-conserving properties of root systems, transfer of fertility from herds to forests, home-grown medicines and methods of contraception — is lost. Ecological and human needs go unfulfilled; societies and cultures disintegrate' (1997: 92–93).

The restoration of the 'embodied materialism' that characterises women's traditional labour is Salleh's project. Though only a sample of it is given here, her gender-suffused approach to political economy is realised as a fully developed system of ecological economics.

SOME PROBLEMS: THE TENACITY OF PRIVATE DESIRES

It is difficult to present a critique of liberal rights environmentalism that is simultaneously short and coherent, because it is a long distance from Wissenburg to Eckersley. The Wissenburg/Achterberg end of the spectrum is likely to be least approved within the environment movement. It can be argued that Wissenburg's bottom line — that the sovereignty of the atomised individual is the non-negotiable essence of political liberalism — can never satisfactorily accommodate all necessary ecological contingencies. The inverse restraint principle, for example, can only work when polluters are identifiable and their effects are precisely determinable. Where offenders are multiple (car-driving pollution generators, for example) and the effects diffuse and cumulative through many generations, it is difficult to see how workable strategies can be derived from it.

The problem comes into sharpest relief when we move, as ecology insists we must, to a consideration of collective entities and the status to be given to relational processes. Liberalism struggles to find a ground for value in units of life consisting of co-dependent relationships, such as species and ecosystems. Wissenburg is blunt and unbudging: 'from a liberal point of view, it is only the life and welfare of individuals (with interests, feelings, etc.) that counts. Species are, like societies for humans, only important in so far as they sustain and promote the lives of individuals' (1998: 179). It is her unwillingness to accept this as the final word that led Eckersley to search within liberalism for a basis for relational processes. Opinion will differ over whether, and to what extent, she has succeeded.

Ted Benton is one who thinks the liberal rights tradition is beyond ecological redemption. Critiquing animal rights theory, Benton argues that 'the intrinsic incapacity of non-human animals to make rights claims on their own behalf is a source of difficulties for the rights view', because 'so central is the value of individual autonomy and the authority of the individual in the judgement of her or his interests in the liberal tradition, that the attribution of rights to beings who by their nature cannot make rights-claims on their own behalf must induce conceptual strain, to say the least' (1993b: 165). Eckersley, who claims to be able to work around Benton's objections, paraphrases his position thus: 'Benton's primary objection to the liberal rights discourse is that it ignores both the social and ecological "embeddedness" and "embodiment" of individuals', for both 'biophysical embodiment and ecological interdependence' must be seen as 'indissolubly bound up with the individual's identity and experience of self' (Eckersley 1996a: 185). Another for whom the liberal rights discourse is a blind alley is Val Plumwood. She finds within liberalism's overdeveloped defence of 'liberal privacy' the 'construction of many varieties of unfreedom — as subordination, discipline and vulnerability in both private and public worlds' (1996: 154). The subordination hidden, and thereby protected by liberal privacy applies most notably to women and nature. Plumwood, thus, finds no possibility in liberalism of 'a public ethic of care and responsibility for nature or the future', and if it precludes the prospect of an enlarged and public form of moral life, 'liberalism can at best yield a narrowly instrumental form of public policy in relation to nature, and is hard pressed to stabilise even enlightened instrumental forms in the tug of war between interest groups' (1996: 155). Her summation is withering: 'ecological failure leads us to conclude that it is not democracy that has failed ecology, but rather liberal democracy that has failed both democracy and ecology' (1996: 147).

Though attracting growing interest, rights-based liberal environmentalism is still not a strand of environmental thought with much of a profile. I think this needs to change. What is at stake here is nothing less than whether hegemonic western values and the insititutions based upon them have a capacity to deliver us from looming ecological devastation. If they do not, the consequences are dire. A first question which should then be asked is what should give — ecological integrity or the cherished and venerable values upon which western civilisation rests? The answer to this may not be as obvious as most people within the environment movement would assume. Here, then, is a debate that urgently needs fostering.

I have noted that little credence is accorded free-market environmentalism within environmental thought, that it is largely regarded as an intellectual con-job, but that, perhaps paradoxically, there seems rather more support for notions of green capitalism and green consumerism. Some radical environmentalists have found this advocacy alarming. Labelling the notion of environmentally friendly capital-enterprise 'greenwash', critics such as Tom Athanasiou (1996) and Sharon Beder (1997a) chart the cynical manipulation of much PR-crafted corporate 'environmentalism'. The consumer end, too, has its critics. They include Timothy Luke, who argues that 'green consumerism, rather than leading to the elimination of massive consumption and material waste, instead revalorizes the basic premises of material consumption and massive waste' (1993: 170; see also 1997; Martell 1994: 69), and that this form of environmentalism 'is virtually meaningless as a program for radical social transformation' (1993: 171). Green consumerism, thus, leaves intact 'an economy and a culture that thrive upon transforming the organic order of nature into the inorganic anarchy of capital' (1993: 170), with 'the well-being and survival of other animal species, plant life forms, or bioregions virtually ignored' (1993: 171).

Though few green thinkers have critically involved themselves with the details of free-market environmentalism, many have explored the perceived shortcomings of free-market environmentalism's overarching paradigm — neo-classical economics — and it is at this point that the green attack upon environmental economics usually begins. Birkeland *et al.* provide an excellent to-the-point summary of the green case against neo-classical economics:

> neoclassical economics has been condemned by the environmental movement for emphasising individual preferences over collective needs, failing to recognise and take into account the economic contributions of women, not respecting environmental constraints, and regarding the 'environmental services' provided by nature as externalities (1997: 125–26).

Birkeland and her team also emphasise the social context in which preferences are formed. Conventional economics refuses to treat as relevant the process whereby preferences come to be determined; in fact, because the individual is constructed as radically sovereign, there is a tendency to make the wholly implausible assumption that there is no broader context for preference formulation. Lindblom (1977; 1982), in his rebuttal of the notion of democracy as a political market, has most prominently made the case for the political construction of preferences. Daly and Cobb similarly describe as folly the core neo-classical assumption that 'individuals all seek their own good and are indifferent to the success or failure of other individuals' (1989: 159), because 'the social character of human existence is primary' (1989: 161), and it is in social relations, in the ebb and flow of daily discourse, that political, cultural — and economic — preferences are formed.

As a neo-classical economics, environmental economics shares with free-market economics a faith in the efficacy of the market, notwithstanding the substantial differences between them, particularly concerning their respective assessments of whether the market is perfect or flawed; environmentalist criticisms of market economics also gather up environmental economics in their critical sweep. In his wide-ranging assessment, *The Market* (1998), O'Neill sides with non-market institutions and against the domino-fall capture by markets of ever-more life domains, whilst Dryzek observes that 'markets have no mechanisms for dealing with the common property and public goods problems which they generate' (1992: 21). Reasoning similarly, Ekins *et al.* argue that an ecologically responsible economics must allocate a more prominent role for the state than is to be found in most market-based formulae for environmental protection (1992: 34–35; 80–81; see also Kinrade 1995).

The status environmental economics assigns individual preferencing has been widely criticised in ecological thought. An influential critique was developed by Mark Sagoff, who argued (1988) that people interact with the world in more complex ways than the simple needs-preferencing individual assumed in neo-classical economics (including its environmental economics variety). In addition to behaving as consumers people also behave as *citizens*. In this context personal needs are modified by responsibilities, obligations, and assessments of what should *rightly* occur: 'most of us seek outcomes, especially with respect to the environment, for intrinsic ethical rather than self-serving reasons. Our personal well-being is not what we necessarily seek. We debate public policy in terms of the values and goals of the community itself' (Sagoff 1993: 5). Michael Redclift agrees that preferences are socially and ethically determined in large part. Rejecting the conventional assumption that all social interaction is 'instrumental', 'designed to maximise the individual's utility' rather than 'constituting value in its own right', he argues that the neo-classical paradigm simply cannot acknowledge that individual behaviour includes important components of thought and action that 'do not correspond to short-term economic advantage' (1993a: 15; see also 1993b: 110–11).

Employment of the concept of 'existence value' in contingent valuation is the main means by which environmental economics has attempted to meet this criticism. This is a device whereby the valuation process incorporates ethical and aesthetic preference expressions. But Sagoff argues that a logical error is committed here, a 'category mistake' involving the collapsing of values into preferences. He explains a 'category mistake' thus: 'if I say the square root of two is blue, I have not uttered a meaningful sentence but an absurdity, since color concepts are not of a logical type that can meaningfully be predicated on numbers' (1988: 93–94). The problem with 'existence value' is that private and public preferences belong to different logical categories':

> public 'preferences' involve not desires or wants but opinions or views. They state what a person believes is best or right for the community or group as a whole. These opinions or beliefs may be true or false, and we may meaningfully ask that person for the reasons that he or she holds them. But an analyst who asks how much citizens would pay to satisfy opinions that they advocate through political association commits a category mistake (1988: 94; see also Keat 1997: 41).

Redclift is, like Sagoff, critical of neo-classicism's equation of the human being with a self-absorbed, rationally calculating *homo economicus*. Such a construction is not value neutral, though a determined effort is made to universalise the ideological biases upon which it is based (1993a: 15). But 'the "rational individual calculator" beloved by economists sits uneasily in cultures other than those which helped develop the paradigm in the first place' (1993a: 16). Hamilton agrees: 'the fundamental problem with environmental economics is that it allows us to be only one thing, private economic agents'. This is, moreover, an aspect of human life 'that many thoughtful people regard as venal, exploitative and self-destructive', and to put it forward as the template for human behaviour is 'ideological folly of the first order' (1994: 77–78).

The most formidable charge against neo-classical economics is that it ignores (in the case of free-market economics), and cannot adequately account for (in the case of environmental economics) entropic constraints. As Schumacher had earlier argued, 'it is inherent in the nature of economics to ignore man's dependency on the natural world' (1974: 36). Norgaard develops this insight, arguing that environmental economics 'assumes that environmental systems are divisible and can be owned. It fails to acknowledge linkages that cannot be matched with a market feedback'. Thus, 'markets fail to allocate environmental services efficiently ... because environmental systems are not divisible'. 'Dividing the indivisible' can only be achieved 'through the destruction of the more highly interconnected portions of ecosystems' (1992: 83; see also 1985). Redclift, considering Norgaard, observes that 'economics ... is epistemologically predisposed towards a reductionist view of resources and their utility' (1987: 41).

The market's claims would seem to be chimerical. Whilst claiming to liberate individuality, it promotes cultural homogenisation. Whilst claiming to be the mechanism that will deliver sustainability, it destroys natural capital by transforming it into high-entropy, system-overloading wastes. Even its promised freedoms, defined as they are in the narrow terms of product choice on the supermarket shelf, are largely illusory. As O'Neill observes, 'the differences between modern centralised bureacratic economies and free-market economies are less notable than the similarities. A spirit of technical rationalism pervades both' (1998: 141).

We have not distinguished systematically between 'pure' neo-classical economics and the revisionism of environmental economics because a running theme within green critiques of environmental economics is that its reformulations have done little to rescue neo-classicism from its ecological ignorance. In sum, writes Tim Hayward, environmental economics 'retains too many assumptions of conventional economics' (1995: 104).

There is, though, much criticism of specific components of environmental economics. We have noted Sagoff's criticisms of contingent valuation in cost-benefit analysis (CBA). Other purported deficiencies are reported in the literature. Most basic is the now familiar criticism concerning the measuring of what is beyond measurement. Redclift maintains that 'economists cannot value what the environment is *worth*; merely its value in monetary terms' (1993a: 14), whilst for Schumacher 'cost/benefit analysis is a procedure by which the higher is reduced to the level of the lower and the priceless is given a price' (1974: 37; see also T. Hayward 1995: 100). (Pearce responds to this criticism by insisting that, in a world of fierce competition for resources, decision-makers will offer no choice; so we must simply get on and do the best job of valuation that we possibly can; Pearce *et al.* 1989: 54.) A similar first-principle objection is that CBA is a measure of human preference in which no calculation is made for the good of entities other than humans (Hayward 1995: 102; Jacobs 1997: 212; for an interesting variation on this argument, see Holland 1997).

Intractable problems with the *procedures* of CBA are also described. Beder (1997b: 92), for instance, argues that 'willingness to pay assumes that the environment does not already belong to the community', whereas a willingness to sell would assume the opposite, and an excision from the public realm would require commensurate compensation. (Goodin makes a similar case in relation to pollution permits, drawing an elegant analogy with medieval 'religious indulgences', in the case of which ecclesiastical authorities sold what was not theirs to sell — God's grace. In the case of the selling of 'environmental indulgences' in the form of pollution permits, Goodin argues that it is 'presumptuous' for humans 'to grant indulgences on behalf of Mother Nature. Forgiveness is the prerogative of the party who has been wronged'; 1998: 240.) Wenz argues that contingent valuation and discounting afford such scope for subjective interpretation that an unchallengeable dollar value is simply unattainable. He points out that different questions intended to establish the value of a human life have produced results varying in the range $136,000 to $2.6 million (1988: 220–21; see also Jacobs 1991: 198, 200–01; 1994; M. O'Connor 1994c: 146). These problems are not merely technical, but stem from the non-negotiable ideological assumptions within which CBA is bedded. They are consequences of a worldview which defines the human as narrowly and exclusively *homo economicus*; a view in which 'public domain concerns are regarded as irrelevant or too difficult', and in which 'any notion that resource allocation should entail political determinations is rejected out of hand' (Rosewarne 1995: 112–13).

Several authors, examining the ideological underpinnings of CBA, defend discursive democratic process as a superior mode of preference formulation and adjudication. Three analyses are presented here.

John O'Neill returns to CBA's problem of *commensurability*: comparing like with unlike. CBA 'treats preferences grounded in scientific, communitarian and aesthetic judgements as necessarily on a par with preferences for, say, this or that flavour of ice cream' (1996a: 101; see also 1993: 110). Comparing like with unlike is not, however, a problem unique to CBA; it is, in fact, a problem with which people are regularly faced. What happens when these sorts of category problems occur? People do *not*, says O'Neill, normally attempt resolution by seeking some common unit of currency: 'they weigh not measures but reasons for and against a proposal. They argue, debate and come to some agreement'. So the economist's 'measuring rod of money' is 'contrived'. Worse, it 'gets in the way of the process of reasoned deliberation' (1997: 82). CBA, thus, constitutes the 'marketisation of politics' (Grove-White extends this beyond politics to all human nature; 1997), in which the validity of reasons for value-preferences is held to be of no relevance, and in which no mechanism for testing the soundness of one's value expressions against others is deemed necessary (O'Neill 1996b: 758).

Michael Jacobs also deems political discourse a superior mode of preference determination to CBA and the market. People see 'environment' as a public good imbued with ethical import (echoes of Sagoff here), which contrasts with the largely non-ethical nature of private goods. And 'environment' also has 'social import': 'valued parts of the natural world are widely regarded as goods to society, over and above the benefits they provide to individuals' (1997: 212). Environmental economics factors these concerns into contingent valuation as 'existence values'; but ethical values are not reducible to needs preferences. The questions that elicit values and preferences are different:

> the question, how much is this worth to me? ... is an appropriate question for private goods because the consequences of my choice fall on me.

> In the case of public goods, however, the consequences fall not only — indeed not primarily — upon me. They fall on others, upon society in general, and they raise questions of right and wrong (1997: 214; see also 1991: 66).

Jacobs extends Sagoff's case and presents a defence of politics; a case for a public and deliberative realm of activity in which preferences are collectively obtained (see also Barry 1996). This is a necessary undertaking because CBA's ideological bedrock — 'the normative political theory that public policy decisions should be made on the basis of aggregating private preference-based choices' — finds no place for deliberative and collective action; for *citizenship* (Jacobs 1997: 215; see also 1995c: 67–68).

Peter Wenz also maintains that decision-making in the public realm is not analogous to the private nature of individual market transactions. In the former it is accepted that the principle of 'equal say' should prevail, as encapsulated in 'one person, one vote'. To allow people to buy additional votes is considered 'seriously unjust'. But 'this is just what CBA *requires.* People's preferences are weighted according to how much money they are willing to spend on the issue. In sum, from the democratic point of view, political injustice is *inherent* in CBA' (1988: 217). Does contingent valuation solve this? No, it does not. Poor people express little interest in paying for clean air, for example, because they continue to assume a low capacity to pay in their 'hypothetical' contributions to contingent valuations. CBA is, thus, 'a form of public decision-making that counts dollars rather than people as equals' (1988: 218). Pearce *et al.* make a virtue of this. They argue that CBA can reflect the strength with which preferences are held, and Pearce contrasts this with the vote, which treats strongly and weakly held views identically (Pearce 1989: 55). But the position articulated by Wenz has considerable currency within environmentalist assessments of CBA specifically and the market conception of justice generally.

Wenz also criticises contingent value and discounting on the ground that they cannot supply an objective dollar value. This leaves the process open to manipulation by those with an interest in securing a given outcome — what Wenz calls the 'fudge factor' (1988: 220–21). Beder, too, argues that CBA is only as sound as the integrity of those applying it, and political and business exigencies frequently find the limits of that integrity, with indirect costs quietly overlooked and indirect benefits determinedly winkled out, so that 'cost-benefit analysis, far from being an objective input into the decision-making process, can be used to justify projects' (1997b: 101; see also Sagoff 1988: 93). It is not, after all, environmentalists who carry out the valuing, but 'economists employed by industry and government' (Beder 1997b: 103).

Finally, it is objected against neo-classical economics ('pure' and environmental) that, despite the tide of opinion to the contrary, the financial incentives approach does *not* beat regulation on the efficiency score. Again, Jacobs is to the fore. 'It is true that financial incentives do generally provide a continuing motivation to reduce environmental damage', he writes, 'but in some cases it may be more important to stop the activity altogether, in which case regulations are more effective' (1995c: 58). It can similarly be argued that the environmental interest is not substantially enhanced if producers merely choose 'to pay the added values to deplete and pollute' (Martell 1994: 70). As Goodin notes, 'the problem with being able to buy your way out of a nefarious activity is that anyone with sufficient ready cash is consequently led to take the nefariousness of the activity insufficiently seriously' (1998: 243). There is, too, no good reason why regulation should require more bureaucracy than taxation; for, to check the validity of claims, taxation requires a system of inspectors, who 'are not less bureaucratic because they are tax inspectors rather than regulatory ones', and 'taxation requires the additional structure of revenue collection, which regulation does not, whilst not dispensing with the legal framework'. Thus, 'the great divide between regulatory and financial instruments is unhelpful and should be abandoned' (Jacobs 1995c: 58).

We have noted environment movement misgivings about the sustainability discourse. Here we will confine ourselves to the principle of substitutability as advanced in the 'weak' version of sustainability that neo-classicism favours. Criticism focuses upon the assumption that, under the 'push' of scarcity, technological enhancement will redress losses of natural capital. Hamilton, for example, argues that prohibitive costs may critically impede technological substitution. The question is not only whether resources will run out, but 'whether technological solutions will be available in time to offset major and irreversible impacts' (1997: 52). Others argue that artefact-for-nature substitutability mistakenly assumes physical stasis, ignoring both the systemic nature of ecological 'goods' and consequent questions of biological resilience thresholds and equilibrium disturbance 'as the residuals generated in production are returned to the system' (Perrings 1987: 8; Common and Perrings 1992). Substitutes will undoubtedly be found for some resources that the market decrees to be scarce, but 'there is no substitute for the assimilative capacity of the biosphere' (Dryzek 1992: 20).

We have seen that Marxist environmentalism is critical of branches of ecological thought that are not couched within class analysis. This branch of environmental thought can, therefore, have no truck with any strand of green economics that is based in the capitalist market: not environmental economics, for example — and not ecological modernisation.

Nor is hostility to ecological modernisation confined to adherents of ecosocialism. The response of the environment movement generally has been decidedly underwhelming. Nevertheless, possibly acknowledging that it is likely to prove tenacious as both idea and strategy, there have been attempts to recast it in more radical mould.

Peter Christoff argues that what he calls 'weak' ecological modernisation reduces the environment 'to a series of concerns about resource inputs, waste and pollutant emissions' (1996a: 485), whilst cultural needs and nature-preservation values, being irreducible to money terms, are marginalised or ignored in calculations. Dryzek puts it well: the natural world is treated as 'a giant waste treatment plant, whose capacities and "balance" should not be overburdened. Denied is any notion that nature might spring surprises on us, defy human management, have its own intrinsic value, and its own open-ended development pathways' (1997: 144). One possible reading of ecological modernisation is that it is a narrowly technical process, devoid of both ideas and consideration of first principles (what Dryzek has termed 'a discourse for engineers and accountants'; 1997: 147), as well as being concerned only with 'increasing the environmental efficiency of industrial development and resource exploitation'. From such a perspective it is 'only superficially or weakly ecological' (Christoff 1996a: 486).

Furthermore, ecological modernisation has been confined to developments within states, and is 'unable to integrate an understanding of the transformative impact of economic globalisation on environmental relations' (Christoff 1996a: 486). It also offers little to non-western countries. 'Ecological modernization is completely silent about what might be the appropriate development path for Third World societies', Dryzek writes (1997: 146); it can, in fact, entrench and exacerbate existing economic inequalities:

> 'weak' ecological modernization in the wealthy countries could easily be bought by transferring risks to poor countries — for example, by locating polluting industries in these countries, or exporting wastes to them, or exploiting their resources in unsustainable fashion. It is not hard to find examples of the cleanest and greenest countries engaging in such actions (1997: 151).

There are also tensions within ecological modernisation over whether this is a process to be pursued democratically or technocratically. Weale, for example, writes of ecological civic virtue and moral reform, envisaging a robust sphere of 'discursive rationality', where dispassionate exploration of alternative courses of action supplants adversarial power-mongering. He prescribes (1992) a highly integrated, corporate state that is simultaneously open, democratic and participatory. Many would pronounce this impossible, perhaps even a contradiction in terms.

Christoff resolves these dilemmas and weaknesses via a 'strong' version of ecological modernisation (1996a: 490–97). It ceases to be a foreclosing-other-options technocratic strategy, and becomes more flexible and less-easily defined, a diversity of development strategies being welcomed under its banner. The transnational aspects of environmental problem-solving are afforded genuine institutional recognition. Enhanced recourse is had to democratic and participatory processes, and changes of a fundamental nature to governmental and economic institutions are freely considered. Ecological modernisation ceases to be, in Dryzek's damning phrase, 'a discourse of reassurance' (1997: 146).

Apart from a contribution from David Pearce, who finds ecological economics 'a hazy and ill-defined subdiscipline' (1998: 4), and who pleads against adopting an alternative economics for fear that 'we lose the influence that can be exerted to conserve natural environments through the use of the neo-classical model', it is ecological economics that receives least within-the-paradigm scrutiny. There is some criticism of Daly's continued commitment to a private market-economy even after his requirements for extra strong sustainability have been met (Daly 1992). As we have seen, it is difficult to see how such an economy could function when deprived of the motivating dynamism of growth. Stuart Rosewarne is one who questions the consistency of Daly's adherence to market individualism, pointing out that the 'laissez-faire economy remains driven by profit and the accumulation of capital, and the contradictions that are inherent within this system will invariably be manifested in terms of environmental degradation' (Rosewarne 1993: 67–68). Moreover, 'this cannot be easily reconciled with Daly's advocacy of a more instititutionally-defined construction of community, where the community defines the boundaries of interactions with the environment' (Rosewarne 1993: 67).

But it is Daly's continuing stress on the entropically determined need for a steady-state economics that has attracted most critical scrutiny. It can be argued that belief in the inevitability and desirability of growth is so deeply entrenched that insistence on the centrality of the goal of a steady-state within the green critique will consign environmental thought, in the eyes of John and Janet Citizen, to the land of the lunatic-fringe. Whatever the ultimate validity of the zero-growth case (some argue), if the practical consequence is to jeopardise ecological achievements that *are* attainable, it would make more sense to significantly reduce the current emphasis on zero-growthism. Certainly, neither governments nor mass media are willing to seriously confront the negative consequences of growth because both function as agents of the forces for whom growth is the point of the game.

Others have misgivings about the critique of growth which are more substantive than merely strategic. Though the pattern varies, people who support environmental imperatives are, nevertheless, more likely to mesh those beliefs with older, established political allegiances, where the goal of growth remains axiomatic, than to align with purpose-built green parties. Whilst the latter tend to see the need for radically new priorities and radically new patterns of behaviour, the former seem wedded to the belief that basic human aspirations and key social structures require an overhaul, but not a revolution. Labour/democratic socialist parties seem particularly adept at providing a voice for people of this cast of mind. H.C. Coombs articulates just such a perspective, noting that 'economic growth has not been a negligible factor in human welfare' (1990: 52), asks: 'need an ecologically acceptable economy be without growth?' (1990: 53–54). He concludes: 'I see no reason why, after the transition to an ecologically acceptable economy was complete, the ingredients of growth would not continue to provide the stimulus to the production of progressively more of the goods and services demanded or their production at progressively lower real cost' (1990: 55).

Assessments along these lines abound. The critique of growth has survived better than most components of 1970s environmental thought, but some shifting of ground is evident, and outright rejections of growth *per se* are now less common than qualified rejections. Thus, Jacobs writes: 'much environmental degradation is not caused by growth, but by a rate of consumption (whether growing, static or declining) which is above the natural regeneration or "sustainable" rate' (1991: 57). Degradation can occur without growth, and the greater need is to evaluate whether consumption of some individual resources is too high, given that in others further growth may be possible. Jacobs concludes that 'zero growth is not a sensible environmental objective', even though it is undeniable that 'current patterns of growth are causing immense environmental degradation throughout the world and will have to be changed' (1991: 58).

Some move still further from the no-growth position. Even within environmental thought some credit is given to the argument that care for the environment can only occur under conditions of economic expansion. Such positions are usually buttressed with reference to the post-1970s record within western countries, where the evidence may suggest that environmental imperatives will be valued and given policy enactment in robust rather than muted economic times.

A general point in closing. Though environmental thought turned early to matters economic, and though, as this chapter attests, there is a vast range of positions within green (and green-ish) economic thought, there is also a tendency to regard the very endeavour of theorising economics as inimical to the 'real' concerns of the planet and its intricately ravelled living processes. A common response is to look wistfully to a greatly simplified, passive economics, for which John Clark's wishfulness can serve as paradigm: 'rather than continuing the system of obsessive, uncontrolled production and consumption, the community will practice true *economy*, the careful attending to and application of "the rules of the household"' (1992: 89). Even stronger responses are in evidence. As Naess has observed: 'especially among activists within the ecological movement, people have been so fed up with unecological policies that the term "economics" itself has become a kind of nasty word', with the result that 'nothing can be expected, think activists, from a study of economics — the economists are to be fought' (1989: 105–06). The judgement cast upon economics by Hazel Henderson, herself an economist, could scarcely be stronger: economics is 'a form of brain damage' (1988: 181).

Such attitudes are understandable. Anne Bell argues that 'language forces us into patterns. Monetary valuation is the language of exploitation and trade-offs. To adopt that language with the intention of countering the very values that it embodies — profit maximisation, economic growth, the human domination of nature — seems problematic at best' (1994: 94). Perhaps more than merely problematic, for there is a fundamental issue here: that the economic view relegates ecological processes to a sub-set of the economy. Beder points out that the economisation of environmental debates can generate 'subtle changes in the way people think about the environment. It is difficult to habitually use terms like natural capital, stock, resources and assets without beginning to see the environment as a set of inputs and outputs of the economic system'; 1997b: 99). A sub-set then, and a non-vital one at that. As Hamilton colourfully puts it, 'the world and its natural riches becomes a supermarket in which the biosphere is preserved and nurtured only to the extent permitted by the self-centred economist within' (1994: 79). Only a paradigm shift would obviate this, and that is not presently in sight.

There is an even more basic issue at stake here, one that may render futile all attempts to construct an alternative economics: the apparent resilience of an economics of private desires. This robustness is manifested not only in the readiness of western peoples to forgo public amenity and the bonds of community in aggressive pursuit of private consumption, but also in the apparent helplessness of successful and stable non-industrialised cultures to resist the private-goods-based capitalist market. It is not easy to refute the claim, made by propagandists of industrial capitalism and its market institutions, that the 'satisfaction' of private desires at the expense of the common realm (including the contextualising biophysical realm) answers to an innate need in, or essence of the human species.

If this is the case — if we are, indeed, under the thrall of runaway appetites and are doomed to an ever-more frenzied pursuit of shifting satisfactions that remain always tantalisingly beyond reach, delivering instead a planet-impoverishing, exponential increase in consumption — what are our choices? The choice has seemed to many to lie with a looked-for transition to one of several forms of socialism. Which brings us to the next chapter.

9

GREEN POLITICAL THOUGHT: THE SOCIALIST TRADITIONS

There are, of course, a great many socialisms: 'gradualist' socialism (labourism, democratic socialism, social democracy), Marxism, post-Marxism (and even the *post*-post-Marxisms of the western intellectual left), utopian and anarchist socialisms, 'actually existing socialism' (China, Cuba), eurocommunism, the 'stone-age socialism' of the Khmer Rouge — and the list goes on. Just three of these will be considered here: 'gradualist' socialism, into which will be conflated, perhaps unwisely, the 'democratic socialism' of trade union-based parliamentary parties and the 'social democracy' of Western Europe (the latter lacking a formalised link to the trade union movement); Marxism, which will be conflated here with post-Marxism, again, a conjunction that many will want to challenge; and socialism's anarchist sub-tradition.

In each instance the focus will be on the interplay between these earlier left-radicalisms and the new green kid on the ideological block: that is, on how socialists have responded to the ecological critique, and how, in turn, greens have sought to variously ally with, and differentiate from the older institutions and ideas of the political left.

DEMOCRATIC SOCIALISM AND THE ENVIRONMENTALIST CRITIQUE

On the surface it would seem that environmentalism should have much in common with democratic socialism. Both have an economic vision which constrains the laissez-faire market. And, throughout the western world, democratic socialist parties have certainly moved to embrace environmental *causes*, though they have often needed convincing first.

Michael Jacobs, pointing to the fact that the market's environmental 'externalities' are disproportionately met by the poor, and that environmental policy brings on significant sectoral and regional employment dislocation (even if its overall impact is to increase employment), nevertheless insists that 'none of these factors constitute reasons for socialists to abandon environmentalism' (1995a: 9). Instead, 'environmental policy should be focused on improving the environments of the poor and the environmental services used by them. Housing and physical regeneration, public transport, the creation and maintenance of open space and so on, should be priorities for public expenditure', whilst 'environmental measures have to be incorporated into wider economic and social programmes' (1995a: 10).

But such expressed intentions on the part of a prominent green democratic socialist are not necessarily reciprocated in the broader environment movement. The green critique holds (at least in part) that there are significant limitations upon the capacity of democratic socialism to fully embrace environmental imperatives, and these limitations stem from values central to the accumulated traditions of democratic socialism.

Democratic socialism has long viewed environmental quality as one of the social goods the socialist is concerned to provide, though 'environment' has a somewhat more tenuous pedigree in democratic socialist thought than such social goods as welfare and education, and may often lose out in the establishment of policy priorities. This is particularly so in the case of labour parties that provide institutionalised influence within the party to trade unions, as these affiliations will almost certainly include trade unions organised to protect the interests of labour engaged in environmentally destructive enterprises. There will be a tendency, in any case, for environmental problems to be defined primarily in terms of 'quality of urban life' issues, in keeping with both the demography of 'socialist' electoral support and the entrenched productivism and human-centredness of its ideological tradition. Thus, in a pamphlet written for the British Labour Party's green group, the Socialist Environment & Resources Association (SERA), Raymond Williams makes this revelatory statement, whilst making a case *for* the proposition that socialists should also be ecologists:

> I hope we are not going to be the people who simply say 'keep this piece clear, keep this threatened species alive, at all costs'. You can have a kind of animal which is damaging to local cultivation, and then you will have the sort of problem that occurs again and again in environmental issues. You will get the eminences of the world flying in and saying: 'You must save this beautiful wild creature. That it kills the occasional villager, that it tramples their crops, is unfortunate. But it is a beautiful creature and it must be saved.' Such people are the friends of nobody, and to think that they are allies in the ecology movement is an extraordinary delusion (nd: 14).

But, sparked by the 'ecological impulse', the 'such people' to whom Williams refers — though not usually 'eminences', and not usually overmuch concerned with the value 'beauty' — are very much the heart of the ecology movement. Extremely few environmentalists would regard the fate of the world's large predators, all of them now facing uncertain futures in the wild, as irrelevant to their project. What this statement most reveals is the extent to which democratic socialists are encaged within the paradigm of productivism (P.R. Hay 1998: 116–17), and the consequent difficulty they have in perceiving the autonomous ideological possibilities within environmentalist thought. Democratic socialists tend to regard 'environment' as mere 'single issue', and the possibilities inherent within an ecological perspective for independent social theory construction — that is, to provide a set of first principles in its own right — as simply uncomprehended. Hugh Stretton, for example — whose *Capitalism, Socialism and the Environment* (1976) was influential and much acclaimed at the time of its publication — provides persuasive arguments for why democratic socialists should be environmentalists. But why, in turn, environmentalists have not seen their obvious political home to be within democratic socialism is something that troubles him:

> movements for environmental reform are unlikely to get far by transcending party politics, as some eco-preachers recommend. They need to get into real politics rather than out of them: to talk about rich and poor, city and country, costs and distributions, prices and taxes, rather than humanity at large (1976: 14).

Most green-leaning people would agree with Stretton's sentiment. But many would dispute the assumption that the parties of democratic socialism hold a monopoly on such issues of 'real politics'.

The problem here is that in democratic socialism we have probably the most *entrenched* human chauvinism of all the major systems of political thought. Conservatism exalts the wisdom of tradition, and is opposed to programmatic reform and the notion that humankind can be greatly improved by collective action. Liberalism has faith in the abstract principles of the market. Marxism places its faith in the scientific laws by which history unfolds. But democratic socialism believes in *no* abstract social process. It merely believes that, through rational and judicious use of the instruments of government the species is capable of greatness, that it can throw off the blights of despair, poverty and exploitation, and that it can do so by its own pragmatic ingenuity, not through the inevitable workings of grand social laws. It is, we might say, *super*-humanist. It came into being, not in response to the intellectual endeavours of a thinker in a garret, but in the real struggle of real people battling for security and a modicum of comfort in the real world. Its passionate commitment to the wellbeing of humankind is, thus, intense, and is manifest in the internationalism of its tradition —

its high-minded vision of a transnational 'family of labour'. It is wedded to the Enlightenment view of history as inexorable progress and believes deeply in the liberatory possibilities of industrial technology (feudalism offered no such possibilities). The notion of a moral subject beyond human beings, as posited by much radical environmental philosophy, is thus far from the reach of democratic socialist thinking.

Given the history and ethos within which the values and myths of gradualist socialism were founded, it is not surprising that 'environment' should be seen as an issue rather than the inspiration for a distinct system of values; that it is an urban issue above all; and that gradualist socialism is resistant to calls to decrease consumption and move beyond industrial society. It is unlikely that democratic socialism will be shifted from its faith in economic growth and technological productivism as the necessary conditions for the liberation of all people and the triumph over want, poverty and exploitation.

The British Labour Party's 1987 *Charter for the Environment* is illuminating in this respect. While it insisted on the necessity for 'sustainable' growth, it nevertheless made not the slightest retreat from a vision of full-on growth: 'our economy must grow, new industries must be created, and old industries revived'. The strong impression gained is that, for the authors of the *Charter*, 'sustainable' was a mere buzz-word, a content-free slogan. It is not defined or explained, and there is little sign, in the *Charter*, of the economic rethinking upon which such 'growth' would need to be based. Thus, the empty commitment to 'sustainable' growth was comprehensively ignored in a second Labour Party 1987 election manifesto, *Industrial Strength for Britain*, which showed no concern for the quality and purpose of production; it simply called for considerably more of it. There should be no surprise about this: where 'environment' has the status of mere issue, it will be allotted its own separate, stand-alone box within the compartmentalised policy process — the assumption being that it does not impinge upon other neatly labelled and self-contained policy areas, nor they upon it. It is perfectly understandable, then, that in the context of the 1987 election the British Labour Party should have called for dramatically faster growth, as the Conservatives failed to maintain the high levels of economic expansion of the early 1980s, whilst simultaneously committing itself to 'sustainable' growth.

Democratic socialist parties stand in uneasy relationship to the new social movements of late modernity. They draw a modest stream of recruits from the new social movements, and this has given them, intermittently, a light green patina. But new social movement recruits have adapted with varying degrees of success to the values and organisational culture of left legislative parties. First wave, or liberal feminists seem to have made the transition most successfully; people recruited from the environment movement, by contrast, have not generally found democratic socialist structures satisfactory vehicles for their aspirations. The labour movement's traditional productivism seems insurmountable (for contrary cases, see Dunkley 1992; Jacobs 1995a). So, too, the capitalist market economy's reliance on continuous expansion in a more general sense may also be insurmountable. A political movement which takes the basic assumptions of contemporary economic existence as the given framework within which it intends to work will, thus, have great difficulty in sustaining an economics of managed, regulated growth, let alone an economics subordinated to planetary health imperatives, perhaps involving only minimal growth, or even no growth at all.

MARXIST HOSTILITY TO ENVIRONMENTALISM

Marxists reacted with varying degrees of scepticism and hostility to the rise of environmentalism in the late 1960s. It was typically viewed as a manifestation of the narcissistic and excessive individualism that characterised the 'counter culture' and made of it a false revolutionary movement. Marxist assessments of environmentalism in this early phase of its development were usually unremittingly hostile, and they followed a standard pattern. In accordance with the principle 'read one read them all', I will consider, as exemplar, a single instance of these writings — a 1970 paper by Sandor Fuchs. The title is as blunt as the argument: 'Ecology Movement Exposed'.

Fuchs argued that the ecology movement was deliberately created by the ruling class of the United States to divert attention from the class-based political issues of the day: Vietnam, American imperialism, domestic poverty and racism. Racism and imperialism were, in fact, promoted by the ecology movement, because it blamed 'scarcity' on population pressures exerted by the world's poor. Fuchs conceded that pollution and depletion are real issues; for this reason it is the ruling class manipulators, not the ordinary, sincere people in the environment movement, who should be exposed (1970: 52). Real problems, then, but the ecology movement's treatment of them is, lacking a class perspective, thoroughly inadequate. And it is more than merely misguided — it is *explicitly* anti-working class. The critique of population is held to be Malthusian: to attribute poverty and environmental degradation to population instead of capitalism is to blame victims, not perpetrators. Fuchs writes:

> the bourgeois ecologists blame people for pollution. Workers are pictured as slobs who overconsume and dump litter on the highways. The only Americans who aren't pigs are the 'eco-freaks' who drive bicycles rather than cars and who put bricks in their toilet bowls to conserve water (1970: 58).

The use of all-inclusive language has the same effect. To eschew class analysis for 'abstract labels' like 'man and technology' (1970: 52) is to divert attention from the differing extent to which labour and capital are responsible for environmental problems, and, thus, away from the class-specific plight of those who are exploited under capitalism — which, according to Fuchs, is precisely the intention.

The ecology movement was created by the ruling class for a second reason: to serve as a conduit through which the loyalties of alienated students who had turned their backs on American capitalism and its political institutions could be reclaimed. Because 'ecology' was an issue that was held to transcend class — the metaphor of 'spaceship earth' connoted 'we're all in this together' — students could reaffirm their allegiance once 'the system' demonstrated its efficacy and good faith by 'solving' the problem. Fuchs is somewhat inconsistent on this point, though. Elsewhere in his paper the environment movement is trenchantly taken to task, not for its promotion of faith in the system, but for its promotion of 'pessimism and defeatism among alienated youth' (1970: 52). This promotion of pessimism damages the cause of revolutionary socialism:

> pessimism comes in strong with the notion that the world is on the brink of disaster. If the revolution isn't around the corner, why fight? If you decide to fight, why do so in alliance with piggish overconsuming workers? And if you're foolish enough to ally with workers to fight for socialism — then you're doing it all for nothing because world-wide socialism with its emphasis on industrialization only exacerbates a crisis of too much production and consumption using limited and fast receding supplies of world resources. What great service these ecology arguments do for the U.S. bourgeoisie! (1970: 58).

Here is what is perhaps the rub. It is the anti-industrialism of the early environment movement that really generated such dedicated Marxist opposition. The environment movement — particularly its sub-scientific counter-cultural manifestation — was already tending to locate itself outside the tradition of Enlightenment progressivism, of which Marxism is one of the foremost articulations. One of the ecology movement's justifications for pessimism, wrote Fuchs, was the prospect of 'imminent mass starvation' (1970: 52). But world hunger is largely a problem of social inequity and of under-utilisation of fertile land, the latter being a technical problem that science will solve (just as problems of population and resource depletion will disappear under socialism's rational distribution and deployment of resources and its lack of a motor for conspicuous consumption). Fuchs castigates the environment movement for lacking a progressivist faith in science. The progress of science 'is just as limitless and at least as rapid as that of population': indeed, 'what is impossible for science?' (1970: 55).

Anti-environmentalist Marxism of the late 1960s and the 1970s recycled these themes. The stress Fuchs placed on environmentalism's perceived role of inducting disenchanted students back into the values of liberal capitalism was somewhat novel. And not all anti-environmentalist Marxist writings concede that there are 'real' problems rather than mere fabrications by evil geniuses, nor do they always insist on a difference in culpability between ruling class manipulators and the movement's ordinary people (though, as one of the quotes used above shows, Fuchs was not uniformly generous to 'eco-freaks'). But the notion that the environment movement is a concocted phenomenon, deliberately set up by agents of capital to defuse revolutionary class consciousness, recurs. So does the attack on environmentalism's failure to develop a social analysis that sees some people to be more responsible for environmental ills than others. So does the rejection of the environment movement's fixation upon population as the root environmental problem. So does the assertion of the rightness of modernist technocratic productivism (few, though, match the purity of Fuchs's ringing triumphalism). And so does the equation of environmentalism with a socially repugnant neo-Malthusianism (for example, Wells 1982).

These first Marxist reactions, though crude in some respects, hit the mark in others. The early environment movement *was* naive and simplistic in its social theorising, it did neglect differentials in reponsibility for environmental problems, and it did produce extremely illiberal political prescriptions. By the end of the 1970s its mainstream was flowing in different channels, these earlier deficiencies acknowledged and (notwithstanding the persistence of some tension) corrected for. As for Marxism itself, from the mid 1970s an unmitigatingly negative assessment of the environment movement becomes somewhat less common. As anti-environmentalist hostility on the part of capital became increasingly pronounced, it became correspondingly harder to maintain a credible environmentalism-as-ruling-class-conspiracy position. Marxist responses to environmentalism became, from this time, somewhat more complex.

MARXIST CO-OPTION OF ENVIRONMENTALISM

In 1974 a paper appeared in English translation by the noted German poet and social commentator, Hans Magnus Enzensberger. This paper was to exert considerable influence upon Marxist responses to environmentalism, and serves as a watershed between the hostility that characterised Marxism's reaction to environmentalism in the late 1960s and early 1970s, and the more diverse and complex interpretations that followed.

Enzensberger rehearses the familiar Marxist objections to first-wave environmentalism, though he sets out these objections rather more cogently than was customary, and subsequent anti-ecological Marxist writings drew heavily upon Enzensberger's paper. Environment movement activists 'are overwhelmingly members of the middle class and of the new petty bourgeoisie'; thus, 'their activities have generally modest goals' (1974: 8). This is consequent upon the absence of class analysis. Enzensberger asks why it is that the environment movement only came into being at the time it did, when the appalling environmental conditions to which the working poor were subjected in the early years of the industrial revolution merited such a movement all those decades earlier. The answer is that 'the ecological movement has only come into being since the districts which the bourgeoisie inhabit and their living conditions have been exposed to those environmental burdens that industrialization brings with it' (1974: 10).

Enzensberger also argues that bourgeois political interests are concealed by the inclusivist nature of the 'population' bogey. Here, too, the Malthusian character of the environment movement — the belief that immiseration is the inevitable lot of the poor because their propensity to breed always outstrips increases in agricultural output — is invoked. The central metaphor used by theorists of population crisis, 'spaceship earth', is a glaring instance of 'false consciousness', the 'ideological purpose' of which is to obfuscate the difference 'between first class and steerage, between the bridge and the engine room' (1974: 15). Because it is blind to social differentials, the environment movement cannot develop an adequate politics of response.

In Enzensberger's paper, though, there is an important variation: for, whilst the environmentalist critique concealed class interests and serves ideological ends, the ecological crisis is nevertheless real. It might be manipulated for ideological ends, but it is not an invention of the ruling class. On the contrary, the ruling class is as much threatened by ecological crisis as other classes, and that is why 'avowed representatives of monopoly capitalism have recently become its spokesmen' (1974: 10). The environmental crisis 'threatens the material basis of production — air, earth and water' — just as it threatens productive labour, 'whose usefulness is being reduced by frequent physical and psychical illnesses'. And to these must be added 'the danger of uncontrollable riots over ecological questions as the conditions in the environment progressively deteriorate' (1974: 11).

This explains why environmental politics does not seek revolution but instead calls for intervention by the state: because 'the state only goes into action when the earning powers of the entrepreneur are threatened' (1974: 11). The state will respond most readily — perhaps exclusively — to monopoly capital, whose 'influence on legislation is decisive', and it is for this reason that those interests, signified by the leading role taken at the time by the industrialists of the Club of Rome, 'attempt to acquire influence over the ecological movement' (1974: 12). Some environmental action escapes this control, but 'a long process of clarification will be necessary before the ecological movement has reached that minimum degree of political consciousness which it would require finally to understand who its enemy is and whose interests it has to defend' (1974: 13).

The distance from Fuchs to Enzensberger may seem none too great, but that is not the case. The environment movement may be tragically insensitive to the crucial variable of social class, but the need to take the ecological crisis seriously renders the technocratic exuberance of Fuchs inappropriate, too. Enzensberger distances himself from the 'irrationalist hopes' in this declaration, attributed to Fidel Castro:

> no one who is conscious of what man can achieve with the help of technology and science will wish to set a limit to the number of human beings who can live on the earth. What characterized Malthus in his time and the neo-Malthusians in our time is their pessimism, their lack of trust in the future destiny of man. *We shall never be too numerous* however many of us there are, if only we all together place our efforts and our intelligence at the service of mankind (cited in 1974: 14).

The left critique of the early 1970s has limitations, according to Enzensberger. It does not reach to 'the heart of the problem' (1974: 18). What, then, *is* the appropriate Marxist response to the environmental crisis?

Enzensberger argues that the socialist project is not merely one of shifting ownership from private to public whilst leaving other relationships untouched. It must lead, says Enzensberger, to:

> totally revolutionized relationships between men and between men and things — that is to say, it must revolutionize the whole social production of their lives. Only such a view of capitalism, i.e. as a mode of production and not as a mere property relationship, allows the ecological problem to be dealt with in Marxist terms (174: 20–21).

This has not been the case within existing socialisms. There, the capitalist mode of production, in which the products of labour are commodified and expropriated, continues to exist. The consequence for the relationship between humans and nature is similar to that pertaining to industrial production in the west; thus, we also get environmental destruction 'in countries where the capitalist class has been expropriated' (1974: 21). As vastly increased technological capacity links with unreformed capitalist modes of production, a 'destructive potential' is unleashed that 'threatens all the natural bases of human life' (1974: 28). For Marxists this means that a more critical assessment must be made 'of the concept of material progress which plays a decisive part in the Marxist tradition' (1974: 22). And it means acknowledging that the conspicuous wealth on display in some countries has been attained through 'a wave of plunder and pillage unparalleled in history; its victims are, on the one hand, the peoples of the Third World and, on the other, the men and women of the future' (1974: 22). Socialism must, then, abandon its dream of a world of material abundance for all:

> the social and political thinking of the ecologists is marred by blindness and naivete ... Yet they have one advantage over the utopian thinking of

> the left in the west, namely the realization ... that every political theory and practice — including that of socialists — is confronted not with the problem of abundance, but with that of survival (1974: 23).

The solution must be a revolutionary one. It is absurd, says Enzensberger, to assume that the powerful social forces that control the political system will voluntarily relinquish their resource-guzzling opulence. He notes the union of humankind and nature envisioned by Marx in the *1844 Manuscripts*, wherein 'the naturalism of man and the humanism of nature' are 'brought to fulfillment' (Marx 1964: 137), and concludes pessimistically: 'if ecology's hypotheses are valid, then capitalist societies have probably thrown away the chance of realizing Marx's project for the reconciliation of man and nature' (Enzensberger 1974: 31).

The environmental politics to which Enzensberger is responding is, of course, the doomsday politics of the early 1970s, and this has coloured the terms of his response (but see Carpenter 1997: 129 for a more recent case along similar lines). The 1974 paper is historically important because it was the first influential Marxist critique of environmentalism to unambiguously concede that the ecology movement has something momentous to say, even if its social analysis has been naive and misguided. Enzensberger's is the first case of substance in which the need to factor global environmental crisis into Marxist praxis is argued.

Thus, 'green' Marxism was born, and two distinct post-Enzensberger developments are discernible. One of these is concerned with theory building: synthesising 'red' and 'green', whilst insisting that the latter ultimately lodges within the former. Andre Gorz can serve our needs here. The other category is less concerned with developing theory than with dissecting the thought of the environment movement in order to explain where and why it fails to pass muster — by which is meant, where and why it resists incorporation within Marxist analysis. This 'genre' is most apparent within British writing on environmentalism, and I have elsewhere suggested why this should be so: 'the strength of the British radical tradition's identification with the working class as both the motor and the end of history' is such that non-Marxist left radicalisms 'must pass through a Marxist filter' (Hay and Haward 1988: 440). Such radicalisms, therefore, struggle for legitimacy. We will look at two British Marxists whose writings on environmentalism exemplify this tendency.

In 1986 Joe Weston edited a collection of essays, *Red and Green: The New Politics of the Environment*, that was, for a time, widely read. In his own contributions Weston attacked the environment movement's concern with 'nature', which he saw as de-radicalising, because concerns such as 'hedgerows, butterflies and bunny rabbits' (1986b: 12) are irrelevant to the working class. The real environmental issues — those that need priority — are the living conditions of the urban poor. Why does the environment movement get it so wrong? The problem is 'ecology':

> for most people in the green movement the term 'the environment' essentially means that which is related to the Earth and its ecological systems. Acid rain, the destruction of tropical rainforests and modern agricultural practices are seen as an attack upon 'nature'. Even non wildlife issues like nuclear energy and air pollution are seen in terms of their effect upon the biosphere — the Earth's ecology (1986b: 13).

The greens, thus, see people as the prime cause of environmental problems, whereas in fact they are its main victims:

> the victim, as the phrase 'ecological crisis' suggests, is seen as being 'nature' — which relegates those suffering poverty, despair and hunger throughout the world to the periphery of their concern. Yet in fact, it is people and not 'nature' who suffer the greatest hardship as a result of ecological damage. 'Nature', after all, will always reappear, albeit in a different form from that which has been destroyed; people however rarely live long enough to make up for the disruption and poverty caused to them when other people destroy their environment for personal economic gain. To think whooping cranes are important — possibly more so than people — one has to be free of the more pressing human problems like that of poverty (1986a: 2–3).

The dominant green view is, thus, characterised by Weston as 'narrow', because it neglects the social environment. It is also based upon an incorrect assumption of what constitutes 'nature'.

We have already looked at the claim that 'nature' has no meaning save as a human construct. Within environmental thought this position has been most forcefully put by Marxists. Marx distinguished between 'first' (or unmediated) nature and 'second' nature, which is the product of the mixing of unmediated nature with human labour. 'First' nature, Marx maintained, has everywhere been extinguished. Weston writes within this tradition. 'Nature', he says 'would not exist if it were not for people who make use of it for social and economic benefits' (1986b: 13); thus, 'our relationship to nature is dependent upon social rather than physical factors' (1986b: 14):

> in their neglect of the social construction of the environment, modern environmentalists have selected priorities which do not reflect the true scale or the true causes of the world's problems. As a consequence their movement is largely irrelevant to the mass of humanity. The problems with which most people are faced are not related to 'nature' at all: they are related to poverty and the transfer of wealth and resources from the poor to an affluent minority of the Earth's population (1986b: 27).

From a false construction of nature it is only a small step to belief in the 'laws of ecology', which are used to postulate a 'web of life' linking all living entities to the ecological demands of the earth, and reducing people to 'little more than links in the vast chain of being' (1986b: 18). This is an extrapolation to 'nature' of pure, unvarnished conservatism. The notion of an ecosystem based upon balance between interdependent parts, translated into social terms, becomes a vision of social harmony maintained through the free competition of independent but balanced interests, a thoroughly anti-radical view of the social process. The term 'industrial society' — the greens' term, socialists employ 'capitalist society' — is a concept taken straight from conservative pluralism, used to convey a notion of freely competing interests not based upon class. Indeed, the term is deliberately used to *distinguish* it from 'capitalist society': it 'entails a rejection of the Marxist focus upon capital and the existence of two great classes, one exploited, the other exploiting' (1986b: 27).

But it is those relations of production, Weston argues, that are *primarily* in need of reform. He turns the heat on greens (such as Capra) who think that structural change is unnecessary, and that we can effect change by moving *en masse* to a new system of spirit-suffused values, which is, in fact, already beginning to sweep the world (1986b: 23–24). To think that values will change first, then structures, is nonsense, says Weston. Values are structurally determined and hence secondary phenomena, and the green faith in an inexorable spread of benign values through the world without reference to the objective conditions that shape them is, to a Marxist of Weston's stamp, idealistic nonsense. It comes as no surprise, says Weston, to find that the sociological evidence indicates that greens are drawn from a narrow segment of the middle class, which is, nevertheless, outside the normal economic relationships of capitalism. Green politics is, thus, an attempt by a specific social group to come to terms with its increasingly marginal status within capitalism.

It might be contended that Weston's is an analysis of unrelieved animus and that it really belongs in our previous category. Weston undoubtedly retreats from some of the middle ground taken up by Enzensberger. He does not buy the contention that the world is in biophysical crisis, and he does not find anything problematic about industrialisation as such. But, at the time of his book's publication, Weston was chairman of the Strategy Committee of Friends of the Earth UK, and he locates himself — theoretically as well as practically — within the environment movement, using the term 'social environmentalism' to describe his position. So, his 'social environmentalism' is a form of environmental politics, albeit Weston complains that 'the green perspective has hijacked environmentalism' (1986a: 2), and albeit he rejects the view that we face an 'ecology crisis' instead of a 'social or economic crisis' (1986a: 2). We *do* face an 'environmental' crisis: a crisis of the living and working environments of the world's poor. Weston is at pains to stress that 'this rejection of green politics does not mean we now believe that natural resources are infinite'. Quite the contrary.

And industrial pollution 'based upon the wasteful misuse of fossil fuels' must be resisted, too. So, 'what we are objecting to is not the issues that the greens take up; it is the issues they neglect as a result of their narrow interpretation of "environment"' (1986a: 2). And, at day's end, the greens 'still confront capitalism ... they are implicitly part of the forces ranged against the capitalist system and should be embraced as such' (1986b: 28–29).

Though Weston is conveniently to the point, his is not the most formidable exposition of the case for subsuming the environmentalist critique within Marxism. That achievement belongs with David Pepper, who has argued this position in several books and papers (most notably 1984; 1993a). Unlike most Marxists, Pepper has read closely the key environmentalist writings. We have earlier cited his criticism of deep ecology and radical ecophilosophy, and we will later draw upon his critique of the strong 'libertarian anarchist' strand (which he sharply distinguishes from 'communal anarchism') within ecopolitical thought. Passing, for the moment, over his 'case against', we will here set up the scaffolding of his 'case for'.

First, though, a digression. Our discussion has focused upon reaction, largely hostile, by Marxists to the radical challenge from environmentalism. But few of the critics, not even Enzensberger, have drawn in any systematic way upon the actual words of Marx himself. In Pepper's analysis, by contrast, Marx himself is foregrounded. Pepper was not the first commentator to do this. Responding to the perceptions of environmental crisis in the early 1970s, Alfred Schmidt's *The Concept of Nature in Marx* (1971) examined the apparent absence of any concern for nature within Marx other than as inherently valueless resource for industrial production: 'when Marx and Engels complain about the unholy plundering of nature, they are not concerned with nature itself but with considerations of economic utility' (1971: 155; for a critical evaluation of Schmidt's position, see Burkett 1997). In this passage Schmidt articulates the case most persistently contended by non-Marxist greens against Marx. And he identified passages in Marx that bear upon the human–environment relationship. In 1978 the latter project was completed comprehensively by Howard Parsons. Parsons's motive was quite different to Schmidt's; he defended Marx's position on nature and technology and rejected 'the radical ecological argument that we should simplify human needs, reduce human population and consumption, and respect nature' (Eckersley 1992: 83). Pepper's thought is in the tradition of Parsons rather than Schmidt. This context established, we can return to Pepper.

It is sometimes assumed, writes Pepper, that, because 'radical green politics are substantially infused by anarchism', they are 'necessarily and largely "socialist"' (1993a: 204). But the two are not the same (1993a: 205–06). Within green political thought the fundamental divide is between 'red-greens', those drawing upon socialism, and 'green-greens', those using libertarian anarchism as their touchstone. The differences between 'red-greens' and 'green-greens' are thrown into sharpest relief when the question is posed: 'what causes environmental degradation?'. For socialist greens the answer is, unequivocally, exploitative class relations (1993a: 207); whilst for libertarian greens 'it is *power relations* between people — relations of hierarchy and domination'. These latter concepts are 'ahistoric', because 'hierarchy and domination are assumed to be possible in all modes of production, whereas socialists would argue that they are not possible in proper socialism' (1993a: 208).

Pepper is at pains to locate his analysis *within* environmentalism. 'Ecosocialism' is urged upon the larger movement; his project is one of 'pushing ecologism *closer* to ecosocialism' (1993a: 217). He does not wish to see anarchism expunged from green thought (1993a: 3); whilst 'greens should make more of an accommodation with reds by dropping those aspects of their anarchism that are more akin to liberal and postmodern politics', reds should also shift and join with greens 'by reviving those traditions in socialism ... of decentralism and of the society-nature dialectic' (1993a: 3). His project is, thus, 'not to save' Marxism, nor is it to judge the legitimacy of green claims by testing them against the 'sacred' word of Marx.

Whilst not wishing to detract from the sweep and rigour of Pepper's scholarship, I do not think that his description of his project accords with his analytical practice. It is difficult to take from Pepper's writings any sense that Marx is other than sacred text. No positions are advanced that cannot be sourced to Marx, and no qualifications upon Marxist precepts, let alone retreats from them, are articulated. The perceived 'ecological sins' of Marxism — that it is deterministic, mechanistic, positivist and aridly scientistic, and blind to non-material needs — are the result of misconceptions about Marxism rather than inherent flaws. By contrast, Pepper calls for the abandonment of some definitive green precepts. That global biophysical processes are under massive stress is nowhere acknowledged, and an assumption to the contrary is strongly implied (for example, 1993a: 100–01). Certainly, there are no absolute limits to growth, these being socially determined (1993a: 101) — though this seems to be contradicted later: 'the ultimate constraints of nature and historical legacy are acknowledged and we can never be entirely free to construct what is "out there"' (1993a: 218). This, indeed, is the tripwire awaiting capitalism: 'the expansionist drive is both inherent in capitalism, and will destroy it, through ecological contradiction' (1993a: 111). Ecologism is portrayed as misanthropic (1993a: 147–49), whilst the fiercest criticism is levelled at ecocentrism (1993a: 221–25), the 'good' of Pepper's ecosocialism being entirely the good of humans. 'Ecosophy' in which 'human desires are not privileged' is contrasted with:

> the Marxist socialist, monist, anthropocentric position of egalitarian development and growth through human labour and scientific ingenuity, to satisfy materially limited but ever richer human needs through democratic, collective, planned production that emphasises resource conservation, non-pollution, recycling and quality landscapes ... there is no possibility of a 'socialist biocentrism', since socialism by definition starts from concern over the plight of humans (1993a: 224; see also 1993b).

Little concern is, thus, evinced for biodiversity issues (1993a: 223). Ecocentrism is, moreover, an expression of bourgeois class interests (1993a: 150); and, 'in its anti-statist, anti-collective, people-must-take-responsibility-for-their-own-lives, individualistic ethos', it is closely congruent with 'Thatcherite liberalism' (1993a: 151). Pepper's environmentalism is, too, located within the tradition of progressive modernism (1993a: 244) and he is fiercely critical of non- or trans-rational tendencies within environmentalism, even arguing that a reverential concern for 'nature' actually alienates humans from 'nature' because — in making 'her' inviolate, worshipful — 'she' is rendered remote, apart, unreachable (1993a: 115).

We have seen that Pepper urges us to refocus on Marxism's 'society-nature dialectic'. In assessing the relevance of Marx's thought to the environmentalist paradigm, his view of the relationship between humankind and other forms of life is fundamental. In none of his writings did Marx provide a comprehensive treatment of the human–other life relationship (nor did he theorise environmental degradation). But there are many incidental references to this relationship and Schmidt and Parsons have done us a great service by pulling these together. For Marx the world beyond the human realm is the 'primary source of all instruments and objects of labor' (Schmidt 1971: 15). It has no status, perhaps no reality, until its 'appropriation' by humans in the creative act of labour. And this act of appropriation is an 'improving' one — a dialectical process which transforms, for the better, both human and non-human realms. Humanity is 'naturalised'; nature is 'humanised'. Marx's metaphor for nature is that of 'man's ... inorganic body ... within which he must remain in continuous interchange if he is not to die' (Marx 1964: 112). The transformation of this 'inorganic body', through human labour and its projection, technology, enables humans to achieve their essential humanness, or 'species being'. Humanity's 'species being' is defined by the consciousness which humans alone possess. Other animals do not reflect upon their 'life activity'; an animal '*is* its life activity', whereas 'man makes his life activity itself the object of his will and consciousness' (Marx 1964: 113). Marx develops this analysis in the chapter of *The Economic and Philosophic Manuscripts* entitled 'Estranged Labor'. It is a vital component of his critique of the alienation of labour under capital. Labour, he says, exists in an alienated state when the worker-creator is divorced from the product of 'his' labour; when the work process itself ceases to be 'spontaneous, free activity'; and when nature, the 'original tool house', is also withheld: 'estranged labor tears from him his species life, his real objectivity as a member of the species and transforms his advantage over animals into the disadvantage that his inorganic body, nature, is taken away from him' (Marx 1964: 114). Crucial to Pepper's argument is a case for Marx holding a position on the question of the human–nature relationship that is organic rather than mechanistic, and monist rather than dualist (1993a: 109–11). This is inherent in the notion of nature as humanity's 'inorganic body'. In explanation of historical change Marx might occasionally refer to 'first' and 'second' nature, but human and non-human are actually linked *dialectically*, which involves relational understandings that closely parallel the general systems paradigms of contemporary physics and ecology: 'dialectics hold that all reality is ultimately motion, not rest ... Furthermore, all change involves not mere movement but an alteration in quality, and there is a universal changefulness in things' (Pepper 1993a: 110). For non-human life, though, this is comprehensively interventionist — a thoroughgoing process of 'improvement' for exclusively human ends. It is hard to escape the impression that, for Pepper, the 'humanising of nature' is likely to manifest itself somewhat more dramatically than the 'naturalising of humanity' might. On this crucial axis, too, the socialist component of the 'ecosocialist' formulation is ascendant over the 'eco'.

As noted above, the work of French theorist, Andre Gorz, is of a different order, less concerned to measure environmental ideas against an immutable Marxism than to draw upon both in the creation of new theory.

Unlike Weston and Pepper, Gorz develops a favourable critique of environmentalism within a broader theory of post-industrial socialism. He argues the need for economic and political decentralisation to enable a restoration of the key political value of autonomy. But this does not extend to advocacy of eco-anarchist self-sufficiency. The self-sufficient commune ('communal autarky') is, for Gorz, inevitably impoverishing. As Frankel puts it, 'the more self-sufficient and numerically limited a community is, the smaller the range of activities and choices it can offer to its members'. It has no realm of intellectuality, nor of production beyond the limits of community need. Thus the community becomes a prison ... only constantly renewed possibilities for discovery, insight, experiment and communication can prevent it becoming suffocating (1987: 59). On the other hand, full self-management is impossible in complex institutions. Gorz, thus, proposes a middle way. He advocates 'a socialist society which utilises the best of advanced technology in combination with local autonomy' (Frankel 1987: 60). Such a change is deceptively revolutionary; in fact, if equality and democracy are to be maximised, old social relations, especially work practices, must be swept away. 'Farewell', proclaims the title of one of his books, 'to the working class'.

Gorz posits everything in a new combination: local autonomy and international integration; high-tech and the accessible, human-scale 'convivial tools' advocated by Ivan Illich (1973); state planning and local control; heteronomous (externally directed) and autonomous work; an end to the capitalist market, but not to private enterprise in service delivery. The virtue is in the balance. For example, Gorz believes that, with the application of high technology necessary drudge work can be reduced to the barest minimum, 'while efficient production and planning can provide the necessary conditions for the conduct of autonomous, creative activity' (Frankel 1987: 59). And the latter, in this 'dual economy' (Gorz 1982: 90–104) must be paramount. Gorz makes much of Marx's notion of a 'realm of necessity'. In Eckersley's apposite summation, for Gorz 'even unalienated, self-managed material labor is a "lower" form of freedom than unnecessary labor and/or leisure activity' (Eckersley 1992: 93). The sphere of individual freedom, where people can pursue non-necessary activities at both individual and communal levels and live and produce according to their dreams and not just their needs, is the end product of human existence, and necessary wage labour is to put in place the conditions that will enable this.

In what way is this analysis 'ecological'? In calling for the deconstruction of work practices, Gorz argues that the old basis of production must be dispensed with because it threatens the ecological system in which all social activity resides — 'the point is not to deify nature or to "go back" to it, but to take account of a simple fact: human activity finds in the natural world its external limits' (1980: 13). The domination of nature also results in the domination of people. Gorz might accord a place for advanced technologies, but these are democratically seated technologies, which 'can be used and controlled at the level of the neighbourhood or community; are capable of generating increased economic autonomy for local and regional collectivities; and are not harmful to the environment' (1980: 19). Unlike Marx, and certainly in contrast to Pepper, Gorz does not hold technology to be structurally irrelevant: 'the ecologist's concern is working at another and more fundamental level: that of the material prerequisites of the economic system. In particular, it is concerned with the character of prevailing technologies, for the techniques on which the economic system is based are not neutral' (1980: 18).

In what way is this analysis Marxist? For a Marxist of Pepper's orthodoxy the answer would be 'scarcely at all', and one can certainly sympathise with such an assessment. But Gorz begins *Ecology as Politics* with a contextualising observation that fully justifies his inclusion here with the Marxist co-opters of environmentalism:

> ecological thinking already has enough converts in the ruling elite to ensure its eventual acceptance by the major institutions of modern

> capitalism. It is therefore time to end the pretence that ecology is, by itself, sufficient: the ecological movement is not an end in itself, but a stage in a larger struggle. It can throw up obstacles to capitalist development and force a number of changes. But when, after exhausting every means of coercion and deceit, capitalism begins to work its way out of the ecological impasse, it will assimilate ecological necessities as technical constraints, and adapt the conditions of exploitation to them (1980: 3).

MARXIST–GREEN SYNTHESES

There is a third category of Marxist response to ecologism, one in which attempts have been made to draw synthetically upon both traditions, without subordinating one to the other.

This can, but need not, involve the construction of a Marxist ecocentrism. Such formulations are not common, with Donald Lee's (1980; 1982) attracting most attention (see also Orton's case, in O'Connor and Orton 1991). Lee offers an ecocentric extension of Marx's metaphor of nature as humanity's 'inorganic body'. This relationship, says Lee, must cease to be interpreted merely metaphorically. Marxism must reject the anthropocentric view of the human–nature relationship which the metaphorical interpretation of this idea has hitherto sanctioned. Instead, Marxists must 'truly see nature as our "body"' (1980: 3) and accept the responsibilities of ecocentric stewardship that flow from such a view. Lee does not argue that this is what Marx originally intended. Marx viewed the human–nature relationship anthropocentrically, but an ecocentric development of the ideas first advanced in *The Economic and Philosophic Manuscripts* can be effected to bring about the reconciliation of the radical green and Marxist critiques that the times demand. Marxism 'must move beyond its present homocentrism, which advocates the "solidarity" of all human beings, to a new solidarity with other living beings; what Marxists advocate with regard to social systems we must extend to ecosystems' (1980: 15; for a riposte, see Plumwood, as Routley 1981).

But most Marxist–green synthesisers are not much concerned with connecting at the purely philosophical level. Here it is difficult for *us* to be synthetic, because these syntheses (and there are quite a few of them) tend to be idiosyncratic to the particular authors and, thus, difficult to generalise. Eckersley (1992: 87–95) attempts such generalisation within a category, 'Humanist Eco-Marxism', that roughly overlaps the one employed here, but only one of the writers considered below is included in it, whilst Enzensberger and Gorz, categorised differently by me, are included. Just two Marxist–green synthesisers are presented here — not as representatives of a nonexistent 'school', but as a sample of the variation such syntheses evince.

The best-known synthesiser of green and Marxist ideas is Rudolf Bahro, a former East German Communist Party member, whose four books from 1978 to 1986 reveal an extraordinary ideological trajectory, a shift from Marxism, to ecoMarxism, to a fundamentalist position within the German Greens (becoming, in fact, the key theorist of the 'Fundi' wing of Die Grünen), to an extreme green fundamentalism, such that in 1985 he split with Die Grünen, which he considered too willing to compromise basic principles (particularly on animal experimentation) and too wedded to an inappropriate legislative path to the green society. Along the way Bahro abandoned any compunction for the dislocation caused to the working class by the radical de-industrialisation which, he says, is the most urgent priority with which we are faced. There can be no compromise on this, for the earth does not have the capacity to supply the 10–15 billion people predicted for the next century at the present level of consumption experienced by 'the North American middle class' (1982: 130). Yet, this is the spectre with which the world is faced. In *From Red to Green* (1984), and *Building the Green Movement* (1986) Bahro talks about 'exterminism', 'the threat of the mass destruction of humankind, as expressed in atomic, biological and chemical weapons', as well as in 'virtually every type of production' (1986: 142–43). 'Exterminism' is elsewhere equated with 'the ecological crisis' (1986: 152; for a critique of Bahro's concept of 'exterminism', see Frankel 1987: 120–24).

Bahro did not stop writing in 1986. But it is impossible to consider the post-1986 Bahro as, in any sense, post-Marxist. Along with industrial disarmament, Bahro argues for an immediate move to basic, self-sufficient communal existence, and for the withering away of state bureaucracies, and an end to the capitalist market. Such a society would be demonetarised, essential to a market-free economy. And life in the communes would be frugally lived (conjoining scarcity with communalism led Bahro to embrace the Benedictine monastery as the model living mode for the green future; 1986: 90–91). Bahro, contra Pepper, accepted the existence of a global limit to growth as beyond dispute; for Bahro, scarcity is — in Ted Benton's apt phrase — 'a trans-historical feature of the human condition' (1993a: 202). Hence the need for that dramatic step away from industrial technology; of all the green visionaries, there is none so implacably hostile to modern and emergent technology as Bahro. Benton, unwilling to concede Bahro's departure from the socialist tradition, writes that Bahro saw 'a cultural revolution of ascetic renunciation of material well-being as a condition of eco-socialist transition' (1993a: 202–03).

Though Bahro makes frequent reference to Marx throughout *Building the Green Movement*, his conception of the political project has by this time (1986) ceased to be Marxist. Far from being the revolutionary subject, the working class has become collaborative with capital in the exploitations perpetrated by industrialism (1986: 79; see also Frankel 1987: 214–15; Redclift 1984: 55). In any case, the environmental crisis is largely spiritual and psychological: it is the psychological foundation of rampant materialism (and its attendant species imperialism) that must be confronted. Dobson puts it nicely; for Bahro 'the proper territory for political action is the psyche rather than the parliamentary chamber' (1990: 143). Bahro envisions change potently generated by a contagion of exemplary action:

> the accumulation of spiritual forces, the association of people who create a common field of energy which confronts the old world with a pole of attraction, will at a particular point in time which can't be foreseen exceed a threshold size. Such a 'critical mass', once accumulated, then acquires under certain circumstances a transformative influence over the whole society (1986: 98).

This theme — the need for psychic or spiritual rather than political renewal — remained prominent in Bahro's subsequent writings (for example, 1994).

I have chosen to include Bahro under this subhead because, though he only fitted here for a short time, his period of green–Marxist ideological fluidity is that which has attracted most interest in the environment movement. In 1982 Bahro could still say, with most ecoMarxists: 'many Greens may initially not think of capitalist industrialism but of the consequences of the industrial system in general. But in this respect they are misled' (1982: 24). Bahro is, however, already talking of the 'external' contradictions between humans and other species and between the industrial world and the rest as more fundamental than class contradictions in the developed west. We must go beyond Marx's view of the proletariat as the motor of history, he wrote then, 'and direct ourselves to a more general subject than the western working class of today. Like the utopian socialists and communists who Marx sought to dispense with, we must once again take the species interest as our fundamental point of reference' (1982: 65). External contradictions aside, the internal contradictions within the shifting planes of Bahro's thought are readily apparent.

Ted Benton, by contrast, continues to self-locate within Marxism, even though, unlike Pepper, he concedes the validity of several key themes within environmental thought, themes that have no obvious grounding within Marxism: he is concerned, in particular, with the status of the higher animals under Marxism (1988; 1993b; Benton and Redfearn 1996; see also Eckersley 1993b: 119–22). He describes the early Marx's view of history as the 'humanization of nature', in which nature is seen as:

> an external, threatening and constraining power [to] be overcome in the course of a long-drawn-out historical process of collective transformation. The world thoroughly transformed by human activity will be a world upon which human identity itself has been impressed, and so no longer a world which is experienced as external or estranged (Benton 1993b: 30).

For Benton such a view stands in apparent contradiction of Marx's metaphor for nature as 'man's inorganic body'. Benton's interpretation of this metaphor goes beyond Donald Lee's. It will be recalled that Lee saw the metaphor as capable of extension beyond Marx's original intention, to justify an ecocentric Marxism. For Benton no such extension is necessary: the metaphor establishes a vital, non-instrumental mutuality between humans and other nature, 'affirming the reality of nature as a complex causal order, independent of human activity, forever setting the conditions and limits within which human beings, as natural beings, may shape and direct their activities' (Benton 1993b: 31). But Marx's vision of the 'humanization' of nature, also to be found in the *Manuscripts*, is 'only dubiously consistent with a notion of the natural world as a purpose-independent complex of causal ideas' (Benton 1993b: 20).

Whilst Lee was content to make a philosophical point, Benton draws a longer bow. Drawing synthetically upon Marxism, animal rights philosophy, and the unfolding practice of ecological politics, Benton constructs a rights-based practical philosophy, which he describes thus:

> the realizability of a view of rights and justice grounded in environmental, social, developmental and bodily needs would presuppose something like the all-round human emancipation which Marx distinguished from more narrowly political emancipation: the bringing of all aspects of social life with a bearing on individual well-being, including our practices of interaction with external nature — under normative regulation. Whatever its current vicissitudes this has always been the central socialist preoccupation (1993b: 209).

Benton holds his Marxism confidently. Marxism is compatible, he maintains, with the most central of green preoccupations — once the contradiction within Marx's own writing is recognised and resolved. Benton also confronts the issue which has so bedevilled much green–Marxist writing — the issue of cornucopian materialism versus emergent scarcity. Pepper, as we have seen, is unwilling to shift from the standard Marxist faith in industrial productivism. Not so Benton. Pepper himself nicely summarises Benton's position:

> Benton ... believes that strands within Marx's analysis led him to under-theorise or leave out limits on the use, i.e. transformation, of raw materials, the fact that they and technologies all ultimately come from nature, and how naturally-given geographical and geological conditions just cannot be subsumed as 'instruments of production' ... Consequently, Benton thinks, Marx overestimated the role of human behaviour and underestimated the significance of non-manipulable nature (1993a: 100–01; for Benton's more detailed case, see 1989).

For Benton, then, there is a bottom line to the exploitation of nature, and scarcity in some form will persist under socialism. Thus, 'whatever moral case can be made for socialism, it cannot, any more, rest on the transcendence of scarcity' (1993a: 209). This reality makes socialism's historical preoccupation with issues of distributive justice more rather than less relevant (hence Benton's interest in a rights-based politics), just as the fact of scarcity makes more rather than less relevant the green preoccupation with questions of justice in relationships between species.

And class remains essential to environmental political analysis: 'class politics has always, and quite centrally, especially at "grass roots" level, been about environmental questions, and the new agenda of environmental politics both extends and is intelligibly continuous with that longer history'. 'Essential to ...' — but not the whole answer, for 'there is also something which transcends class politics in the new agenda of environmental politics'. Benton returns to the old, and difficult project of coalition building between labour and environment movements, though not, significantly, by rendering one a subsidiary of the other. The environmentalist constituency is 'a further possible element in a new coalition of the left, binding together social movements organised on the basis of class interest and ones deriving from moral perspectives' (1997: 44).

Questions concerning a long-term relationship between the old Marxist and somewhat more recent green radical traditions remain vexed. Considerable obstacles exist, though there are people attempting to dismantle these obstacles. At the forefront of this project is the journal *Capitalism, Nature, Socialism* and its editor, James O'Connor. Via a series of books and articles, O'Connor has rehearsed many of the issues discussed in this chapter (for example, 1989; 1991a; 1991b; 1992a; 1992b; 1993; 1998; Faber and O'Connor 1989a; 1989b). *Capitalism, Nature, Socialism* has also carried a series of essays on 'Marxism and Ecology' (for example, Burkett 1996; Kovel 1995; Leff 1993; Luzon 1992; Perelman 1993; Rudy 1994; Skirbekk 1994) and regular symposia on 'The Second Contradiction of Capitalism' — 'the absolute general law of environmental degradation under capitalism'. Remaining committed to Marxism, and rejecting philosophical ecocentrism, O'Connor nevertheless sees much congruence between the social movement aspects of environmental politics and Marxist thought — unlike Pepper. 'Workers, oppressed minorities, communities, environmentalists, and others engaged in identity politics and politics of place today are', he writes, 'struggling to subordinate exchange value to use value and the production for profit to production for need' (1998: 332).

One consequence of these developments has been the emergence of a less ideologically defensive ecological Marxism, one more given to forthright criticism of its parent tradition. 'By elevating the ignorance of nature and life to new heights and adopting the mechanistic view of the world', writes ecoMarxist Jean-Paul Deleage, '"official Marxism" has endorsed the most absurd choices in the field of development' (1994: 46). Thus, 'the dismal ecological record of "existing socialism" is equally disappointing as that of capitalism', and 'a Marxism capable of understanding ecological economics and an ecologically sound socialist policy must take this truth into account' (1994: 47; see also J. O'Connor's tabulated 'comparisons and constrasts between traditional socialism and ecological socialism'; 1998: 334–37).

We should also note that there are neighbour-to-Marx socialisms that have cross-fertilised with ecological thought. Martin Ryle, for example, styles himself an ecosocialist, and looks rather to the 'libertarian socialism' of William Morris than to Marx. Of course, Marxists have long claimed 'ownership' of Morris (for example, Arnot 1934; E.P. Thompson 1976), but Morris has also been claimed by the environment movement (for example, Gould 1974; A. Taylor 1997). His communitarianism and his insistence upon an 'ethical' socialism lived in harmony with nature, rather than a 'utilitarian' industrial socialism within which pride in craft and human dignity are impossible, are seen to chime with certain values prominent within environmentalism. There are problems with Morris — gender relations for example, and a pronounced anti-urbanism (Ryle 1988: 22–23) — but, for Ryle, Morris provides the 'minority tradition' within which the socialist and ecological cases can be united (1986: 36). He is particularly drawn to Morris's distinction between 'real' and 'sham' needs and Ryle himself distinguishes between irreducible and constructed needs (1988: 68–73). If some are irreducible, of course, needs are not infinitely contextual. Unlike Pepper, then, Ryle recognises ecological constraints. Thus, 'the prospect of unlimited ... proliferation of needs is untenable' (1988: 70–71), and as capital is blind to this truth, it follows that 'an ecologically sustainable society is incompatible with the economic imperatives of capitalist production' (1986: 36).

It is, nonetheless, the anarchist tradition within radical socialist thought, the tradition to which Bahro and Gorz eventually gravitated, that has most attracted green thinkers. As John Young observes: 'environmentalists of the left are now seeking to re-emphasise this minority tradition because of its ideological compatibility with such goals as decentralisation, bioregionalism and small-scale economic activity' (1991: 158). Time to turn thither.

CLASSICAL ANARCHISM AND THE ENVIRONMENTALIST CRITIQUE

There is a school of sociological literature concerned with something called the 'ecological' (or 'alternative', or 'human exemptionist', or 'new environmental') paradigm, which is usually juxtaposed against something called the 'dominant paradigm' (for example, Catton and Dunlap 1980; Cotgrove 1982; Cotgrove and Duff 1980; Dunlap and

Van Liere 1978; Milbrath 1984; 1989). The word 'paradigm' commonly refers to a body of fundamental assumption held at a sub-ideological level (and perhaps even providing common ground for what, at the level of ideological disputation, seem irreconcilable opposites); though this meaning is wider than the one employed by Thomas Kuhn, who, in *The Structure of Scientific Revolutions* (1970), popularised the term. 'Paradigm' served Kuhn as a conceptual tool to explain the periodic replacement of basic assumptions concerning the status of scientific knowledge and the nature and operating processes of the universe, with radically different basic assumptions.

The status of the term 'paradigm' is contestable, and it is not now as evident in environment movement analyses as it was through the 1980s. From various sociological treatments of the environment movement and several tabulations of the rival paradigms to be found in the 1980s literature, I have produced a composite in the following table.

NEW ECOLOGICAL PARADIGM	DOMINANT SOCIAL PARADIGM
1 High valuation of nature	**1** Low valuation of nature
2 Environmental protection valued over economic growth	**2** Economic growth valued over environmental protection
3 Generalised compassion	**3** Compassion reserved for only those 'near and dear'
4 Science and technology not always good	**4** Science and technology a great boon to humankind
5 Limits to growth	**5** No limits to growth
6 A new society, with an emphasis on: • participation and openness • the public sphere • post-materialism • simple lifestyles • co-operation	**6** Contemporary society is fine, as is its emphasis on: • hierarchy and efficiency • the market sphere • materialism • complex and 'fast' lifestyles • competition
7 A new politics, with an emphasis on participation, consultation, devolution and direct action	**7** Contemporary politics is fine, as is its emphasis on centralisation and economies of scale and on decision-making by technical experts, and by delegation/representation

Note how pronounced the contrast is between the New Ecological Paradigm (NEP) of 1980s environmentalism and the Hardin brand of 1970s authoritarian environmentalism. Note, too, that the NEP corresponds remarkably closely with the values of classical anarchism.

That the small, non-hierarchical political structure conducing to high participation in decision-making should be regarded as more appropriate than centralised authoritarian structures is only to be expected. This is, after all, the social mode deemed appropriate to benign technology, it being argued that large, centralised technologies have a greater capacity for environmental destruction, and that such technologies can only be sustained in large, centralised political systems. Others insist that such small-scale systems not only *conduce to* the implementation of environmental values but *are of themselves* the appropriately ecological mode for human living — larger systems being inevitably de-powering and alienating. The small, non-hierarchical social unit, on the other hand, corresponds to what is to be found in nature. Kirkpatrick Sale, anarchist and bioregionalist, enjoins us to:

> regard the behaviour of all primate species except the human. Nowhere are found organisations that extend beyond the small troop — nowhere a collective of troops, an amalgam of like species, nothing that would resemble the structure of a state. Within some troops there are various temporary leaders and sometimes a 'pecking order' so to speak, with so-called alpha males establishing dominant positions in sexual affairs — that is far more common among baboons than among more humanoid gibbons, orangutans, chimpanzees or gorillas — but even then there are no multi-troop leaders, no chieftains or kings for all the cities in an ecosystem, nor are there extensive collaborations or alliances (1985a: 15).

As we have seen, some environmentalist thinkers have serious misgivings about this recourse to biological determinism; nevertheless, here is a particularly evocative expression of the view that 'small is beautiful because small is natural'.

The other argument noted above — that small-scaleness is to be preferred on ecological grounds because of the technological consequences that flow therefrom — seems ultimately of greater significance.

The question of technology is crucial in explaining the dramatic shift *away* from a politics favouring an environmentally benign centralised authoritarianism and *to* a politics of devolution and human-scaleness. It may even be that 'the technological question' is the single most important site of contest upon which the environment movement divides; a more internally contentious matter, for instance, than ecocentrism. Scratch any movement activist, even one who is a- or anti-theoretical, and invariably the appropriateness of an ecocentric perspective is conceded (often impatiently, with a comment of the type 'I'd have thought that was self-evident'). But there is little agreement about what an ecologically appropriate attitude to technology should be.

There *is* agreement that big energy- and resource-guzzling industrial technology is not to be countenanced. But from that basic level of agreement there is sharp divergence. Some within the movement — not, I think, a very great number — argue that we must substitute pre-industrial technological modes for industrial technology and its mega-institutions. Some seek similarly low-tech but decidedly post-industrial solutions. Others see in the development of information-intensive, post-industrial technology the capacity to transcend, and thereby consign to history, industrial technology as we know it. The best-known anarchism within environmental thought is bioregionalism, and the foremost theorist of bioregionalism, Kirkpatrick Sale, explicitly connects a vision of environmentally benign information technology to his advocacy of devolved political structures (1985a: 22).

It will be clear that environmentalist thought has much in common with anarchist formulations. O'Riordan sees 'the self-reliant commune modelled on anarchist lines' as the logical organisational consequence of the ecocentric perspective (1981: 307; see also Dobson 1990: 117). But — with the conspicuous exceptions of Murray Bookchin (who, in *Our Synthetic Environment*, established an umbilical connection between ecologism and the anarchist tradition as early as 1962), and anarchist bioregionalism — no 'parental' relationship can be said to exist between classical anarchism and current environmentalism, and the early anarchists have been rediscovered by an action-focused movement, in the mid-1970s and thereafter, seeking a complementary theory. The social principles outlined here are also to be found in the various European green parties, and this helps account for the ambivalence felt within those parties about the taking of the legislative path, and the self-flagellations that those parties indulge in over this matter as a consequence. 'A green movement', runs the argument, 'must be built from below'. It must 'build the new society within the shell of the old'. It is not obvious that such a strategy has much to do with, or is even compatible with a strategy centred upon the winning of seats in legislatures.

The green movement has, thus, adopted a politics of *cultural* transformation, in sharp contrast to Marxist praxis and the emphasis there on cataclysmic revolutionary struggle wherein production relations are captured in struggle and radically transformed. Here, by contrast, is a politics of quiet revolution; a politics that plays down the importance of change in economic spheres in favour of change in cultural spheres. Old institutions are not confronted head-on. They are quietly subverted and, where that is impossible, they are rendered irrelevant: one simply works around them, puts in place informal processes that render them impotent.

Clearly, for a politics that emphasises such grass-roots political process, the classical anarchists provide the best precursors. Though some of the classical anarchists provide very little sustenance — the grim William Godwin, for instance — most are of some present-day interest (Proudhon, Bakunin, Fourier, Owen). One stands far 'greener' than all these, however — the Russian anarchist prince, Peter Kropotkin (1842–1921). His work, in particular, has been taken up, for he was unique in emphasising 'the need for a reconciliation of humanity and other nature, the role of "mutual aid" in natural and social evolution ... and his vision of new technics based on decentralisation and human scale', in addition to the more standard 'hatred of hierarchy' (Bookchin 1982: viii).

Kropotkin's greatest work is *Mutual Aid* (first published in 1902). Kropotkin was a geographer by profession, and much-travelled. His experience in the field led him to view the evolutionary process as a struggle against environmental adversity, and co-operation as one of the most effective strategies in this struggle, rather than the dog-eat-dog competition that was, in the view of the dominant neo-Darwinists of the time, evolution's sole shaping principle. In *Mutual Aid*, George Woodcock has observed, Kropotkin 'related the social responsibilities which he found characteristic of men when they were undisturbed by coercive institutions with the sociality he found so widespread among animals' (1974: 84). Woodcock also points out that, 'since he was concerned mainly with what happens within species, Kropotkin did not explicitly delineate the complex pattern of mutual dependencies which the modern ecologist sees when he looks at the natural world', but we can take as implied that Kropotkin's law of mutual aid was a universal one. Certainly it applied to us — human social development is based on 'the same principle as that which ensures the survival of other species' (1974: 84).

Kropotkin's other great work, *Fields, Factories and Workshops* (first published in 1899), also resonates with the concerns of today's environmentalists. In this tract Kropotkin related anarchism, through concrete proposals for industrial and agrarian reform, to the need to preserve and conserve in the interests of a 'richer' existence. Much classical anarchist thought was technologically exuberant, but Kropotkin is a notable exception. He, thus, provided a most convenient hook for a prominent strand of twentieth century environmental thought.

The marriage of anarchism and ecologism is most fully developed in Bookchin's work. *The Ecology of Freedom* (1982) adapts classical anarchism to ecological social principles, and is one of the classic works in the present-day environmentalist *oeuvre*. *Toward an Ecological Society* (1980) details prescriptions for the establishment of ecological communes. Still earlier, *Post-Scarcity Anarchism* (1971) established the relevance of contemporary anarchism to the central questions of the emerging ecological critique, moving anarchism away from the technocratic faith that so characterised much nineteenth century anarchism. In literature, novelist Ursula Le Guin did likewise. Her wide ly read science fiction novel, *The Dispossessed* (1974), significantly located her experimental anarchist community on a resource-poor moon rather than upon its lavishly endowed planet, a clear break from the cornucopian optimism of earlier anarchist formulations. Human wealth, in the view of this new ecological anarchism, stems from less tangible sources than extravagant productivism; it is to be found in the satisfaction to be derived from rich relationships and the development of individual and social potentialities: 'if abundance must rely on infinitely expanding productivity and exhaustion of nature as a resource, it obviously can never be achieved. But for anarchists, abundance is to come from the development of social needs and from the desire for a creative and joyful existence' (J. Clark 1984: 149).

Bookchin excepted (he is considered separately below), the importance of ecology for anarchist thought is most succinctly articulated by John Clark (1984: 191–99). He notes the technological implications of ecological crisis:

> a belief that human beings are beings within nature, rather than above it or apart from it demands the development of what Illich calls 'convivial tools', Schumacher labels 'intermediate technology', and Bookchin (perhaps most adequately) describes as 'liberatory technology', or 'ecotechnology' ... such technologies offer the possibility of achieving human goods without morally indefensible domination of other species and of the biosphere (1984: 196–97).

But corporate capitalism and its main modern alternative, state socialism, have built themselves upon large, ecologically destructive industrial technologies. 'A new technological practice aimed at ecological regeneration and founded in a respect for nature must be accompanied by a new political practice', which will 'oppose centralization with decentralization, hierarchical control with self-management, atomistic individualism with community, and domination with cooperation' (1984: 198). For Clark, then, only political envisionings within the anarchist tradition can provide the necessary technological basis for a 'respectful' relationship between humans and nature.

The most significant meeting-point between classical anarchism and current environmentalism is the emphasis each places upon *scale* — though, in emphasising scale as a key social variable, the modern environment movement looks less to the anarchist classics than to one of its own modern classics, Fritz Schumacher's *Small is Beautiful* (1974). Like Schumacher himself within the body of his argument, most environmentalists have preferred to speak of 'human scale' rather than 'small' *per se*, arguing that scale is a matter of what is humanly comprehensible and manageable from a participationist perspective, and, though this will usually translate as 'small', there will be instances when larger social units are possible and desirable.

THE 'SCALE' VARIABLE: 'SMALL IS BEAUTIFUL' AND THE CRITIQUE OF URBANISM

Whilst the emphasis on scale has been important in environmental political thought, on the related question of social *mode* — the *character* of the small communitarian unit — there has been much less agreement.

Environmental thought contains a strong anti-urban theme, perhaps most explicit in the writings of Theodore Roszak. 'Of all the hypertrophic institutions our society has inflicted upon the person and the planet', he writes, 'the industrial city is the most oppressive' (1981: 254).

Against this, Roszak advocates the small, economically self-contained 'monastic' commune which is juxtaposed against the 'urban-industrial colossus':

> its natural focus is the small community, or, at largest, an anarchist network of communities — the form which most monastic orders finally assumed as each house, having reached an optimum size, sent out its members to find new land and to make a new, small beginning ... my interest here is precisely those social forms that disintegrate bigness, seeking to replace it with socially durable, economically viable alternatives (1981: 302–03).

We have seen that it is possible to locate environmental monasticism within the political traditions of the right. Paehlke notes that 'the environmental expression of the themes of self-management, self-reliance, autonomy, and decentralization' partakes of key concerns of the new right: 'both seek a return to a wholesome — and perhaps mythological — past. Self-sufficiency in energy, shelter and food has obvious appeal in economically unstable times, especially in societies dominated by large international corporations and bureaucratic governments' (1989: 155). But Roszak emphasises conviviality and the development of human potentialities hitherto untapped under the stultifying institutions of industrial capitalism, rather than the tight moral integration envisioned by rightist advocates of self-sufficiency (and characteristic of the closed monasticism of Edward Goldsmith). It is an anarchist tradition, which, its individualist strands notwithstanding, is appropriately located within the socialist family of ideas.

Roszak's preferred option, a thriving rural and village life serviced by small autonomous towns, is one commonly expressed in environmentalist literature, but it may not command overwhelming support. Others have expressed unease over the danger of what Marx bluntly termed 'the idiocy of rural life', and have portrayed the flight from cities to rural communes as a mere temporary phase for most, stemming from a disillusionment with the city, to which the refugee is eventually lured back by a yearning for urban stimulation. We should also remember that the small-scale communitarian unit of social organisation favoured by most environmentalists is also possible in the city, and that it is not necessarily the case that to de-urbanise would be ecologically sound (on this, see Paehlke 1989: 156–57, 244–50). There have been attempts, more or less plausible, to overcome rural/urban dichotomy. Timothy Luke's solution can stand as exemplar. He argued that such an end should be sought via 'deconstruction' to a 'demo-communitarian' small city of 30,000–50,000 people, a process that would 'ecologically fuse the rural and urban', and 'deconcentrate urban resources, bringing urban commerce, art, society, letters into balance with "rural" crafts, culture, community, customs', and 'rehumanise nature', whilst 'renaturalising humanity' (1983: 28).

At a more abstract level, though, the congruence between ideas of decentralised social, economic and political structures and other key ideas within the environmentalist value-set is very strong (Paehlke 1989: 154–57; J. Young 1991: 36), and particularly was this so through the 1970s and 1980s. Much of this envisioning has taken place within the context of economic thought (we looked at this in chapter 8), but it has relevance beyond economic thought. Again, historically, the key figure here has been Fritz Schumacher.

Schumacher (1974: 53) notes the existence of two apparently mutually exclusive human qualities that social structure is expected to foster: 'freedom', for which is needed small, autonomous organisational units; and 'order', for which large-scale, possibly global units are required. Though these needs are accorded equal importance in the abstract, and the importance of global structures 'to the indivisibility of peace and also of ecology' (1974: 54) is explicitly stated, it is clear that, for Schumacher, this level of social organisation is of secondary importance to the need for structures small enough to enable the flourishing of personal autonomy and forms of human activity that are genuinely goal-achieving rather than merely symbolic. Whilst 'for every activity there is an appropriate scale', society's current 'idolatry of giantism' has meant that the balance is wholly wrong (1974: 54) and its restititution necessitates the promotion of scaled-down structure.

Before mass transport and communications systems, movements of people and goods occurred at the social margins, and most needs were locally met. We are accustomed to seeing the increased complexity of present-day society as a force for system resilience, but Schumacher argues that this is not so; that, in fact, 'structures are *vulnerable* to an extent that they have never been before' (1974: 57). Increased fluidity expands human freedom 'only in some rather trivial respects', whilst threats to that freedom are posed by the escalation of insecurity, found most notably in large countries, where 'footlooseness' is a major problem, along with its attendant social dysfunctionalities, including the unbalancing of the rural and the urban as 'vitality is drained out of the rural areas' (1974: 58).

These social dysfunctionalities are all attributable to the denial of autonomy, meaning and self-achieving social action within giant organisations. Schumacher returns repeatedly to the importance of social scale in determining the possibility of fulfilled human existence. 'There is no such thing as the viability of states or of nations', he writes, 'there is only a problem of viability of people: people are viable when they can stand on their own feet'. And 'what is the meaning of democracy, freedom, human dignity, standard of living, self-realisation, fulfilment? Is it a matter of goods or of people? Of course it is a matter of people. But people can be themselves only in small comprehensible groups' (1974: 62).

Many of the themes that were to dominate green thought through the 1980s are prefigured in Schumacher's work in the early 1970s. His emphasis on the indivisibility of a larger, ecological whole which is nevertheless coexistent with the need for maximised possibilities for self-directed action, freedom, autonomy and democracy came to be distilled into the best-known green movement maxim: 'think globally, act locally'. His critique of organisational 'giantism' as intrusive, technocratic and inimical to social and ecological wellbeing has become a central theme within environmental thought. In his stress on the need for embeddedness, on the need for a social and ecological context of belonging, he anticipated the environment movement's concern for the integrity of place. His critique of the imperialist nature of urban/rural relations has remained a prominent thread within green thought. And his identification of the importance of meeting local needs from local resources anticipated the school of thought which became known as bioregionalism, perhaps the most broadly influential application of anarchist ideas within environmentalism.

BIOREGIONALISM

Bioregionalism provides environmental thought's most emphatic endorsement of the scale variable. For Kirkpatrick Sale, scale is the *key* environmentally relevant variable:

> the issue is not one of morality, but of scale. There is no very successful way to teach, or force, the moral view. The only way people will apply 'right behavior' and behave in a responsible way is if they have been persuaded to see the problem clearly and to evaluate their own connections to it directly — and this can be done only at a limited scale (1985b: 53).

'Bioregion' is an old and familiar concept: it has existed in biogeography at least since Odum and Moore's work in the 1930s. But, if we restrict ourselves to Sale's usage, 'bioregion' is an area 'naturally defined by its location, geology, soil structure, wildlife and the human communities as well as the cultures that arise within its confines' (1984a: 167). And to these areas of ecological soundness, political systems should accord. Though there are variants of bioregionalist thought that do not draw such utopian conclusions (for example, Raberg 1997), in the dominant version within 1980s green thought, such bioregions should be largely autonomous and live within the constraints of their resources. Whilst by no means the only prominent voice in the construction of utopian bioregionalism, Sale is usually taken as its exemplar because he has provided the most fully fleshed bioregional political theory. As Eckersley has observed, this is somewhat atypical, for 'much of the bioregionalist literature is poetic, inspirational and visionary and more concerned with cultivating a bioregional consciousness and practice than with presenting a detailed political analysis' (1992: 168).

In terms of continuities and discontinuities with the canon, the latter characteristic places bioregionalism across the divide from the technocratic modernism of most nineteenth century anarchism, and links it with the anti-modernist concern for life-reverencing spirituality and with interest in indigenous interactions with the environment. 'What is essentially represented by the term "bioregionalism" has been the common knowledge and practice of native peoples', writes David Haenke: 'it really has to do with living in spiritual and physical harmony with all aspects of the lives of the ecosystems in which we make our homes. Some native peoples, when they and the lands where they live are left alone, still do this' (1984: 3). There is interest in both pre-European regionalisms in North America, and 'the pre-nation-state kingdoms of Europe' (J. Young 1991: 136). More generally, 'Gaean' spirituality is given high priority in Sale's thought (1985b: 3–22, 183–92), whilst Jim Dodge accords 'spirit' the status of one of the three defining components of bioregionalism (the other two are 'natural systems' and 'anarchism'): 'what I think most bioregionalists hold in spiritual common is a profound regard for life — all life, not just white Americans, or humankind entire, but frogs, roses, mayflies, coyotes, lichens: all of it' (1981: 9).

Not surprisingly, bioregionalism has made common cause with deep ecology. Sale, for example, has frequently written in defence of deep ecology, being 'struck by how similar' the two are (1985a: 21; see also 1988). Bioregionalist anthologies usually include some deep ecology content (for example, Seed 1990), whilst deep ecologists have, in turn, saluted bioregionalism. And in his assault on deep ecology, Bookchin specified Sale and spiritualist bioregionalism as prominent instances of wrong-headed 'eco-lala'. But bioregionalism has also received support from quarters critical of deep ecology. Jim Cheney, for example, in his influential paper 'Postmodern Environmental Ethics: Ethics as Bioregional Narrative' (1989), argues that bioregionalism, through its recourse to 'storied residence ... bioregional truth ... or ethical vernacular', can '"ground" the construction of self and community without the essentialization and totalization typical of the various "groundings" of patriarchal culture' (1989: 134). Andrew Dobson provides an interesting perspective on this, one that captures the place of bioregionalism in the green scheme of things: 'Greens will always advocate decentralized political forms' even if they stop short of bioregionalism *per se*; so, 'if Greens depart from bioregional principles, the place they eventually arrive at will bear a family resemblance to such principles' (1990: 117). If, then, not all greens explicitly self-identify as bioregionalists, the central preoccupations of bioregionalism are so near the heart of the environmentalist paradigm that few greens will adopt a stance in hard opposition to it.

Bioregionalism's pre-modernism might not be crucial to this central locus, but 'the guiding principle of bioregionalism ... that the "natural" world should determine the political, economic and social life of communities' (Dobson 1990: 119) surely is. (Perhaps bioregionalism's social anarchism is also crucial. We have seen that this is made explicit, bioregionalism's theorists deploying 'anarchism' unselfconsciously and locating their movement within the wider stream of anarchist history.)

Natural regions, says Sale, occur at four levels. These are, moving from largest to smallest:

- the 'ecoregion' ('a huge area of perhaps several thousand square miles' that takes its definition from 'the broadest distribution of native vegetation and soil types'; 1985b: 56);
- the 'georegion' ('identified most often by clear physiographic features such as river basins, valleys and mountain ranges'; 1985b: 57);
- the 'vitaregion' (territories of life);
- the 'morphoregion' (territories of shape, 'identifiable by distinct life forms on the surface — towns and cities, mines and factories, fields and farms — and the special land forms that gave rise to those particular features in the first place'; 1985b: 58).

Dwellers within these regions are enjoined to live according to the distinctive regional rhythms; to be self-sufficient and accepting within those constraints.

Four principles guide this endeavour. The first two of these, 'knowing the land' and 'learning the lore', concern the discovery of the home range and its potential and the development of a respectful intimacy with it, including its human folklore, history and indigenous technologies, as well as its natural processes (1985b: 44–46). At this point the bioregionalist project clearly intersects with the environment movement's wider concern for the authenticity of place. 'What is your homeland?', asks bioregionalist Peter Berg. 'Well, it's these plants and animals and natural systems that are in this life-place, in this bio-region ... I feel it is *my* home. And by knowing about it, by learning about it, I begin to authenticate my identity with it' (1990: 24). Here, the concept of 'reinhabitation' is important, a concept that connotes not only protection but rehabilitation:

> if the life destructive path of technological society is to be diverted into life sustaining directions, the land must be reinhabited. 'Reinhabitation' means developing a regional identity. It means learning to live-in-place in an area that has been disrupted and injured through past exploitation. It involves becoming native to a place through becoming aware of the

> particular ecologic relationships that operate within it and around it (Berg and Dasmann 1978: 200).

Sale's other two bioregionalist principles are: 'developing the potential' of a region towards self-reliance by 'using all the biotic and geological resources to their fullest, constrained only by the logic of necessity and the principles of ecology' (1985b: 46); and 'liberating the self' from 'distant and impersonal market forces, remote governments and bureaucracies, and unseen corporations dictating consumer choices — while within the bioregion both economic and political opportunities would be inevitably opened up' (1985b: 47).

Though there are, as we have seen, several bioregionalist tiers, the logic of his four key principles leads Sale to prescribe the local community of 5000–10,000 people as the desirable political unit, though this may vary according to the level of resource abundance at the community's call. The communities within a morphoregion would constitute a confederation of polities, autonomous and, in Sale's preferred vision, anarchist. Concerning this latter characteristic, though, there is a sense in which Sale holds his anarchism lightly, for whilst he is precise when it comes to prescribing scale, he is conspicuously less so on the question of political mode. On that, he argues that autonomous communities will 'go their own separate ways and end up with quite disparate political systems — some democracies no doubt ... but undoubtedly all kinds of aristocracies, oligarchies, theocracies, principalities, margravates, duchies and palatinates as well' (1984b: 233). Despite the likelihood of political disparity, the sovereign communities of a morphoregion would trade with each other, network information, and share some facilities — 'a university, and a large hospital and a symphony orchestra' are cited as examples of institutions that might only be feasible on a morphoregional scale (1985b: 74–75). Sale sums his vision thus: 'the bioregional mosaic would seem logically to be made up of communities as textured, developed, and complex as we could imagine, each having its own identity and spirit, but each of course having something in common with its neighbours in a shared bioregion' (1985b: 66).

MURRAY BOOKCHIN AND 'SOCIAL ECOLOGY'

Finally, there is the work of the most explicitly anarchist of ecological theorists, Murray Bookchin. As we have seen, Bookchin labels his thought 'social ecology' (though not all deployments of the term follow his usage; see Hill 1999), in which emphasis is placed upon the twin pathologies of 'hierarchy' and 'domination' and the need to create conditions conducive to the development of socially and personally competent self-hood. Bookchin links his 'social ecology' to a somewhat contentious evolutionary teleology. A combative man given to vituperative criticism of those with whom he disagrees — there is nothing remotely pluralist in his response to the multi-facetedness of ecological thought — Bookchin excites equally strong and varied reactions to his own work. John Clark has described Bookchin's thought as 'the first elaborated and theoretically sophisticated anarchist position in the history of political theory' (1984: 202), Bookchin's masterwork, *The Ecology of Freedom* (1982), being 'a work of sweeping scope and striking originality' (J. Clark 1984: 215; see also 1982). Bookchin, on the other hand, has written of Clark: 'I, for one, want neither to deal with him nor his supporters, who are graying the world in the name of greening it'; 1997: 172).

Bookchin identifies hierarchy and domination as the source of *all* political pathologies — including of environmental degradation, which is a consequence of the domination by some people of others (ecofeminists think similarly). He is, thus, at odds with those who see the domination by some people of others as a consequence, rather than a cause, of environmental exploitation (the position usually attributed to deep ecology). The origins of social hierarchy are to be found in prehistory, at the point when 'the conceptual apparatus of domination' (Eckersley 1992: 148) arose to justify the nascent push for control on the part of the old over the young, and men over women (Bookchin 1987b: 7–8; 1990: 46–66, 76–80). This quest for domination required control over resources, and these were to be found in the realm of 'external' nature, which therefore had, in turn, to be dominated. The mental framework that gives rise to pecking orders within human societies generates the repressive technologies of domination of human over human, and of human over nature.

Though the conceptual apparatus for domination was developed in prehistorical times, Bookchin sees, in early tribal societies, largely non-hierarchical, egalitarian and co-operative ways of life in which human communities were integrated within a larger community of nature. He speaks of 'the extent to which their sense of communal harmony was also projected onto the natural world as a whole' (1990: 48). In Bookchin's view, societies transfer the working assumptions of their social structures onto nature, and so the individualised conception of society predominant within market capitalism generates a view of nature which is commensurately one of a-social individuals preying upon each other in cut-throat competition. By contrast, 'in the absence of any hierarchical social structures, the aboriginal vision of nature was also strikingly nonhierarchical' (1990: 48).

But Bookchin eschews idealisation of primitive societies, setting the positive features noted above against perceived limitations and failings (for example, 1982: 67). He instances, too, some modern systems that have gone some way towards applying the principles he promotes: the Paris Commune of 1870, and the Spanish collectives of the 1930s, for example. But the historical exemplar to which he returns in almost all his writings is the *polis*, the city-state of ancient Greece (and the principles of political thought to which it gave rise). The exemplary position of the *polis* in Bookchin's writing is well summed up by Clark:

> what he admires most [about the *polis*] is its success in nurturing a highly individualized social self by cultivating a balance between values like civic duty, developed personality, and freedom. A responsible self and an active citizenry were consciously created through the system of assemblies, councils, courts, and other institutions in which direct democracy prevailed (1984: 226).

Bookchin maintains the classical anarchist preoccupation with the cultivation of citizenship. Social institutions are, above all, educative. They are to be judged according to their capacity to produce citizens — socially and personally integrated 'selves'. Their essential purpose is a liberatory one, the production of self-managing beings who can also participate confidently, effectively, and *directly* in a richly textured social existence. Bookchin tilts against externally imposed structures, and against 'representational' structures that preclude unmediated political involvement. There can be no authentic selfhood arrived at through those means. 'The people' are neither the de-humanised proletariat of Marxism, nor the manipulated, homogenised consumers of market capitalism. They comprise a community of responsible, self-directing, socially skilled selves.

At the level of structure Bookchin argues for 'affinity groups' (tenants' associations and local environment and resident action groups, for example) and the decentralised ecological city. Only in small groups of socially skilled, similarly minded individuals can self-managed activity be effectively practised. Such groups, 'linked together by proliferation and combination' (1980: 48), can form the basis for a revolutionary politics of 'libertarian municipalism', a politics based upon community assemblies. Despite its apparent simplicity, this notion is revolutionary because it seeks to confront the state with 'a communitarian society oriented toward meeting needs, responding to ecological imperatives, and developing a new sensibility based on sharing and cooperation' (1992: 94) — not mere 'political strategy', but 'a kind of human destiny' (1992: 94).

The small primary group, linked to other groups within a framework of libertarian municipalism, can also provide the infrastructure to bring technology and production into line with human need and aspiration. Bookchin rejects the position (common to Marxism and liberal democracy and implicit in much classical anarchist thought) that technology is instrumentally neutral, capable of being used for liberation or domination according to its deployment. He also rejects the opposite view — that technology has become a repressive juggernaut, with all other social structures and processes subordinate to it. 'Liberatory' technology will be generated when autonomous selves come together in spontaneous co-operation within affinity groups. Whether we develop liberatory technology or stay with technologies of dominance depends on whether we have the will to develop the self-directing social existence in which liberatory, democratically fashioned technology can flourish. Latterly, Bookchin has directed his attack more toward the 'anti-Promethean' tendency in ecological thought, behind which lies 'a very privileged disdain for human intervention as such into the natural world'. Humans are, he writes, 'biologically unique organisms precisely in that they have the nervous system and anatomy to intervene into first nature and "manage" their future — to innovate, not merely to adapt to a pre-given environment' (1997: 170–71; see also 1995).

In none of the above does Bookchin move beyond the staked terrain of classical anarchism. In bedding his thought within an ecological and organicist paradigm, however, he takes anarchism into significantly different territory.

As long ago as the mid-1960s Bookchin was arguing the far-reaching implications of ecological insights for social and political thought. Each of his social principles is sourced to a justification from ecology. 'Ecology', he writes, 'denies that nature can be interpreted from a hierarchical viewpoint' (1980: 271). The living components of nature exist interdependently, 'more like variations in the links of a chain than organized stratifications of the kind we find in human societies and institutions' (1982: 29). A society in which freedom and citizenship are enjoyed, then, is one grounded within the ecological processes of nature.

Bookchin's is an 'organicist' view of nature; hence this passage, written in 1980, which reads somewhat strangely from a man who would soon be directing much of his writing into raillery against the 'spiritualising tendency' in green thought:

> the sun, wind, waters, and other presumably 'inorganic' aspects of nature would enter our lives in new ways and possibly result in what I called, nearly a decade ago, a 'new animism'. They would cease to be mere 'resources', forces to be 'harnessed' and 'exploited', and would become manifestations of a larger natural totality, indeed as a respiritized nature, be it the musical whirring of wind-generator blades or the shimmer of light on solar-collector plates (1980: 93).

From other viewpoints within ecological thought his theory of nature becomes most controversial when this organicism becomes overtly teleological, with human beings the privileged agency of nature's telos. Bookchin writes of 'this sweeping drama in which we split from blind nature only to return again on a more advanced level as nature rendered self-conscious' (1980: 95–96).

He is not the only ecology movement thinker to argue this position. James Lovelock has also done so; though the similarities between the two fade once we move beyond the bald claim for humanity as nature rendered self-conscious. Bookchin's natural teleology is cased within a theory of evolution in which he posits a *human-led* process: that is, rather than species finding their own evolutionary destinies, there is an ordained evolutionary vector and it is for humans to make that vector explicit and direct movement along it. For Bookchin, 'subjectivity' is nascent within all life; all strives toward a coherent end, because all life partakes of 'Logos' — immanent reason, 'the organizing and motivating principle of the world' (1982: 10). Humanity's task is to 'make the implicit meanings in nature explicit' (1982: 316); in Eckersley's well-crafted paraphrase: 'there must be an infusion of human values into nature, because humans are the fulfillment of a major tendency in *natural* evolution' (1992: 155). That 'major tendency' is intellectuality, the reason inherent in nature rendered self-aware, and charged with the task of shaping an ever-more complex and varied ecology.

SOME PROBLEMS: HITCHING A RIDE ON A FALLING STAR?

The most severe critics of socialist forms of environmentalism have been adherents of rival socialist environmentalisms. In the case of Marxism, for instance, the most withering condemnation has come from Murray Bookchin.

Bookchin's rejection of the ecological credentials of Marxism is contextualised within a broader attack upon Marx's capacity to ground a genuinely radical politics. He dismisses the notion that Marx's thought can be distinguished from practical misapplications of it, arguing that the flaws of Marxism in practice are readily sourced to Marx himself. Marx's thought has been miscast as revolutionary praxis; it is much more appropriately seen as 'bourgeois sociology', and Marx's 'largely bourgeois image of reality' (Bookchin 1990: 130) explains why in practice Marxist ideas led to state capitalism rather than to a genuinely liberatory socialism. Marx welcomed industrial production and the centralisation of economic and state power as 'historically progressive' (Bookchin 1990: 135); in Ariel Salleh's telling phrase, Marx 'endorsed the capitalist stage as a way station towards a world of infinite possibility in the growth of human needs, provided for gratuitously by Mother=Nature' (1997: 74). That these developments were to pave the way for a sociality wherein — against the logic that had created and applauded the technocratic system of centralised productivism — a re-humanised social life would somehow come to pass, came to be seen as less real, less *logical*, than the perpetuation of 'progressive', rigidly and hierarchically controlled, industrial productivism.

And Marx 'reduces' people to an agglomeration of economic beings, transmuting all other needs and interests into the economic. 'At every point', Bookchin laments, 'we are impeded by economistic categories that claim a more fundamental priority' (1980: 209). But it is within the non-economic dimensions of life, says Bookchin, that liberatory activity must take place, because claims for the working class as the spearhead of revolutionary change made on the basis of economic abstractions are now plainly spurious, and it is for humans as humans, not as economic cyphers, that liberation must be sought:

> the working class has now become completely industrialized, not radicalized. It has no sense of contrast, no clash of traditions, and none of the millenarian expectations of its antecedents. The proletariat as a *class* has become the counterpart of the bourgeoisie as a class, not its unyielding antagonist ... the working class is simply an organ within the body of capitalism, not the developing 'embryo' of a future society (1990: 133; see also 1994: 5–6).

This is the larger context within which Bookchin attacks Marxism's ecological credentials. The reduction of humans to 'mere economic resources' leads to the objectification of nature, to its treatment as 'mere natural resources' (1990: 136). More, it leads to the depiction of nature as alien and hostile. Ecological factors are seen to 'inhibit the growth of industry and the mining of the natural world'; nature is 'a cruel "realm of necessity" and an ensemble of "natural resources" that labour and technology has to subdue, dominate, and rework' (1990: 136). Thus, the ecological viability of Marx's thought is constrained by 'the "progressive" role he imparts to capitalism's "success" in supposedly achieving the technical domination of nature' (1980: 293).

Bookchin's critique focuses mainly upon Marx himself. Whilst dismissive of the neo-Marxist thought within which environmental Marxisms are broadly located, he does not (as I have here, and as Eckersley does) distinguish between varieties of environmental Marxism. One ecological Marxist, however, is singled out for the Bookchin 'treatment'. Andre Gorz's *Ecology as Politics* is dismissed as an incoherent 'pastiche', an attempt 'to refurbish an orthodox economistic Marxism with a new ecological anarchism' (1980: 291). But Gorz, claims Bookchin, understands neither 'ecology' nor the anarchist tradition from which he 'freely pilfers' (1980: 290). He is rather 'the tombstone to an era when revolutionaries took their ideas seriously; when they criticized their opponents with ruthless logic; when they demanded clarity, coherence, and insight' (1980: 290).

Responses to Marx and Marxism from anti-, non- or a-socialist perspectives within green thought also proliferate. Robyn Eckersley's is one of the more prominent of these. In a 1988 paper she compares Marx unfavourably with John Muir (taken by her as paradigmatic of a nature-preservationist ecocentrism). The 'problem' with Marx, she argues, is that he saw the fulfilment of humanness to lodge within the perpetual act of expressing oneself upon (transforming) nature (the 'external world'):

> we realise our humanity by transforming nature through our technology and productive activity. To conquer nature, rather than be subject to it, was seen to be our historical calling. Any form of reverence for nature, such as that displayed by primitive cultures, was considered to be childish and backward, for it hampered humanity's development (Eckersley 1988: 145).

The instrumental status of nature is reinforced in Marx's theory of value, wherein 'value' is created in the act of mixing one's labour with 'the external world'. Nature itself, though, 'is value free, it makes no normative claims upon us, it is raw material to be bent and transformed as an instrument of human labour' (Eckersley 1988: 145). The notion that humanity is 'at one with nature' ('man's inorganic body'), because we have 'made it over' as an 'expression' of ourselves, says Eckersley, is 'what Marx meant when he spoke of the "humanization of nature"' (1988: 145).

For Eckersley, then, 'despite their shared contempt for *laissez-faire* capitalism', the Marxist and ecological perspectives are ultimately irreconcilable. To privilege the human-defining act of 'making over' the natural world leads to a 'technological dream world' that will not only undermine the earth's biophysical processes but will fail to even realise its own more modest liberatory ends, as it will 'inevitably serve to alienate and enslave the masses' (1988: 146). The latter point is important. Eckersley rejects the left's charge that the placing of humans within wider ecological frameworks, and deriving ethical import therefrom, amounts to a rejection of the moral claims of the working class, or that it constitutes a form of misanthropy. This charge 'merely betrays the Left's narrow moral universe and outmoded picture of the human and non-human worlds as two mutually exclusive zones'. For Eckersley, a concern for nature, its components and its processes does *not* displace a concern for humans; it merely contextualises that concern: 'the Marxist circle of compassion (concern with social and economic justice) is but a subset of the Green circle of compassion (concern that all life forms be able to "live and blossom")' (1988: 147).

This position is developed at length in her 1992 book, *Environmentalism and Political Theory*. In some respects, Eckersley is kinder to Marx in 1992 than she had been in 1988. In 1988 she cited Marx to his great disadvantage, from the *Grundrisse*: 'for the first time, nature becomes purely an object for humankind, purely a matter of utility; ceases to be recognized as a power for itself; and the theoretical discovery of its autonomous laws appears merely as a ruse so as to subjugate it under human need' (cited in 1988: 145). In 1992 she deemed it a more complicated matter, noting that 'there are frequent passages in *Capital* where Marx observed how the dynamics of capital accumulation have led to the exploitation of the labourer and soil alike' (1992: 81; Foster draws upon these same passages to argue that Marx had 'a deep-seated understanding of the central ecological crisis of capitalist agriculture and of capitalist production generally'; 1997: 292). Engels, too, Eckersley observes, 'showed a keen awareness of the many unintended ecological dislocations brought about by the labouring activities of humans' (1992: 81).

But, the ambiguities notwithstanding, there is no getting round Marx's contrasting realms of freedom and necessity, 'the former corresponding to the mastery of social and natural constraints and the latter corresponding to subservience to social and natural constraints' (Eckersley 1992: 81). And this theme has endured within Marxism. It even endures within environmental Marxism. Gorz, for example, remained within its thrall. Eckersley instances his view of 'true freedom' as a leisure only attainable through the application of 'high technology or a high energy and material throughput'. This view, moreover, is misguided, for the enjoyment of 'ample creative leisure' does not require 'the technological subjugation of nature' (1992: 94; a similarly constrained view of freedom as 'free time' is attributed by Eckersley [1992: 212], following Schmidt 1971 and W.J. Booth 1989, to Marx himself).

Eckersley concludes that Marx cannot be enlisted to support the liberation of the non-human world, for 'Marx's notion of freedom as *mastery* achieved through struggle, as the subjugation of the external world through labour and its extension — technology — can be achieved only at the expense of the nonhuman world' (1992: 95). Such damning comment upon the 'profligate' productivism of Marxism is a prominent theme within green literature (for example, S. Bell 1992: 218–19; J. Clark 1989: 250–57; Jung 1983: 169; Redclift 1984: 6–8; Salleh 1997: 76). There is no great diversity, however, among environmentalist critiques that find Marxism unable to measure up on ecological grounds. Bookchin and Eckersley occupy different positions within environmental thought (and have aired their differences publicly), but it will be noted that the commonalities within their separate critiques of Marx are more salient than are the differences.

Anthony Giddens also sees the Promethean and 'nature as instrumental' themes within Marx's thought as being more fundamental than the tentatively ecological passages to be found in his early writings (1991: 165–66), but he extends his critique beyond Marx. He sees the transmutation of all facets of human life into aspects of production to be a position common to both reformist socialism and Marxism, and an important reason why neither is appropriate to a time in need of a coherent and effective radical politics: 'the welfare state confines itself largely to economic matters, and leaves other issues aside, including emotional, moral and cultural concerns. Thus, in general, socialists have been ill-prepared to cope with issues of life politics, although these are actually intrinsic to questions of welfare' (1994: 77). In his later work, Giddens's critique of both the Marxist legacy and the reformist tradition of socialism stresses stultifying bureaucratisation as the inevitable concomitant of the central economic co-ordination that is thought in both traditions necessary for the rational mobilisation of resources (1994: 62). In a green movement increasingly committed to realising the elusive promise of democracy, the old processes of rational bureaucratic planning, in which socialism placed such faith, can be expected to come under increasing attack (see Ehrenfeld's essay, 'The Overmanaged Society' for a forthright articulation of this position; 1993: 49–64).

Several similar analyses could be adduced, but these can suffice. It is rare, though, to have the door emphatically shut against Marx. Where most critical green analysis typically concedes Marxism some value is in Marx's explanation of how industrial capitalism works. I have myself argued that Marxism 'best interprets the nature of political power under advanced capitalism' (P.R. Hay 1988: 55), and even Bookchin, rarely inclined to generosity towards those he deems to have gotten it wrong, applauds Marx's 'extraordinary insights into the commodity relationship and accumulation process' (1990: 134) and his 'compelling demonstration that the very law of life of capitalist competition, of the fully developed market economy, is based on the maxim, "grow or die"' (1980: 293). Such acknowledgements have little persuasive effect, though, because, Eckersley maintains, 'it is not necessary to adopt a Marxist perspective in order to acknowledge the many ways in which the profit motive and the dynamics of capital accumulation have contributed to our currrent environmental ills' (1992: 76–77).

Many green activists are conspicuously Marxist. Most others, though, remain singularly unmoved, even suspicious of their Marxist collaborators. There are good reasons why this should be so. Marxists invariably seek to 'capture' environmental organisations for service in what is deemed the 'wider' revolutionary struggle. They are mostly unable to move beyond an analysis of the dynamics of capitalism as a *relation* of production to an analysis of the dynamics of industrialism as a *mode* of production. And Marx's historical materialism so ties history to the determinism of vast inexorable forces that it problematises the individual's activism. As I have written:

> when, as a Marxist, I subscribed to the notion that history was a ponderous and inexorable affair, that structural forces moved it along, and that the only action that mattered was the action of great slabs of humanity moving in mass, I could find no personal impetus for activism. What was the point? My puny little flailings would be of no account in the tides of history. Even my very support for materialist analysis was apparently not objectively defensible, but the product of the structural forces at work upon me (P.R. Hay 1992: 232).

Lastly, and perhaps most importantly, as *prescription* Marxism is now thoroughly discredited, the experiments carried out in its name having been systematically dismantled. Moreover:

> some of this discredit inevitably, though perhaps unjustly, spills across to analytical insights. To take up Marxism, even in part, is to risk the taint of obsolescence, even though it may well be possible to argue that, even if Marxism is to have no lasting legacy in the world of exemplary practice, the theoretical insights of its critique of capital may be as vital as ever (P.R. Hay 1992: 229).

I am, nonetheless, slightly out of step with some of the critics discussed above, and have written of the promise that seems to me to inhere within red–green dialogue (1988: 55). The ecology movement still struggles to construct a coherent praxis, and here the left has much to offer. 'It may be', I have argued, 'that some people within the green movement *essentially* over-value Marxism; overwhelmingly, on the other hand, it is *instrumentally* under-valued' (1992: 231). This is because Marxism best explains how capitalist society functions and reproduces; this being so, I would contend, against Eckersley, that it can provide vital guidance for the intermediate or strategic dimension of political envisioning, in which, by contrast with end-state and tactical thinking, the ecology movement noticeably lags.

Just as anarchist Murray Bookchin has mounted the most sustained attack on Marxism within ecological thought, so has Marxist David Pepper mounted the most forthright attack on green anarchism, including Bookchin's variety thereof. In Pepper's view, 'Bookchin replaces the concepts of class and exploitation by "hierarchy and domination" as ahistoric universals' (1993a: 138), whilst 'some of the social ecology rhetoric' is 'mystifying, if not pretentious' (1993a: 168). Nevertheless, though he perceives points at which Bookchin's thought is vulnerable, and though Bookchin regards himself as within the libertarian rather than Marxist tradition, Pepper clearly exempts Bookchin from most of the follies he attributes to green libertarianism. He even occasionally styles Bookchin a neo-Marxist (for example, 1993a: 147; given Bookchin's unrelenting assault upon the ecological credentials of Marxism one would like to know his reaction to such a tag!).

Eckersley's is a more thoroughgoing critique. She finds much of value in Bookchin; his is 'a radical ecological humanism' that nevertheless fails to deliver its emancipatory promise (1992: 160). In the first place, she argues, Bookchin's account of historical pathologies is too reductionist. She criticises the depiction of social hierarchy as the font of all social injustice and ecological disjunction. Social hierarchy 'can either enable or oppress, depending on circumstances'; it can even 'facilitate personal self-realisation'. She instances a monastery where members voluntarily submit to 'the authority of the abbot or Zen master as a means of facilitating their own self-realization'. It is more plausible, then, to see social hierarchy as 'merely a necessary (as distinct from sufficient) condition for social domination' (1992: 150). Nor is it the ultimate source of ecological degradation. Many hierarchical societies 'have lived in relative harmony with the non-human world' (1992: 151), whilst 'it is at least theoretically possible to have a society that is free of social hierarchy but that nevertheless dominates the non-human world through large scale technologies in order to minimize "necessary labor" (such as Marx's communist society)' (1992: 152). (Bookchin's own emancipatory project is different, of course; it employs 'democratically manageable technologies'.)

Eckersley is also critical of Bookchin's teleological view of evolution, the shaping of which is in the charge of evolution's stewards, *homo sapiens.* Eckersley forbears to say so, but these days the view of evolution as teleological attracts near-universal derision from evolutionary biologists. Moreover, Bookchin's view of evolution does not conduce to any sense of crisis. With humans 'playing an active and creative role in the evolution of the planet' (Eckersley 1992: 158; see also 1989), the 'progress' of life on earth is in the best of hands (for example, Bookchin 1982: 343). Bookchin has, thus, shown little concern for wild-area preservation or species extinction, though as we have seen, these may be *the* prime motivations behind a green commitment.

Bioregionalism also has its detractors, Bookchin himself not least among them. Bioregionalism is, in his view, mystical and much tainted by the soft mush of deep ecology (see his criticisms of Kirkpatrick Sale; 1988: 6). Pepper finds bioregionalism to merit less condemnation than more purely libertarian forms of green anarchism, but its romantic anti-urbanism is depicted as extraordinarily naive. 'The whole concept of "natural" regions' is 'increasingly untenable as it overlooks the reality of an urban oriented (if not based) population'. This being the case, 'where there still is regional differentiation it is based far less on watersheds than on economic function, and other criteria such as language and religion. To suggest re-identifying and reinstating old cultural, let alone physical, regions appears grossly unrealistic' (1993a: 187). Other recurring criticisms (for example, Allaby 1986; A.L. Booth 1997) are that bioregionalism, or at least Sale's version of it, sees no need for distributive mechanisms to assist less well-endowed bioregions; that, given the consequential discrepancies, it is not likely that bioregional communities will live together in peace and harmony and that war and instability are more likely outcomes; and that bioregional political systems are likely to succeed only if they are rigidly conformist and intolerant of dissent.

Challenging critiques also come from Eckersley and Plumwood. Eckersley notes, with Pepper, that 'linguistic, religious, and cultural boundaries do not necessarily follow bioregional lines', and that, moreover, 'ceding complete political autonomy to the existing local communities that inhabit bioregions will provide no guarantee that development will be ecologically benign or cooperative' (1992: 169). Plumwood argues similarly. She agrees that people living in near proximity to the ecological consequences of their activities 'should, other things being equal, be well motivated to make decisions that are ecologically benign', but the example of her own small community suggests that 'the proximity to local nature does little to guarantee the first condition of the bioregionalist, the transparency to inhabitants of ecological relationships and dependencies'. This is because 'other cultural factors' and 'the intractability of local economic and social relationships' can obscure these relationships and needs (1998: 567). It may be, for example, that the small political units advocated by bioregionalists will be insufficiently flexible to generate alternatives to ecologically impoverishing economic processes. In addition, a focus on the bioregional dimension cannot ensure cognisance of the more broadly regional and even global ramifications of local activities, and, 'given that contemporary ecological effects are rarely likely to be contained within a single political community', here is a dilemma for bioregionalism at the level of first principle, for it is axiomatic that 'those most affected by decisions should have a proportionate share in making them' (1998: 567).

As well as these criticisms of specific eco-anarchist formulations, there have been more broadly presented critiques of this, through the 1980s the most prominent strand within ecopolitical thought. According to Pepper, libertarian anarchists, the 'green-greens', fundamentally misconstrue the motive force of historical change. They are '*idealistic* — seeing social, economic and historical development as fashioned essentially from the introduction of new *ideas*, and from the development of old ideas and values, particularly at the level of the individual'. Against this, Pepper sets Marx's historical materialism, a perspective which holds that 'the predominant ideas which influence social change are not autonomously derived from abstract thinking and reasoning processes — rather they themselves are related strongly to the mode and organisation of economic life and to concrete events which have occurred in history' (1985: 16). From green anarchist idealism, Pepper argues (1993a), stems a multitude of sins: biocentrism and 'nature worship'; noble savage 'primitivism'; naive anti-urban and anti-industrial romanticism; an incapacity to appreciate that 'nature' has no objective reality but is socially constructed; alienating and reactionary mysticism; an almost narcissistic concern with the moral and spiritual wellbeing of the individual which conduces to political quietism instead of dedicated activism; a misplaced belief in hierarchy rather than class relations as the motor of ecological breakdown and social injustice.

Others have found problems with green endorsement of the anarchist commune. We have seen that Robert Paehlke perceives an ominous trajectory to this component of the green paradigm. A challenging case is also put by Michael Kenny, who finds strengths in communitarianism, but much that is misguided as well. Kenny argues that communal adherence to a 'right' ecological way is so solidaristic that it is difficult to see any capacity for accommodating the diverse values and interests that democratic vitality demands. 'Defence of minority rights could be submerged beneath the *gemeinschaft* logic of ecological communitarianism', whilst 'the sustainable world conjured up by ecologism can appear one-dimensional, dull and monolithic' (1996: 22). A narrowly prescriptive conception of the social good 'puts greens at odds with a range of other critical and radical theoretical currents, including feminism and the arguments of radical democrats', and 'it is hard to envisage ecologists connecting their strong political goals with the prevailing desire for social and cultural diversity and pluralism' (Kenny 1996: 23).

Eckersley also finds eco-anarchism generally wanting. 'Leaving it all to the locals', she argues, 'makes sense only when the locals possess an appropriate social and ecological consciousness' (1992: 173). In fact, 'while unilateral action by "right minded" citizens in local bioregions is to be encouraged, it will have minimal effect for as long as recalcitrant neighboring local communities and regions continue to "externalize" their environmental costs' (1992: 174). There is an overwhelming case, then, for some form of overarching authority, since that is the only way much of the green agenda can be delivered. This strongly accords with empirical observation: 'historically most progressive social reform and environmental changes have tended to emanate from more cosmopolitian central governments rather than provincial or local decision making bodies' (1992: 173). And the delivery of a global environmental agenda cannot be effected without national and transnational agency. The urgency of ecological crisis means that the environment movement simply cannot ignore the institutions of the liberal democratic state and immerse itself in the utopian speculation of eco-anarchism. She would agree with Kenny's observation that the idea of 'returning to a small-scale *gemeinschaft* living pattern cuts against the grain of modern society in so many ways that it is hard to imagine the profundity and depth of the cultural revolution which this shift would necessitate' (1996: 29).

And there's the rub. The luxury of end-state theoretical speculation exists for Bookchinites and ecoMarxists because they share an optimism that is a major stumbling block for large sections of the environment movement. Neither are imbued with the sense of urgency that helps shape Eckersley's position. Bookchin and the ecoMarxists envisage a future in which progress to an ecologically and socially benign future is being, or will be achieved. This is because both partake unproblematically of the Enlightenment belief in progressive human betterment; it is upon this axis, indeed, that Bookchin parts company with most other green-movement thinkers within the anarchist tradition. On almost every measure Pepper and Bookchin are poles apart — except that each remains staunchly progressivist.

The green movement in its mainstream has not been progressivist. It is suspicious of western technological 'achievement' and critical of the trend of historical unfolding. 'Progress' is neither inevitable nor to be defined along a one-dimensional axis of human material betterment. The green movement has stood outside and to some extent in opposition to Enlightenment progressivism. And it has a sense of crisis, a sense that there is a very short time in which the fundamentals of social existence must be turned around if lasting ecological damage is not the unavoidable consequence.

But a change is afoot, and it is this sense of urgency that has engendered it. Whilst it is unlikely that the environment movement will adopt the progressivist perspective urged upon it by Bookchin and Pepper, there has been a pronounced shift away from the utopian anarchism that marked green political thought in the 1970s and 1980s towards a less radical — some would say more realistic — politics of the here-and-now. For the same reason the 1980s focus on radical ecophilosophy has moved to political and economic theory, and the shift from first principles of environmental ethics to practical philosophies of politics and economics has been accompanied by a shift to the politics of the moment rather than to the politics of some remote future. These changes have muddied the waters as to where environmentalism stands on the 'progressive-modernism versus its opponents' axis, with an emergent view discernible that this is an increasingly irrelevant bifurcation (though the negative legacy of modernism remains with us) to the construction of a viable new political economy. The emerging politics will draw upon the modernist tradition as it will draw upon its persistent antithesis; it will favour neither the rational nor the spiritual, seeing both perceptual modes as valid, if not hegemonically so. And in the realm of political theory change is taking the form of a new interest in the terms and conditions of green democracy. This is where much of western environmentalism's intellectual energy is now directed, and it is where we now turn.

10
SEEKING *HOMO ECOLOGICUS*: ECOLOGY, DEMOCRACY, POSTMODERNISM

How should we engage politically, we who urge the primacy of ecological imperatives? Variations on this question have largely directed green inquiry in recent times. Identifying (and, as deemed necessary, constructing) processes and structures that would conduce to an ecological being-in-the-world has become the priority. Here we consider recent theories of how to 'be', politically, at a time of worsening ecological crisis. Two foci have emerged during the 1990s, and they point in different directions.

The first seeks to contextualise ecological values within theories and practices of democracy (real and potential). It is a development from, and in part a reaction against the utopian eco-anarchism of the 1980s, and takes the form of wary engagement with 'real' western democracy and the tradition of liberal democratic thought. In the process it has also involved — almost as by-product — a reconsideration of environmentalism's hostility to the value heritage of progressive modernism.

The second is found in the rub of ecological social thought against the new theoretical orthodoxies that loosely and unsatisfactorily coalesce under the rubric of 'postmodernism'. Here, the tendency is to concentrate on the condition of the individual *qua* individual rather than individuals in combination. Ethical holism is rejected, as is much programmatic environmental thought, on the ground that these tend to be *totalising*: they impose unacceptable external constraint upon 'subjectivity'. Postmodern environmentalisms are, thus, somewhat more modest than those against which they react.

CRITIQUING 'ACTUALLY EXISTING DEMOCRACY'

We have seen that the 'problem' of democracy and ecology was an early preoccupation within green thought. The inability (or unwillingness) of liberal democracy to resolve environmental problems was early recognised, and drew two predominant responses. In the 'classical' period of environmental thought, what was seen to be the 'corrupt' capture of democratic institutions by partial interests dedicated to subversion of the general good led to an ecopolitics of centralised authoritarianism, as exemplified in the 1970s writings of Hardin, Ophuls and Heilbroner. In the 1980s ecologically benign dictatorship was, in turn, rejected in favour of its very opposite — utopian, small-scale communitarianism, based upon principles of 'strong' participatory democracy. In the 1990s this, too, largely fell from favour as green theorists turned a more complex critical eye upon the theory and practice of 'actually existing' democracy.

It must be said that activist prescription has not followed the lead of theory, the orthodox grass-roots position remaining one of support for communitarian democracy (for example, Lappe 1999), and such principles continue to shape activist organisational design (on activist structures and processes, see Pakulski 1991: 25). Some theorists also still espouse the decentralist, 'strong democracy' principles ascendant in the 1980s. 'The negation of an unrepresentative, centralized state is a decentralized participatory democracy', Alan Carter writes. Furthermore, 'the urgent need to inhibit the environmentally hazardous dynamic' is 'the most compelling reason' to implement such principles as 'decentralization, participatory democracy, self-sufficiency, egalitarianism' (1993: 54). A more usual 1990s response is to retain the emphasis upon participation, whilst seeing no innate incompatibility between this and the institutions of liberal democracy; the argument being that these institutions, their shortcomings notwithstanding, remain the ones most likely to produce requisite environmental outcomes. Thus, Robert Paehlke argues that 'the fate of environmental protection and the quality of democracy will be very much intertwined in the future'. This is because 'environmental dangers have reached such complexity and magnitude that they now intersect with questions of social justice, world peace, and global economic development', and 'the simultaneous handling of all these challenges will require both intelligence and enhanced democratic institutions'. Paehlke makes some practical suggestions to guide this 'enhancement': to 'introduce an environmental role within all governmental subdivisions', and to 'expand the environmental powers and roles of municipal and regional governments'. With such measures in place 'an effective pluralist democratic system is the best source of balanced, participatory initiatives' (1998: 159; see also 1988; 1990; 1996. Marius de Geus also argues against 'radical decentralisation', advocating instead 'co-ordinated and well-designed reforming steps'; 1996: 209). But there are also those who, remaining sceptical of the greening capacity of liberal democratic institutions, argue for more radical democratic alternatives, though rejecting the communitarian utopianism of the 1980s, and perhaps even envisaging 'strong' democratic alternatives arising through evolution from existing liberal democratic regimes.

This, though, is the business of the next sub-section. The question here is, why has a critical focus even arisen? Why is it that, from the environmentalist perspective, current democratic arrangements seem to fail so comprehensively?

At the checklist level several reasons are advanced to account for this. One is that short electoral timespans significantly disadvantage ecological imperatives. Not only do environmental problems emerge over a long timespan, but, more to the point, their resolution is also unlikely to be effected within a timescale that enables a government to claim election-influencing credits therefrom. Furthermore, the successful resolution of an environmental problem typically results in an *absence* of visible consequence; electorally potent credits are expected to be more tangible, on the other hand, to have a recognisable economic or physical manifestation. Some also argue that, although there is a clear public interest in the maintenance of environmental health and ecological integrity, invariably this will run counter to certain environmentally irresponsible special interests, and these interests, being more than normally influential, will do their utmost to undermine the public's interest in environment-regarding decision-making. A variation of this argument points to manipulation by the mass media, arguing a community of economic interest between the media and capitalist enterprises engaged in environmentally destructive activities. Yet another argument focuses upon the adversarial nature of liberal democracy: one in which institutions are constructed to favour trade-offs between 'stakeholders' rather than a disinterested search for a general interest.

Several theorists have examined the perceived failure of liberal democracy in greater depth. Even those who argue the need to stay with liberal democratic discourse tend to envision a thoroughgoing institutional makeover. Eckersley, for example, wishes to stay with liberal democracy, but not necessarily with the nation state, within which its institutions have evolved (1992: 183). Inevitably, though, the most damning critiques of liberal democratic institutions have been made by those who seek to move beyond them.

For Plumwood, as we have seen, the difficulty experienced by liberal democracy in responding to environmental crisis stems from the 'liberal' part of the equation. She notes that the mediation-between-interests concept of liberal democracy 'is unable to create stable measures for the protection of nature, or to recognise basic ecological priority' even though 'ecological well-being is not just another interest-group concern but ultimately a condition for most other interests' (1998: 569–70). This is because 'the liberal-individualist model stresses a view of politics as the aggregation of self-interested individual preferences', and is thus unable to adequately accommodate a wider, non-individuated interest, even when that interest is of surpassing importance. Environmental 'goods' are collective goods, and in the liberal-individualist discourse such goods 'are not well treated'. Here, 'an unquantifiable, highly diffused, generalisable and perhaps not easily detectable ecological harm is pitted in a political contest against a quantifiable economic benefit accruing to a highly concentrated and influential group'. Even where some ameliorative gains are won under this process of interest brokerage, they 'rarely halt the overall progress of ecological damage', and thus 'it is very difficult to maintain environmental values over the long haul' (1998: 570; see also Lafferty and Meadowcroft 1996). And finally, Plumwood argues, liberal democracy — and again this stems from the *liberal* insistence that its institutions are neutral conduits within which individual interest-claims are adjudicated — rejects the legitimacy of any claim for intercession to rectify the uneven social distribution of 'ecoharms'. Such an unjust distribution stems directly from the correlation between inequality in individual access to the economic market and inequality in political influence within liberal democratic structures (1998: 570–73).

Plumwood's analysis closely accords with John Dryzek's. Dryzek (1996b: 14) rejects Eckersley's view that the need is less for liberal democratic institutional change than for a change in the cultural assumptions within which these institutions operate. And this is so, according to Dryzek, even though 'more progress has been made in dealing with environmental problems in the world's democracies than in any other political system', and even though 'there is little doubt that liberal democracy resolves ecological problems more effectively than the market with which it coexists, and which passes on to liberal democracy the environmental problems it generates but cannot resolve' (1992: 22). Concerning the environmentalist agenda, 'there are good reasons why dominant political mechanisms cannot adopt and implement that programme, or even substantial chunks of it'; this is because a complex system will embody imperatives which 'constitute values that the system will seek. Other values will be downplayed or ignored' (1996b: 15). The most significant of the 'imperatives' within which liberal democracy functions are those of the capitalist market, and they make their presence felt by 'punishing' any government seen to threaten market profitability:

> any state operating in the context of such a system is greatly constrained in terms of the kinds of policies it can pursue. Policies that damage business profitability — or are even perceived as likely to damage that profitability — are automatically punished by the recoil of the market. Disinvestment here means economic downturn. And such downturn is bad for governments because it both reduces the tax revenue for the

> schemes those governments want to pursue (including environmental restoration), and reduces the popularity of the government in the eyes of the voters (1996b: 15; see also 1992: 22–23; 1996c: 12. Dryzek draws upon the influential analysis of Lindblom 1977; 1982).

There is no conspiracy at work here; this outcome is built into the logic of the system, and it will occur regardless of the intentions of system actors. Thus, 'the first task of any liberal democratic state must always be to secure and maintain profitable conditions for business' (1996b: 15). In this way, liberal democracy comes to have as much stake in endless economic growth as the market. 'If growth ceases', writes Dryzek, 'then distributional inequities become more apparent, and the "political solvent" of economic growth is no longer available. This fear of economic downturn means that liberal democracies are imprisoned by the market's growth imperative' (1992: 23).

There is a second important component of Dryzek's case against the greening capacity of liberal democracy. In addition to political structures, ecological imperatives must struggle to obtain policy recognition against the imperatives of a market economy and an adminstrative state. The problem is one of rationality, Dryzek argues. On the face of it, there is nothing new here. We have already noted that, through the 1980s, environmental thought self-located in opposition to the dominant tradition of Enlightenment rationalism, with only Bookchin and the ecoMarxists rooted within the dominant tradition of progressive modernism. But it was also noted that the shift in focus within environmental thought to questions of 'actually existing democracy' has somewhat complicated this configuration, and Dryzek's work has been pivotal here. He, too, stands within the progressivist tradition, arguing that the problem lies not with rationality as such, but with certain dominant modes of rationality. Rationality is not unitary; not indivisible. Each of the three intersecting loci of power — the economic, the political, the administrative — corresponds to a specific rationality. These are largely (though not completely) compatible with each other, though they have very little fit with an ecological rationality.

The rationality of the market is a specifically economic rationality, one which identifies growth and profit as the goals of rational behaviour. By contrast, Dryzek identifies the dominant rationalities within the administrative realm as technical and narrowly instrumental; they are rationalities of means rather than ends (the latter are tacit, assumed, unexamined). Regarding environmental imperatives, the problem with instrumental rationality is that 'its orientation is a manipulative one', which 'translates all too easily into human arrogance in dealing with natural systems'. In the logic of instrumental rationality then, 'nature consists of nothing more than objects to be manipulated and dominated' (1996c: 28), and it is unthinkable that this process could *cause* crisis, because it is the very mechanism for achieving human *betterment.*

Dryzek would agree with Douglas Torgerson:

> although environmental concern achieved considerable acceptance, the perception of crisis was resisted and rejected by the administrative sphere; indeed, to perceive a fundamental flaw in this whole project of industrialization would be to question the *raison d'être* of this institutional complex. Accordingly, those articulating this sense of crisis were ridiculed — and ridiculed fairly easily since the rationalistic imagery of order and progress was at hand to help in portraying as emotional and irrational those who perceived a crisis (1990: 134).

Moving from the market and administrative realms to political structures finds no improvement. The political rationality of liberal democracy 'is that of facilitating arrival at effective collective decisions' (Bartlett 1990: 83), a rationality in which 'solutions to policy problems are normally reached on the basis of compromise between interests' (Dryzek 1996d: 113), and these can be 'quite arbitrary or destructive when viewed in an ecological light'. In fact:

> rarely will ecological and political rationality coincide. It was politically rational for the Clinton administration to backpedal on its commitments to reform grazing on the public lands of the American West, for fear of losing a few key states in the 1996 presidential election; but it is ecologically disastrous (1996d: 113).

With the instrumental rationalities of the political, administrative and market realms sharing enough common ground to be mutually sustaining, it is difficult to see how a non-instrumental rationality might compete. Though sufficiently positive to advocate alternatives, Dryzek pessimistically observes: 'in this light it is not hard to imagine a future in which the ecological irrationality of the capitalist market is compounded by that of liberal democratic politics, and that all will be revealed when eventually some (global?) ecological limits are reached' (1996d: 113; see also Blake 1992: 199).

Another view of the shortcomings of liberal democracy — one that complements those already outlined — is provided by Ulrich Beck. For Beck the technological subversion of democracy is already complete and therein lie the seeds of a differently fashioned democratic politics, one that goes far beyond the liberal version, which was, in any case, limited from the start (1992: 191). But there is no inevitability about such a development. The continuous production of life-menacing hazards, which Beck sees as the defining characteristic of technocratic modernity — 'risk society' — results in 'totally new types of challenges to democracy':

> it [risk society] harbors a tendency to a legitimate totalitarianism of hazard prevention, which takes the right to prevent the worst and, in an all too familiar manner, creates something even worse. The political 'side effects' of civilization's 'side effects' threaten the continued existence of

> the democratic political system. That system is caught in the unpleasant dilemma of either failing in the face of systematically-produced hazards, or suspending fundamental democratic principles through the addition of authoritarian, repressive 'buttresses'. Breaking through this alternative is among the essential tasks of democratic thought and action in the already apparent future of the risk society (1992: 80).

For Beck, a new 'ecological democracy' is 'a future to which some current social trends point, the achievement or prevention of which can be affected by conscious collective awareness and choice' (Dryzek 1996d: 116). The question is, then, 'completely open in both theory and practice' (Beck 1995b: 151).

A new-minted democracy is not inevitable — but why is it even necessary? What is the nature of the 'risk society' that, in Beck's view, has so emphatically destroyed all but the outward semblance of liberal democracy?

For Beck, the institutions of industrial society — created as the vehicles of progressive, problem-solving modernity — have become problem-causing. 'Large-scale hazards of civilization', he writes, 'came into the world as opportunities, with the blessing of science and technology' (1995b: 76); thus, it is 'mainly industry, in conjunction with science, that is involved in the creation of the risk society's risks' (Lash and Wynne 1992: 3). Beck describes this as a 'defection from the side of problem solution to the side of problem causes' (1997: 52).

Two aspects of this development conspire to bring on 'risk society'. One is the sheer, unprecedented magnitude of contemporary hazard. Beck defines a 'risk society' as one faced with the possibility, 'hidden at first, then increasingly apparent', of the destruction 'of all life on this earth' (1995b: 67). He elaborates the distinction between 'risk now' and 'risk then', thus:

> since the middle of this century the social institutions of industrial society have been confronted with the historically unprecedented possibility of the destruction through decision-making of all life on this planet. This distinguishes our epoch not only from the early phase of the industrial revolution, but also from all other cultures and social forms, no matter how diverse and contradictory these may have been in detail. If a fire breaks out, the fire brigade comes; if a traffic accident occurs, the insurance pays. This interplay between beforehand and afterwards, between the future and security in the here and now, because precautions have been taken even for the worst imaginable case, has been revoked in the age of nuclear, chemical and genetic technology. In all the brilliance of their perfection, nuclear power plants have suspended the principle of insurance not only in the economic, but also in the medical, psychological, cultural, and religious sense (1998: 330).

The insurance metaphor is not fancifully chosen. Beck uses it to validate his case according to the standards of common sense: 'a very rough distinction is provided by the principle of private insurability. Wherever insurance companies step out of the picture or refuse to enter in the first place, based on their internal standards of economic rationality, the alarm systems for uncontrollable effects light up' (1997: 52).

The second causal component of 'risk society' is the failure of institutions to respond effectually to hazard on this scale. Part of the problem lies in the failure of risk calculus, as implied in the insurance metaphor:

> nuclear, chemical, genetic, and ecological mega-hazards abolish the four pillars of the calculus of risks. First, one is concerned here with global, often irreparable damage that can no longer be limited; the concept of monetary compensation therefore fails. Second, precautionary after-care is excluded from the worst imaginable accident in the case of fatal hazards; the security concept of anticipatory monitoring of results fails. Third, the 'accident' loses its delimitations in time and space, and therefore its meaning. It becomes an event with a beginning and no end; an 'open-ended festival' of creeping, galloping, and overlapping waves of destruction. But that implies: standards of normality, measuring procedures and therefore the basis for calculating the hazards are abolished; incomparable entities are compared and calculation turns into obfuscation (1998: 330).

Another part of the problem stems from the risk-denying logic of the institutions themselves, which remain trapped within their nineteenth century problem-solving ideologies. This leads to a systematic 'normalisation' of hazard shocks, in which hazard is absorbed within a self-serving belief that 'lessons have been learned' and the institutions of modernity are back in control. Beck has labelled this 'organized irresponsibility' (1995b: 63–65). It is a situation which 'ultimately turns that which controls the production of hazards — law, science, administration, policy — into its accomplices' (1995b: 160). One consequence is that 'historically new hazards' are legitimised as economic and political institutions abrogate the functions of guidance and control by telling themselves (and us) that existing mechanisms can cope (1995b: 80). Doubts about these assurances bring on a crisis of trust. As Christoff points out, in the construction of risk society reality and perception are equal partners: risk society is 'a society which has shed the romantic gloss of productivist progress and replaced it with scepticism about the benefits of science and technology, and an anxiety about the ecological and social consequences of its actions' (1996b: 166–67). The crisis of perception, then, is a crisis of trust in regulatory administration, in the science/technology which has generated hazards on such a scale but has proven inept at finding solutions (Dryzek 1992: 155–82) — and in liberal democracy.

And so we return to where we started. Liberal democracy, having ceded technical authority to technologists (and to spurious military imperatives; Beck 1997: 77–81), has abnegated responsibility for the most important realm of decision-making: that concerning priority-setting in science and technological research and the subsequent deployment of technologies in the real world. 'Democracy in industrial society', Beck writes, 'rests on the fiction that technological decisions of industry cannot nullify and modify the foundations of social coexistence and cooperation. Consequently they do not require special articulation and consent' (1997: 41). It amounts to a crucial 'closing off of the scope for decision-making in the parliament and the executive' (1992: 188), and we are left with a 'truncated democracy', one in which 'questions of the technological change of society remain beyond the reach of political-parliamentary decision-making' (1998: 342) — and one in which there is declining public faith.

TOWARD AN ECOLOGICALLY INFORMED DEMOCRACY

It is one thing to point out the ecological failings of liberal democracy. It is another thing entirely to devise democratic structures that can be sourced to ecological principles. The ecocentrism of radical ecophilosophy is particularly difficult to translate into democratic political practice. 'A democratically-attained majority', I have written, 'is only a register of human preferences' (P.R. Hay 1994b: 17), and it seems likely that it must ever be so (though, as we see below, attempts are being made to overcome this difficulty). And Albert Weale (reviewing Eckersley's *Environmentalism and Political Theory*, wherein a defence of both ecocentrically based and rights-based liberal democracy is presented) insists that 'within democratic institutions ... choices will necessarily be made solely by humans. Democracy is a discursive practice and other species, let alone types of non-living entities, are not able to participate in that practice' (1993: 342). John Barry is as forthright, arguing that a concern for trans-species relationships displaces 'the primacy of the political' in green action, and that, 'from a strictly ecocentric point of view democracy is superfluous at worst or an optional extra' (1994: 371).

Though Eckersley rarely uses the term 'ecocentrism' in her later writings, she has remained committed to a view of 'centres of agency' within nature (for example, 1993c), whilst developing the view that a radical green commitment is, in significant part, compatible with liberal democratic imperatives. It would be wrong to conclude, however, that hers is not a radical democratic theory. It is that. On this account it may not be entirely fair to take her, as I now intend to do, as exemplar of the growing number of theorists arguing that the realisation of environmentalist aspirations should be pursued through the medium of liberal democracy (such a 'category' includes contributors as diverse — even, in some cases, as mutually exclusive — as de Geus 1996; Gundersen 1995; B.M. Hayward 1996; Mathews 1996; Mills 1996; Paehlke 1988; 1990; 1996; 1998; and Saward 1993; 1996).

The institutions of liberal democracy are, Eckersley argues, adaptable to ecological (and more authentically human) ends, but much needs to change. As currently constituted, the liberal democratic state systematically discounts ecological concerns, because it 'represents the existing citizens of territorially bounded political communities', and thus denies the interests of what Eckersley calls 'the new environmental constituency' — species and unborn humans whose interests are significantly affected by political decisions, 'but who cannot vote or otherwise participate in political deliberations' (1996b: 214). Liberal democracy also undervalues public interest advocacy because it assumes a politics based on the interplay of conflicting selfish interests. Thus, environmental advocacy groups are not seen as voices for long-term *generalisable* interests, but as 'sectional' or '*vested*'; their claims, therefore, are reduced to legitimately compromised interests, 'systematically traded-off against the more immediate demands of capital' (1996b: 215).

Eckersley calls this 'the bias of liberal democratic framing devices' (1996b: 215). Her most challenging case against liberal orthodoxy concerns its basic claim to 'procedural neutrality', a claim which, if it stands, must pronounce that, 'whatever the ecological consequences, people must be "free" to make ecologically bad decisions' (1996b: 212). But Eckersley's explication of 'the bias of liberal democratic framing devices' insists that liberal democracy is neutral in neither theory nor practice, being constituted in deference to the sectional interests of industrial society and against the interests of 'the new environmental constituency'.

All this is very damning. But it will be recalled that Eckersley finds in the rights discourse a promising means for extending protection to non-human entities. She brings this to bear on liberal democracy, and finds a lifeline for it therein.

The attraction of extending the rights discourse to ecological interests lies in the non-negotiability of claims that have been transformed into *rights* and removed thereby from cost-benefit calculus and the brokerage of sectional interests (1996b: 216). In 1996 Eckersley published two papers on this subject. In the earlier of the two she argues (as we have seen) that 'an extension of the rights discourse to ecological interests' can only work up to a point, because 'it becomes considerably strained and unworkable' in relation to 'non-human and non-domesticated constituents of the biotic community'. When it comes to 'the human obligation to the broader biotic community', she prefers 'the systematic application of the precautionary principle and more general sustainability planning' (1996a: 193).

In the later paper Eckersley focuses upon human environmental rights, and, whilst warning that 'environmental rights are not offered as a panacea for the green movement or for democracy', here she is much more positive. 'Autonomy' and 'justice' are identified as liberal democracy's base principles, and their customary formulation has it that 'the liberal principle of autonomy respects the rights of individuals to determine their own affairs; the liberal principle of justice demands that this respect be accorded each and every individual' (1996b: 222). But Eckersley finds this interpretation flawed; 'an arbitrary and self-serving way in which notions of "inherent dignity" and "value" have been reserved exclusively for humankind' (1996b: 212). Ecological embodiment gives the lie to it, and institutions and procedures must be adjusted accordingly. She posits a reconceptualised 'autonomy' — 'the freedom of human and non-human beings to unfold in their own ways according to their "species life"' (1996b: 223) — with the following result:

> if we are to give moral priority to the autonomy and integrity of members of both the human and non-human community, then we must accord the same moral priority to the material conditions (including bodily and ecological conditions) that enable that autonomy to be exercised ... humans, both individually and collectively, have a moral responsibility to live their lives in ways that permit the flourishing and well-being of both human and non-human life. This more inclusive notion of autonomy would necessarily involve the 'reading down' or realignment of a range of 'liberal freedoms' in ways that are consistent with ecological sustainability and the maintenance of biodiversity (1996b: 223).

The problem is that, in the liberal tradition, rights are ascribed exclusively to individuals. But 'individual interests do not always coincide with the interests of ecological wholes', and 'it is the broader network of social and ecological relationships that should be the proper field of concern' (1996b: 226; Melucci had earlier argued for a 'democracy of everyday life', in which was envisaged 'the generation of certain "rights of everyday life" such as ... individuals' biological and affective dimensions, and the survival of the planet'; 1988: 258–59). This would not, Eckersley stresses, obliterate individual rights; it would rather call for 'a mediation between these two mutually constitutive and co-evolving realms' (1996b: 227). This could be done by establishing a notion of individual rights in such a way that 'such rights belong to individuals not only as individuals but also as members of social and ecological communities: infringement of individual rights would also be an infringement of collective social and ecological interests' (1996b: 227). As individual rights cannot, in any case, 'be abstracted from their social and ecological context' (1996b: 228), not only *can* this be done, it *should* be done. Some progress to this end has already been made through the legal device of the 'class action', wherein an infringement of an individual right is also seen to be an infringement of a collective right.

Despite her earlier disclaimer, such a formulation would seem to go quite far towards providing a basis for rights within the 'broader biotic community'. But, though she may have overstated the obstacles to applying her rights discourse to non-human nature, a 'democratic deficit' certainly remains. In a more recent paper, Eckersley (2000) has considered the problems of non-representation and externalising risks to those who do not participate, propounding an 'inclusive and ecumenical' 'democracy-for-the-affected' — *for* the affected because the morally relevant constituency in democratic deliberation will, on most issues, be much greater than the sum of those actually participating. The 'demos', thus, becomes a 'community of fate' gathered together, not by common nationality, but 'simply by the potential to be harmed', its boundaries shifting according to the issue (2000: 119). This poses, Eckersley acknowledges, 'complex moral, epistemological and institutional challenges', and the gauntlet thrown down to liberal democratic theory and practice is to meet these challenges (2000: 120). She herself offers signposts to this end.

John Dryzek, as we have seen, is much less optimistic about the extent to which liberal democracy can be harnessed to ecological ends. He is the most prominent champion of the ecological possibilities seen to inhere within 'deliberative', 'discursive' or 'strong' democracy, wherein is emphasised a degree of participation and inclusivist discourse that may well be beyond the capacity of 'merely' representative institutions.

Dryzek defines the process of ecological democratisation as 'any enhancement of democratic values in an ecological context that does not sacrifice ecological values, or any enhancement of ecological values that does not sacrifice democratic values' (1996d: 108). The two in tandem is to be preferred of course, but movement on either dimension brings closer the goal of ecological democracy. Democratic values are enhanced if participation is increased, if the range of issues subject to democratic decision is increased, and if there is movement from 'symbolic' to 'substantive', from 'ignorant' to 'informed', and to greater actor competence. Ecological values are enhanced if 'the content of politics becomes increasingly sensitive to human interests in a clean, safe and pleasing environment', or if 'the content of politics becomes less anthropocentric and more biocentric' (1996d: 109).

Dryzek sees no impediment to incorporating the latter discourse within his rationalist framework. In our interactions with nature we find 'a variety of levels and kinds of communication to which we might try to adapt'. The need is to seek 'the kinds of interactions that might occur across the boundaries between humanity and nature', and 'in this spirit, the search for green democracy can indeed involve looking for progressively less anthropocentric political forms. For democracy can exist not only among humans, but also in human dealings with the natural world' (1996b: 18). But he is dismissive of the anti-rationalist mainstream of environmental philosophy wherein biocentric discourse is usually encountered, arguing that, 'provided our notions of rationality are expanded in the right direction, human dealings with the environment are indeed best governed by rational standards' and that 'a regressive emphasis on spirit is therefore unnecessary' (1990a: 199; see also 1997: 155–71). He also criticises 'romantic' ecophilosophy as providing no steps to a practical politics, 'no effective theory of transition' (1990a: 201), because its concentration on change at the level of the individual rather than social structure is deeply misguided: 'why does social structure matter? The main reason is that macro consequences (in terms of policies, institutions, and events such as revolutions) are rarely if ever a simple extrapolation of micro causes'. This given, 'even if there were large-scale conversions of individuals along the lines sought by green romantics, it is quite possible that nothing at all would change at the macro level' (1997: 170).

The principles outlined above neither prescribe nor proscribe, on the face of it, any particular institutional modes; but, as we have seen, Dryzek has little faith in the institutions and processes of 'conventional' liberal democracy when it comes to recognising the priority claims of ecological rationality. It may be that there is still potential for further democratic enhancement in liberal capitalism, though 'certain areas, notably the economy and national security, remain off-limits', and this limitation 'applies with particular force to ecological democracy so long as economic and ecological values stand, or are seen to stand, in a zero-sum conflict' (1996d: 119). All this given, 'civil society is attractive as a far less constrained place in which to seek deeper democratization'. The problem here is that, whilst civil society becomes 'the location of authentic egalitarian discourse' and a site in which appropriate recognition is accorded ecological rationality, 'the administrative state would still grind out policies insensitive to the play of civil society' (1996d: 119).

Dryzek's solution is to so construct civil society that it becomes not merely adjacent but *oppositional* to the state: it can be the forum of protest against state policy, and it can also nurture activities that involve 'reclaiming from the state control over particular areas of life' (1996d: 120). To avoid co-option, it is important that oppositional spheres within civil society keep their distance from the state. Rather than 'seeking a share in state power', such a politics must generate 'autonomous forms of association oriented towards, but separate from, state institutions'. Dryzek envisages the creation of self-resourced 'public spheres', constituted around 'particular risks, social problems or policy issues', and active at all levels from the local to the global (1996d: 120). Needless to say, their activities will be informed by an understanding of ecological rationality and a recognition of its priority over other rationalities.

Much of Dryzek's work seeks to establish the principles for discourse within public spheres. Here he draws upon the critical theory of Jurgen Habermas, particularly the latter's concept of *communicative rationality*. Habermas distinguishes between instrumental and communicative rationality — norms for non-manipulative discourse between free and competent subjects (1984; 1987). 'Communicative rationality', writes Dryzek, enjoins intersubjective discourse that is 'free from deception, self-deception, strategic behavior, and domination through the exercise of power' (1990b: 14). He makes much of Habermas's 'ideal speech situation', wherein discussion:

> is thoroughly unconstrained; there are no restrictions on who may participate, or on what kinds of arguments may be advanced, or on the length of deliberations. The only resource available to participants is argument, and the only authority is that of the better argument. It is crucial, too, that all participants possess roughly equal degrees of 'communicative competence' (Dryzek 1987: 201).

As Dryzek notes, 'communicative rationality constitutes the model for a democracy that is deliberative rather than strategic in character' (1996b: 20; see also 1990b).

Habermas himself (1981) dismissed the notion that communicative reason could mediate trans-species relations, on the ground that only humans can claim subjectivity, and that, given the communication deficit of non-humans, the latter can never be equal partners in free discourse (see Eckersley 1990; 1992: 97–117 for a rejection, on this account, of the value of Habermas to green thought; for a development and qualification of this position, see T. Hayward 1995: 44–48; for a critical treatment of Habermas as environmentalist on other grounds, see Goldblatt 1996: 112–53).

Dryzek, however, sees considerable merit in 'rescuing communicative rationality from Habermas', and the key to this is the distinction between 'subjectivity' and 'agency' (1996b: 20):

> the key would be to treat communication, and so communicative rationality, as extending to entities that can act as agents, even though they lack the self-awareness that connotes subjectivity. Agency is not the same as subjectivity, and only the former need be sought in nature. Habermas treats nature as though it were brute matter. But nature is not passive, inert, and plastic. Instead this world is truly alive, and pervaded with meanings (1996b: 20).

This is not a matter of verbal communication, which is, of course, impossible; but non-verbal trans-species signals are there for the reading: 'the content of such communication', Dryzek writes, involves 'attention to feedback signals emanating from natural systems' (1996b: 24). Acknowledgement of agency in nature means that 'we should treat signals emanating from the natural world with the same respect we accord signals emanating from human subjects' (1996b: 21). Linking such a view of an 'ecologised' communicative rationality back to human discursive processes, Dryzek writes:

> communicative reason can underwrite a particular kind of democracy in purely human affairs — one that is discursive or deliberative in character,

> whose essence is talk and scrutiny of the interests common to a group of people ... But of course non-human entities cannot talk, and nor should they be anthropomorphised by giving them rights against us or preferences to be incorporated in utilitarian calculation, still less votes. However, there are senses in which nature can communicate (1996b: 22).

As a human process informed by signals from nature, Dryzek's ecological democracy can be viewed as 'a regulative ideal'. Just as the regulative ideal for liberal democracy is 'fairness and efficiency in preference allocation', for deliberative democrats the regulative ideal is 'free discourse about issues and interests'; for that variant of deliberative democracy Dryzek labels 'ecological democracy', the regulative ideal is 'effectiveness in communication that transcends the boundary of the human world', in which case 'the practical challenge when it comes to institutional design becomes one of dismantling barriers to such communication'. Dryzek counsels against blueprinting, merely noting that 'the design of such a democracy should itself be discursive, democratic and sensitive to ecological signals' (1996b: 24), and that the key 'watchword' for this is 'appropriate scale', because some feedback signals register at the local level, and some are global: 'an ecological democracy would, then, contain numerous and cross-cutting *loci* of political authority' (1996b: 26).

How are these multi-scaled loci to be co-ordinated? Not by the state, because Dryzek is concerned with 'the possibility of democratisation apart from and against established authority' (1996b: 27; see also 1996a). It is here that the notion of the public sphere becomes so important. Under the assault of the market and technocratic liberalism the very notion of the public sphere — 'political bodies that do not exist as part of formal political authority' (1996b: 27) — is under attack. These 'political bodies' must be defended. Their institutional unconstrainedness provides a flexibility and fluidity absent from the institutions of state, liberal democracy and the capitalist market; because of this, they are uniquely placed to respond to whatever the issue is, and at whatever is the appropriate level (1996b: 28). And, of course, the notion of the public sphere is also uniquely fitted to realise the principles of communicative rationality and ecological democracy.

Though the concept of the public sphere is integral to Dryzek's thought, its foremost emerging champion is the Canadian, Douglas Torgerson. In *The Promise of Green Politics: Environmentalism and the Public Sphere* (1999), Torgerson provides a penetrating exposition of the linkages between instrumentalist and end-state thought in ecopolitical theory. One of the book's most prominent arguments is that green thought (and practice) should value more highly the enterprise of politics itself. The impulse to cut democratic corners under the imperatives of crisis should be resisted. There should, instead, be a shift of emphasis from green goals to green process, and it is here that the idea of the public sphere becomes so important. Torgerson argues for a 'green public sphere', an 'ethically concerned communication-community ... informed by an ecological ethos' (1999: 124), and characterised by openness, fluidity and diversity. To some extent this has already occurred:

> what environmentalism has most significantly created in the prevailing context is a manner of speaking about the environment that was previously not possible — a range of discursive practices, expressive of green concerns, that allows environmental problems to be recognized, defined, and discussed in meaningful ways. The green movement continues to construct a green discourse and to shape a forum for communication, a green public sphere. Even with its many internal differences ... the emerging green public sphere poses a challenge to the once comfortable framework of industrial discourse (1999: xi).

Here is a plea for greens to adopt open dialogue between a plurality of voices to be accepted as an end in itself. It is a plea I will endorse when offering, in closing the book, some thoughts by way of conclusion.

We have noted Beck's case against liberal democracy. What does he envisage in its place? Beck argues that, faced with the enormity of contemporary hazard, liberal democracy has, along with the other institutions of modernity, abandoned its *reflexive* (self-aware, self-critical) capacity. This position has been developed by Tim Hayward, who identifies parallel tensions within ecological and Enlightenment thought. Hayward argues that each has a tendency to 'the merely dogmatic and arbitrary', producing the sort of error 'that creeps in with the assumptions made before thinking even begins' (1995: 40). In Enlightenment thought this has led to the certainties that conduce to domination. But the Enlightenment tradition began as a fear-nought quest 'to get at the truth about *reality*' (T. Hayward 1995: 52). It needs to reconstitute this critical or reflexive tradition; and 'if the Enlightenment project is understood as one of critique rather than of domination it is not necessarily anti-ecological' (1995: 40). Indeed, 'if ecology is to challenge conventional wisdom, then in so doing it partakes of the *critical* legacy of Enlightenment thought itself' (1995: 52). Thus, like Dryzek, Hayward remains within the Enlightenment paradigm, and so does Beck, despite his attack upon the institutional corruption of modernity. He, too, seeks a revitalised modernity, one in which the self-critical faculty is rehabilitated. To achieve this, institutions without a critical capacity will need to pass away or be downgraded in significance. Beck seeks, in particular, an end to 'the fixation on the political system as the exclusive centre of politics' (1992: 187).

A reflexive democracy will require, if it is to come about (remember that these developments are not inevitable), broad-based social critique extending beyond institutional elites, and a consciously chosen agenda of political action, gathering up the policy spheres abandoned by liberal democracy, science and technology. The vacuum left by discredited scientific authority provides space for extending the democratic scope. Thus, a 'discursive' politics replaces a politics of discredited expertise, its seat no longer within large, inflexible institutions, but within the looser and more mobile structures of civil society. Like Dryzek, then, Beck is a theorist of 'deliberative' or 'strong' democracy.

As Dryzek describes it, 'reflexive modernization as portrayed by Beck entails democratization in society beyond the state, though there is no necessary hostility to the state, whose own democratization is at issue too' (1996d: 117). What is needed is 'public dispute on technological alternatives' (Beck 1998: 342), leading to 'different organisational forms' for science and politics. Within the public sphere 'dissenting voices' and the 'alternative experts' of a 'public science' are prescribed for the revitalisation of the institutions of the state, though Beck clearly envisions this revitalisation stopping short of returning those institutions to the positions of dominance they have hitherto enjoyed. 'Only a strong, competent public debate "armed" with scientific arguments is capable', Beck states, of 'allowing the institutions for directing democracy — politics and the law — to reconquer the power of their own judgement' (1998: 343). Here, Beck envisages rather more constructive interpenetration of civil society and state than does Dryzek, who enjoins the former to keep its distance from the latter. 'My only real objection to Beck's account', Dryzek writes, 'is that it holds no defence against the competing developmental construct of risk *management*' (1996d: 116). Only radical separation can, for Dryzek, guarantee against co-optative strategems.

Beck nevertheless advocates what might be called a process of 'saturated' democratisation. Acknowledging the debt owed the women's movement for stressing the political dimension of the personal sphere (1997: 41), he argues for the democratisation of all civic spheres — family, business, industry, labour, science. Noting 'the impossibility of limiting the basic rights of the citizen to a single field (that of the state and politics, for instance)', Beck sees a transformation of society as the consequence of 'collisions and syntheses of logics specific to particular fields' (1997: 46).

This shift in the focus of democratic activity from the formal institutions of liberal democracy to groups in the public sphere Beck describes as a shift from politics to what he calls *subpolitics* — groups and movements located 'outside the political or corporatist system' who also demand the right 'to appear on the stage of social design'. The realm of subpolitics includes the public sphere and 'citizen initiative groups' that openly construct themselves in opposition to formal structures, but also 'professional and occupational groups, the technical intelligensia in companies, research institutions and management' (1997: 103). The dice are not loaded, then: anything might yet happen. Beck reminds us that 'subpolitics is not open to only one side. This opportunity to fill a vacuum can always be seized and used by the opposite side or party for the opposite goals' (1997: 101). The point is, though, that 'we look for politics in the wrong place, with the wrong terms, on the wrong floors of offices and on the wrong pages of newspapers' (1997: 99).

Beck refers to the subversion of the formal institutional sphere by a self-critical, thoroughly politicised public sphere as 'reflexive democracy'. But what is there, in the ideas we have considered to date, to merit their inclusion in a book on ecological thought? There is a twofold answer. In the first place, the hazards that Beck sees to be so destabilising of trust in science, bureaucracy and liberal democracy manifest primarily in various ecological disequilibria. And, as an unsurprising consequence, it is the environment movement that has most obviously and promisingly led the challenge of the subpolitical realm against institutionalised technocracy. Beck does think that environmentalists get some things wrong; for instance, given his location within the progressivist tradition, he is predictably critical of strands of environment movement thought that posit a realm of nature that is separate from, and not defined in relation to the social (1995b: 40). But pointing out 'threats to nature' also threatens 'property, capital, jobs, trade union power, the economic foundation of whole sectors and regions, and the structure of nation states and global markets' (1998: 337). In effecting the transition from risk society to reflexive democracy, the politics of ecology is the obvious place to start. Thus, Beck is as wont to use the phrase 'ecological democracy' as 'reflexive democracy', and 'the call for ecological democracy' is 'elementally important and urgent' (1997: 93).

We have considered some of the more prominent attempts to re-theorise democracy from within the environmentalist value-set, but we have also noted that this is currently the epicentre of green intellectual preoccupation, and contributions extend across the gamut of positions outlined at the beginning of the chapter. The central point of dispute is, of course, the question of representation versus unmediated participation. At the Paehlke/Eckersley end of the spectrum are those such as Mike Mills, for whom the formal machinery of government is adaptable to a degree that it would be folly to reject it. And this even applies to ecocentric and holistic values. Democratic polity 'cannot ensure green outcomes, but it can ensure a green political process', because 'anything which works to diversify the political community, to expand its moral constituency, to open up the number of political opportunity structures for that constituency's interests, and which encourages tolerance and compassion in decision making will, to a greater or lesser extent, promote some aspect of green democracy' (1996: 112).

Andrew Dobson (1996b) goes still further: he has developed actual prescriptions for representing 'the environmental constituency'. He notes the inadequacy of existing institutional arrangements when it comes to the democratic goods of responsibility and autonomy — in respect to environmental despoliation that spills across jurisdictions, and the autonomous interests of other species and future generations of humans. In this light Dobson considers both a case for transnational parliaments (theoretically sound, practically difficult) and the use of proxies to represent the interests of unborn humans and other life-forms. The case for proxies is not the same in each instance. It is easier to make a case for proxy representation for unborn humans; nevertheless, as animal-rights theorists have shown, 'the human/other species divide is porous: if humans have rights (including political rights) then at least some animals have them too' (1996b: 135). Dobson foreshadows and rejects opposing arguments, calling instead for a 'proxy electorate' (the 'sustainability lobby'), from which potential representatives would seek election, and to which they would subsequently account, electorally, for their defence-of-an-interest record.

There is, though, a sense in which the representation versus participation bifurcation is a false one. Dobson, a proponent of electoral change to enable the proxy representation of environmental interests, nevertheless finds much in the discursive tradition to recommend it, in particular that it enables green and democratic thought to be intrinsically rather than merely instrumentally linked, via the promise it holds for the realisation of autonomy (1996c: 134–46). He sees no contradiction in this. Participation is transformative (1996c: 141): one becomes an educated, 'other regarding', environmental citizen through the act of participation, and this has transformative flow-ons for formal structures. Even Eckersley, the most prominent green critic of Habermas-based discursive ecological democracy, is really only a critic at the level of discursive democracy's perceived opposition to rights-based democracy. She, too, wishes to expand participative opportunities within the polity, observing that there is 'much to be gained from developing more deliberative forms of democracy in state institutions and civil society — both as a supplement or in some cases an alternative to liberal democracy' (Eckersley 1996a: 178). And she has devoted a recent paper to effecting 'a thoroughgoing greening of discursive democracy' (1999: 25).

This notion that effective and responsible citizenship is the outcome of high rates of political participation, though not a specifically green position, has attained particular prominence within green democratic thought, and goes far towards bridging the representation/strong democracy divide. Thus, John Barry (1996), a proponent of discursive theory, nevertheless agrees that the aim of green democracy is to cultivate a certain form of citizenship, one entailing responsibilities and values that reach beyond the institutional machinery of democracy, so that representative democratic institutions 'can be supplemented with more discursive institutional forms and greater citizen involvement in political and non-political spheres' (1996: 122). Peter Christoff also stresses the need for a concept of ecological citizenship ('*homo ecologicus*'), one that would require us 'to defend the rights of future generations and other species just as we are morally obliged, and increasingly legally required, to consider and protect the rights of those humans who cannot be defined as "morally competent"' (1996b: 159). This means that humans 'must assume responsibility for future humans and other species and "represent" their interests and potential choices according to the duties of environmental stewardship' (1996b: 159). Strong democracy is needed to deal with ecologically complex problems as well as to foster the ethic of ecological citizenship that Christoff envisages; but, whilst in favour of a state-transcending, strong democracy, Christoff argues for the continued relevance of constitutionality and legally based processes (1996b: 158).

As a further complication, cracks have also opened *within* ecologically informed strong democracy. There are other versions of deliberative democracy besides those of Dryzek and Beck. Tim Hayward's is reached after exhaustive critical journeying; even then, he holds his support for deliberative democracy conditional upon a successful reconciliation of the apparently rival needs of participatory theory and the reality of social pluralism (1995: 202–06). Others build upon Iris Marion Young's (1990; 1995) criticism of deliberative democracy's notion of free debate via 'ideal speech' on the ground that power differentials, even intangible ones, always privilege some participants over others. Young wishes to treat a range of communicative forms, not just critical argument, as equally valid, and to allow for difference at the level of the group. Bronwyn Hayward (1996) is one who finds much of value to green political thought in Young's theories (Eckersley, on the other hand, believes that Young's argument does not threaten the fundamentals of discursive democracy and is assimilable within it; 2000: 124).

This branch of environmental thought continues to burgeon at bewildering pace; we can expect further development, further fragmentation.

ENVIRONMENTALISM MEETS POSTMODERNISM: A DIFFICULT SYNERGY?

Postmodernism is notoriously difficult to define. Oelschlaeger points to 'a multiplicity of meanings within the natural and social sciences as well as architecture, literature, art and the humanities', and concludes that, 'although some common threads run across postmodern thought, such as the opposition to essentialism (the idea that there are timeless, universal truths), it cannot be claimed that a postmodern paradigm exists' (1995: 1). Zimmerman also finds postmodern thought 'difficult to define', but sees, among its core ideas, 'a denial of both objective truth and an independent reality to which true statements supposedly correspond', and a rejection of 'modernity's notion that history has any direction, much less a progressive one' (1996: 132).

I am on a hiding to nothing here. Whatever I say will attract howls of protest. But I will nominate these as the core 'truths' of the postmodern perspective:

- that ideas, linguistically encoded, are all that there is, or at least all that anyone can hope to apperceive — no objective reality is to be sought behind the concept. 'There is nothing outside the text', Derrida has famously written (1976: 158), because 'the thing itself is a sign' (1976: 49).
- that ideas, 'truth claims' and pedagogic stances not only serve to mask relations of power; in fact they *constitute* that power. Dominant ideas, thus, mask/constitute the interests of dominant elites.
- that there is no 'final and changeless truth' that cannot be sourced to the exercise of power and authority. Foucault writes: 'truth is a thing of this world; it is produced only by virtue of multiple forms of constraint. And it induces regular effects of power. Each society has its own regime of truth, its general politics of truth: that is the type of discourse which it accepts and makes function as true' (1980: 131).
- that the focus of resistance should be *for* the encumbered, disempowered, un-free human and *against* totalising systems of power and the 'grand narratives' of modernity (science, Marxism, Christianity, liberal capitalism, progressive modernism ... environmentalism) that sustain them.
- that there is no simple and immutable subjectivity — all human identity shifts through time, and even at any one time, across contexts. The human is a 'de-centred' subject.

Postmodernist ideas made their powerful impact within western thought in the final three decades of the twentieth century, which was, too, the period in which the ecological crisis flooded public awareness and spawned its own prominent theoretical corpus. That ecologism and postmodernism largely avoided serious collision until the 1990s is to be marvelled at. As Gare puts it:

> it is arguable that the postmodern condition, associated as it is with a loss of faith in modernity, progress and enlightenment rationality, reflects people's awareness that it is just these cultural forms which are propelling humanity to self-destruction. Proponents of postmodernist politics usually endorse ecological resistance as a new form of politics. But the fragmentation of experience, disorientation and loss of overarching perspectives and grand narratives associated with postmodernity are threats

> to the efforts of environmentalists who are struggling to proselytize a global perspective on environmental destruction. Clearly postmodernism and environmentalism are of great significance to each other. Yet little effort has been made to relate the discourse on postmodernity with the discourse on the environmental crisis (1995: 1–2).

Gare's final statement is less true now than it was in 1995. There are now several postmodern environmentalisms — as well as several postmodernist rejections of same. Even the founders of postmodernism, though having little to say directly about environmentalism or environmental crisis, have been brought into comparative environmentalist context. Eric Darier, for example, has explicated the perceived green credentials of Michel Foucault (1926–84). Whilst Foucault personally 'detested nature', his writings 'are having profound, albeit indirect, effects on environmental thinking' (1999a: 6). Darier instances Foucault's concept of 'biopower', a celebration of biological science as 'the entry of life into history' (1999a: 12); his ideas of space and place (1999a: 23–25; see also Chaloupka and Cawley 1993; Heyd 1999; Quigley 1995; 1999); and his 'non-ethics', in which the focus is on endlessly reconstituted subjectivities 'which constantly rework humans' relations with themselves, with other life-forms and with the world generally' (1999a: 27; 1999b)

In Foucault's case one has to search for environmentalist resonances. But this is not the case with the poststructuralists Gilles Deleuze and Felix Guattari. Patrick Hayden has made a convincing case for Deleuze as 'ecological postmodernist', attributing to him 'an ecologically informed perspective that emphasizes the human place within nature while encouraging awareness of and respect for the differences of interconnected life on the planet' (1997: 185). As with all the 'high postmodernists', Deleuze rejects the immutability of 'invariant essences'; but, whereas 'the term *naturalism* is rarely if ever encountered in the writings of poststructuralists, and even then usually appears only as an object of hostile interest' — 'naturalism' being deemed synonymous with 'essentialism' — Deleuze 'promotes a type of naturalism that highlights the diverse connections between human and nonhuman life' (Hayden 1997: 186). Much of the focus of Deleuze's work (see especially 1990) is upon human interaction with the natural world, and the key to the compatibility between such a focus and the postmodernist rejection of invariant essences will be apparent in this passage: 'each individual is an infinite multiplicity and the whole of nature is a multiplicity of perfectly individuated multiplicities' (Deleuze and Guattari 1988: 254). Nature perpetually shifts and re-forms — in much the same way that human subjectivity is constantly on the move, constantly shifting ground and re-forming. Deleuze, in short, 'ecologises' the postmodern position on subjectivity. Nature 'produces itself through new combinations of its heterogeneous elements' (Hayden

1997: 185), though it is not an unchanging essence conferring value.

Ecological concerns are also foregrounded in Guattari's work: one of his books, indeed, is entitled *The Three Ecologies* (1990). The ecologies he delineates — social, environmental and mental — correspond to the three linked crises of social and ecological breakdown and increasing regimentation of thought and action. Resistance must be both discrete to each domain and also across the domains, and it is needed at both macro level (where connections between problems become comparatively important), and micro level (where problems and conditions unique to each domain will be more in evidence). Guattari, like Deleuze, advocates 'the creative proliferation of new value-systems, alternative modes of subjectivity, and innovative human and nonhuman relationships and forms of alliance' (Hayden 1997: 204).

Much of the most important work of Deleuze and Guattari has been collaborative, and in their joint work the 'ecologism' of their thought is further developed. In *What Is Philosophy?* they propose a notion of 'geophilosophy', arguing that 'thinking takes place in the relationship of territory and the earth' (1994: 85). De- and re-territorialisation occur as new combinations of process and matter emerge and disappear. The earth is both a constant and a context for flux. It is 'the fundamental yet never fixed plane of immanence on which the constitution of multiplicities takes place' (Deleuze and Guattari 1988: 34). Here is an ecological view that retains the postmodernist stress on difference, impermanence, and particularity. It does not amount to, nor permit holism. Nature is not transcendent: it is an intricacy of symbiotic interconnection which constantly unravels. There are no intrinsic values, no immutable essences. And activism must recognise this. It must understand local problems and possibilities, make its pitch accordingly, and avoid abstracting from the local. A macropolitics *is* needed, but it will emerge from the interaction of a multiplicity of micropolitics.

Deleuze and Guattari would seem to have cleared the decks for fruitful ecology-postmodernist collaboration, yet few environmentalists have shown anything but deep suspicion of postmodernist ideas. Zimmerman is an exception and and we consider him below. Raymond Rogers (1994) draws upon Foucault, Derrida and Baudrillard in an ecological critique of modernity that strongly affirms radical environmentalism, but he also draws upon thinkers who are anathema to postmodernism, such as Marx. Stephen Rainbow also affirms the relevance of postmodernism to the green project, its 'tolerance of ambiguity and diversity' a necessary corrective to the 'totalizing schemes' of the Marxist strand in ecopolitical thought. It is postmodernism's 'viewing the world as a plurality of heterogeneous spaces and temporalities' that is most attractive to Rainbow. He argues, not for a 'green society', but for a 'greener society, where the green agenda is a pivotal component of the ongoing political process which incorporates a variety of inputs from the varied groups and interests in society'. In contrast to some of the postmodernist theorists noted below, Rainbow even evokes deep ecology founder Arne Naess as a 'radical pluralist' (1993: 92–93).

Others could be adduced. But the postmodernism-onto-ecology grafting that seems to have had the largest impact within environmental thought is Jim Cheney's 1989 essay, 'Postmodern Environmental Ethics: Ethics as Bioregional Narrative'. We saw earlier that Cheney finds, in postmodernism's insistence that voices from the margin should be defended against the falsely universalised ideologies (the 'privileged discourse') of dominant elites, a trend of thought that runs adjacent to, and can be brought to bear upon environmentalist understandings of place. Place is the ground whereon occurs 'the construction of understandings of self, community, and world' (1989: 117), and these constructions body forth in the myth, 'storied residence' and 'ethical vernacular' of 'bioregional narrative', which is place-particular, known to be so, and, thus, avoids the trap of abstracting to universals (1989: 124–34; echoes of Deleuze and Guattari here). Places lend themselves to postmodern metaphors like 'textual communities' because they are unique. They cannot be generalised beyond themselves because place-meaning is subjective to the observer.

But for those who insist that environmental thought 'bend' to fit postmodernism's own non-negotiable 'truth claims', Cheney's postmodernist bioregional narrative will still not do. The marginal voices he champions are, says Quigley, 'politically important', but not because of any insights intrinsic to these voices: they are politically important only 'because they are abused by power'. At day's end, Cheney's is still another 'traditional environmental posture', in which recourse to 'vague' principles 'establishes yet another hierarchical power arrangement' (Quigley 1995: 184).

Quigley's voice is of the type most usually found at the intersection of postmodern and ecological thought. It is a voice concerned less to explore possible synergies between two vital schemes of thought than to prove one (environmentalism) irredeemable by demonstrating how it fails to measure up against the precepts of the other (postmodernism) — presumably on the assumption that the postmodernist critique has already consigned nonconforming paradigms to history's dustbin. Thus, the task is not to seek symbioses; it is to push erroneous environmentalism aside, clearing the decks for a postmodern environmentalism that is discontinuous with what has gone before. Darier (1999a; 1999b); Quigley (1995; 1999) and Van Wyck (1997) fit this bill. All see the essentialising of 'nature' — nature as immutable essence; as the objective ground of value — as irreparably contaminating ecological thought.

Darier rejects ethics-in-nature, though he does put a green ethics in its place, an 'ethics of resistance', which derives from a process of 'constant self-critique'. How is this 'green'? Well, it is not, necessarily; it is only so conducively:

> it is through practical opposition to a new landfill site or an incinerator that individuals and communities may start questioning the conditions which have led to a 'garbage crisis'. Household recycling can be one technical alternative which transforms individual subjectivity from 'wasteful' consumer to recycling or Green consumer. However it could also lead one to re-question the entire process of consumerism, and why and how individuals are seduced by it (1999b: 234).

Peter Van Wyck has a slightly narrower target. He tackles deep ecology, and he devotes an entire book to it (1997). At the end of the demolition he posits a 'weak ecology', or 'an ecology of weakness'. It is 'weak' in the sense that it avoids 'strong' prescription. In language vague and poetic Van Wyck describes his 'weak ecology' as 'self-critical', and 'from its self-critical position it can produce possibilities not imaginable from the positions of the shallow and the deep'. These possibilities are not specified. To do so would, of course, transgress the open, non-prescriptive nature of the project: 'the agency implied by a weak ecology consists in working through the softness and fluidity of boundaries in order to reconstruct lives and meaning, identities and affinities, that make for a future of possibility'. Then, though, the territory becomes slightly more familiar: 'it positions theory within an ethics of responsibility, of care', which requires that one be 'constantly aware of the responsibility implied in the production of any knowledge' (1997: 134).

Extensive use is made by Van Wyck of the work of Donna Haraway, a virtually unclassifiable theorist — which is how she would want it — who straddles feminist, postmodernist and socialist thought. Haraway's ideas have some currency within environmentalism, though her status therein is decidedly ambiguous.

Haraway is best-known for her cyborg metaphor. In science fantasy the 'cybernetic organism' is a human–machine hybrid, and Haraway uses it as a device to signify dissolution of boundary categories, particularly the culture–nature binary. The cyborg is 'the offspring of both biology and technology, organism and machine' (Van Wyck 1997: 112), such that 'the one can no longer be the resource for appropriation or incorporation by the other' (Haraway 1985: 67). Though she sees herself as 'a-modern' rather than postmodern, her vision for human possibility seems distinctly postmodern. It lies in 'taking the risk of imaginatively exploring multiple identities that transgress the boundaries between the human, the natural, and the mechanical, thereby opening up unexpected alternatives to the deadly path forged by technological society' (Zimmerman 1994: 356).

It will be clear that the cyborg is us, our unavoidable becoming. Haraway urges us 'to cease clinging to our identity as gendered, organic and human, and to participate in the dangerous, boundary-crossing technological play that is our destiny' (Zimmerman 1994: 363), because, conversant as she is 'with the trends of high-tech capitalism', she knows that 'there is no way to stop the unfolding of astonishing new technological configurations' (Zimmerman 1994: 356). And, if we join the game, we may be able to subvert the long-standing 'appropriation of nature as a resource for the productions of culture' (Haraway 1991: 150). Hers is not an injunction to disengage, then; indeed, old-fashioned concepts like 'social justice' are prominent within her work (as Harvey points out, 'the effect of the postmodern critique of universalism has been to render any application of the concept of social justice problematic'; 1996: 342). But the nature she defends is not the nature of radical ecology. This nature is not a purity to defend against human depradation; it is not an 'essence to be saved or violated' (Haraway 1992: 296). Nature is, rather, constructed in dialectic with people (literally — she is not talking semantics here). Though this does not put 'all agency firmly on the side of humanity' (1992: 304), Haraway's 'social nature' nevertheless dissolves an attitudinal opposition prominent within environmentalism: cyborgs 'are sympathetic to the holist protest against the technocratic orientation of the world' (J. Bennett 1993: 254), but they are *not* on that account either holists or technophobes, because 'taking responsibility for the social relations of science means refusing an anti-science metaphysics, a demonology of technology' (Haraway 1985: 100). There is no going back. To confront ecocide, we, too, must be technologically adept cyborgs.

I have drawn upon Michael Zimmerman in constructing a sketch of Haraway's ideas. Zimmerman's is an interesting and, until recently, anomalous position in this debate, for he partakes of both radical ecological thought and postmodern critique in constructing his preferred position, and his 1994 book, *Contesting Earth's Future: Radical Ecology and Postmodernity* is the most exhaustive examination yet of real and potential intersections between these two supposed incompatibilities. It is true that, to reach an accommodation with postmodern thought, Zimmerman found it necessary to retreat in large part from the deep ecology position he had hitherto held. He wishes to 'encourage radical ecologists to take into account the political dangers posed by movements seeking to improve humanity's relation to "nature"' (1994: 7). Even though he remains 'a friend', 'postmodernism has disclosed to me the limits of radical ecology' (1994: 374; somewhat paradoxically, it also seems to have kindled in him a grudging support for the institutional fabric of modernity; 1994: 6–7).

What Zimmerman takes from postmodernism is its rejection of fixed essences. He champions a spirituality that affirms both nature and human life 'but absolutizes neither' — that works towards 'maintaining the biosphere on which all terrestrial life depends' whilst accepting that 'nothing permanent is "out there" in the external world since nothing permanent is "in here" in the internal world' (1996: 139–40). But Zimmerman is finally unwilling 'to abandon the idea that there is some direction to cosmic history, including human affairs'. Though he assimilates postmodern influences, he declares for the 'perennial philosophy': belief in a progressive evolution to higher states of consciousness, a process in which the universe generates 'ever more complex forms of awareness'. 'There is a purpose', he writes, 'to the unfolding infinite complexity of the universe' (1994: 373). Zimmerman's perennial philosophy may be 'significantly influenced by critical postmodernism and radical ecology', but here is a teleological faith that few postmodernists could accept.

Zimmerman's position, or at least variations upon it, seems lately to be gaining adherents — though some of these 'qualified constructivists' arrive at their positions via ideational pathways that others who reach a similar position would not wish to follow. It may even be that constructivism is not necessarily irreconcilable with ecocentric philosophy: Mick Smith, for one, has explicitly argued that 'deep ecology has much in common with ... some varieties of constructivism' (1999: 359). Others have also argued persuasively for a rejection of *extreme* constructivism but in favour of a *mild* constructivism, one that can accommodate the 'real' of nature (for example, Peterson 1999; Soper 1995; 1996; Spretnak 1997). Thus, Soper distinguishes between 'nature-endorsing' and 'nature-sceptical' positions, arguing that ecologist and constructivist 'are committed to projects which are in principle mutually supportive' (1996: 23), whilst Spretnak contrasts 'deconstructionist postmodernism' with an 'ecological postmodernism' that 'replaces groundlessness with groundedness, supplanting freedom *from* nature with freedom *in* nature' (1997: 72).

We should finally note the postmodernist sally into environmental thought that has most effectively hit its mark. Probably because they are not much known or read within environmentalist circles, attacks by postmodern theorists on radical ecology have made little impression. But when a distinguished professor of environmental history brought out the proceedings of a 1994 symposium of humanities academics on 'Reinventing Nature', and it became apparent that, under the guidance of that professor, discussion had taken a decidedly critical and postmodern turn, contest was joined. The professor was William Cronon and the book *Uncommon Ground* and it is Cronon's own contributions that have attracted most attention. In his 'Introduction' he sets out the standard postmodern take on nature, 'a profoundly human construction'. This is 'not to say that the nonhuman world is somehow unreal as a mere figment of our imagination', but the way we understand the world 'is so tangled with our own values and assumptions that the two can never be separated', and we can never directly know 'the "nature" we seek to understand and protect' (1995a: 25).

As this perspective is foreign to the environment movement, 'concern about the environment often implicitly appeals to a kind of naive realism ... assuming that we can pretty easily recognise nature when we see it and thereby make uncomplicated choices between natural things, which are good, and unnatural things, which are bad'. Environmentalism's moral force 'flows from its appeal to nature as a stable external source of nonhuman values against which human actions can be judged without ambiguity' (1995a: 25–26). In his other contribution Cronon applies this case to the political use to which the concept of wilderness is put, arguing for an end to the conceptual split between sanctified wilderness and the less exalted nature around us. It is likely that Cronon's case excited particular attention because he wages his argument in conceptual territory that is familiar within ecological discourse, is of a nature to stir passions, and is of clear and direct relevance to both policy and activism. And it is clear, too, that this is not the argument of a nature-alienated sophist, for Cronon writes with passion for 'the wilderness that dwells everywhere within and around us' (1995b: 89).

SOME PROBLEMS: 'GOVERNMENTALITY' VS ECOLOGICALLY CONSTITUTED DEMOCRACY

It would seem that ecological political theory, with its present focus on the ecological citizen *collectively* constituted in democratic interaction, has little in common with postmodern environmentalism. There, the focus remains on freeing individual 'subjectivities' from the constraints that are perceived to inhere when humans gather — or are gathered — into real and abstracted collectivities. Dobson notes that 'green politics has much more obviously called into question the modernizing strategies of advanced industrial societies than it has the central tenets of modernity itself' (1993: 208), and the 1990s focus upon democratic theory has sharpened consideration of where environmentalism stands in relation to progressive modernism. There is a trend — not followed by all — to now situate environmentalism at a (highly!) critical *insider* position in relation to modernism. Postmodernism, by contrast, views the institutions, ideologies and values of progressive modernism as freedom-denying and totalitarian. Darier's observation says it all: 'the challenge to environmental activism is not to establish a binding "ecological rationality", with even more powerful instruments of control and management, but to acknowledge human freedom' (1999b: 238).

Foucault posited 'governmentality' as a major obstacle to the realisation of human freedom. This is a notion with shades of meaning but it most commonly designates a discourse of legitimacy for the state, its institutions and its interventions (Foucault 1991: 102–03; for a critique of 'ecological governmentality' as constituted by environment movement demands upon the state and the responses to these demands from state and non-state actors, see Rutherford 1994; 1999a; 1999b). Though this is usually presented as non-pejorative analysis, it is clear that we are supposed to disapprove of 'governmentality'.

Postmodernism's prioritisation of human freedom/subjectivity and its suspicion of collectivist organisation seems to conduce to a liberal anarchism. In the same way this suspicion may also bias postmodernism against the public sphere as a site of possibility, this being a discourse of collective action for the socially construed betterment of the human condition — in the process of which participants constitute themselves as free and autonomous citizens — rather than one focused upon individual subjectivity. We must confine ourselves here to reactions to postmodernist ideas that are specific to the environment movement. It is clear, however, that these will dovetail with the recurrent left-progressivist criticism that postmodernism cannot explain how ecological and social pathologies are produced in the first place (for example, Norris 1990), and that, stripped of its anti-modernism, it shares such affinity with liberal discourse (for example, Sayer 1995: 231–34) that it can legitimately be viewed as being the equivalent in the realm of culture to free-market liberalism in the realm of economics. Environmental thought, by contrast, is split over the role of the state — green democracy theorists particularly so — but many will agree that 'government is important', because 'the goals of sustainability and social justice require new forms of economic intervention, regulation, taxation and public spending. Moreover they require these not just at a national level but internationally, and indeed locally' (Jacobs, M./Real World Coalition 1996: 111).

Turning specifically to theories of green democracy, some critics of ecological discourse have argued that green politics is intrinsically authoritarian: that it could only achieve its ends via a degree of coercion and regimentation that would be unacceptable in advanced, open societies (for example, McHallam 1991; Passmore 1993). This is not a negligible position; it is passed over with reluctance here because, apart from Hardin's increasingly marginal voice, it has little currency within environmental thought itself. Therein, the possibility of democracy is largely assumed, and debate centres around the terms and conditions a green democracy should take.

Two dissenting voices have prominently opposed the project of theorising a specifically green democracy.

In 1992 Robert Goodin argued that what is special and basic within green thought is its 'theory of value'; its case that value lodges intrinsically within nature. The 'green theory of agency' is subsidiary to this core theory of value; indeed, Goodin argues that much of the theory of agency is seriously flawed, in that it fails to further the agenda demanded by the green theory of value. Such preferred political modes as decentralisation, local control, flat or destructured organisations, and radical 'alternative' lifestyles have hindered rather than assisted the furtherance of green value. 'It cannot be true, at one and the same time', Goodin writes, 'that green issues ought to be of pressing concern because only concerted action is capable of solving them, but that the best way of responding to those issues is to devolve all decision-making powers down to very small political units who lack any way of acting in concert with one another' (1992; 168; see also T. Hayward 1995: 181, 193). In fact, there is no reason why any form of agency should, at the level of principle, be preferred to any other. The choice of agency should be determined by what delivers: 'it is more important that the right things be done', he insists, 'than that they be done in any particular way or through any particular agency' (1992: 120). Goodin personally favours parliamentary institutions; but, as Doherty and de Geus note, that he does so is entirely serendipitous, for 'parliamentary democracy may be desirable in itself', but that is 'separate from the issue of valuing and defending the natural world' (1996: 8).

The project of theorising a specifically green democracy has also been rejected by Michael Saward. Dobson notes, almost incidentally, that Goodin and Saward reach 'roughly the same place' from 'somewhat different' approaches (1996c: 133). In fact, Goodin and Saward proceed from *diametrically opposed* assumptions. Saward *would* concur with Goodin's observation that 'to advocate democracy is to advocate procedures, to advocate environmentalism is to advocate substantive outcomes' (Goodin 1992: 168). The difference is that, for Goodin, the 'substantive outcomes' take such clear priority that a green commitment to democracy must be merely contingent, whereas for Saward the demonstrable procedural superiority of democracy — liberal democracy — over all other decision-making modes is such that he advises greens to treat their core values more 'flexibly'. He urges a 'revision of the green approach to knowledge', abandoning 'foundationalist myths of intrinsic value', and to 'embrace uncertainty' and 'constant self-interrogation' (Saward 1993: 77).

Why should instrinsic value be jettisoned? Because if value is *intrinsic* within nature, it must take precedence over any merely contingent value, which is how democracy, being procedural, must be accounted: 'something of intrinsic value has, in the first instance at least, more value than something valued for contingent reasons. Such values cannot be overridden by something for which only instrumental justifications can be offered' (Saward 1993: 66). The same difficulty pertains to the extreme urgency with which environmentalists imbue ecological crisis. To insist that environmental problems are of such unsurpassed urgency logically suggests that any effective agency is justified, including non-democratic ones. This is Goodin's position, of course, but Saward recoils from it in horror. He presses the need to accord priority to democracy rather than green imperatives, mounting the familiar defence of the former: no one possesses permanently superior insight; only decisions that accord with majority views are defensible and legitimate; minority rights must be protected, and allowed the freedom to pitch for majority status. Thus, governments can only pursue environmental goals if 'a majority of citizens votes for them' (Saward 1996: 79), and 'greens have little comeback if a majority does not want green outcomes' (1996: 93). And there is little point trying to fashion a specifically green theory of democracy, because democracy is a 'self-sufficient political concept', and, whilst there is a 'natural compatibility between liberalism and democracy', this 'does not obtain between ecologism and democracy' (1993: 69). Time, then, to desist from the construction of hybrid frameworks that attempt to fit ecological with democratic principles.

In one sense the cases of Goodin and Saward merit presentation before those of, say, Dryzek and Eckersley, as the former are less responses to the latter than *vice versa*. Eckersley, in particular, has cast her thought specifically to show, against Goodin and Saward, that it *is* possible to logically connect democracy and green values 'at the level of principle', through her demonstration that the key democratic principle of autonomy has been arbitrarily applied within liberal democracy and that, consistently thought through, this concept provides the necessary in-principle link between democracy and green value (1996b: 223). And, too, Goodin and Saward both seem to be moving closer to their critics. In a 1996 paper Goodin develops a case for participatory democracy 'within familar liberal terms of equal protection of interests', arguing further for discursive democracy as the best means of linking 'those familar propositions about green democracy to the greens' other more fundamental values' (1996: 835). Similarly, of Saward's two papers cited above, the 1996 paper is noticeably less 'hard-line'. Though it may not be apparent from our earlier discussion, Saward *is* within the green paradigm; he *does* want environmentalist outcomes. In 1996 he argued that inherent within the 'self-sufficient' concept of democracy is a 'democratic right to environmental health', and — whilst he sternly warned that this is the limit of the extent to which core green values can institutionally interpenetrate with democracy — 'within this right lie a range of major environmental concerns' and, thus, 'several green concerns, far from being in conflict with democratic ideals, are in fact integral to the democratic process' (Saward 1996: 94). He also seemed to soften his stance on the need to always privilege 'pure' democracy over green imperatives and hybridised conceptions of democracy, conceding that 'there is no necessary prescription that democracy must "win" — or win fully — when principles conflict in practice', and contenting himself with a plea that 'to the extent that green outcomes take precedence over strictly democratic outcomes, it ought to be recognised and acknowledged that democracy is being diluted' (1996: 93).

Other points of contest also exist. Eckersley, for instance, is unconvinced by Dobson's elaborately wrought formula for institutionalising non-human interests in representative fora (1999: 28). For his part, Dobson is critical of Eckersley's formulation of green democratic rights, arguing that, whilst a 'right' to 'freedom from harmful ecological actions on the part of other human agents' is undoubtedly important, it is not a *democratic* right, in the sense that it is '*constitutive* of democracy, such that, if violated, a democracy cannot be said to exist'. Similarly, Eckersley's 'right to exist' is 'excessively general'; a right that is 'as preconditional for authoritarian as for democratic practices' (Dobson 1996c: 142).

Most disputation, though, has focused upon discursive democracy, with the criticism most persistently put that it fails to account for tenacious power differentials. 'Given social inequities and resource, knowledge and power disparities among different social classes and groups', Eckersley writes, 'it is unclear how the abstract norms of free and impartial public discussion will provide a check against the power and interests of elites' — and even small communities tend to be elite-dominated (Eckersley 1996b: 218; see also 1996a: 179). Dobson also finds deliberative democracy's ideal speech situation 'naive', as the notion of communication between equally communicatively competent participants, interacting free from 'domination, strategizing and (self-)deception', is nowhere evidenced in real-world democratic practice (Dobson 1996c: 141).

Such analyses echo Iris Marion Young's influential critique of deliberative democracy. It will be remembered that Young argued (1990; 1995) the need to find ways to transcend cultural and social differences in communication, a need that may require equal status to be accorded to communicative modes other than the disinterested rational argument promoted by such proponents of discursive democracy as Dryzek. And, once the complexity and entrenched nature of power disparities is conceded, it becomes much more difficult to maintain the case, as Dryzek does, for a generalisable environment interest, because such a case overstates the degree of commonality between interests.

A second perceived weakness of discursive democracy concerns its capacity for adaptation to the complex institutional realities of the times in which we live. Christoff, for example, though arguing for *more* deliberative democracy within institutions with an international scope, nevertheless notes that at present deliberative democracy is 'poorly suited to the determination of issues that affect an international constituency' (Christoff 1996b: 156), and Eckersley makes a similar point when she notes the absence of 'a robust model of co-ordination between "deliberative communities"' (Eckersley 1996b: 217).

A third criticism observes that, even in Dryzek's reworking of Habermas, discursive democracy only delivers ecological rationality 'as a potential by-product of communicative rationality' (Eckersley 1990: 759; see also Dobson 1993: 198; Krebs 1997). Eckersley notes that it is merely assumed that 'debates and decisions about generalisable *human* interests will effectively incorporate non-human interests', yet 'species that have no apparent use to humans could quite reasonably be considered dispensable and their habitats destroyed on behalf of generalisable human interests' (1996a: 179).

We should also recall that Dobson and Eckersley — who have subjected discursive democracy to its most sustained scrutiny within ecological thought — also have much to say in its favour. For both, the strongest defence of discursive democracy is its usefulness as yardstick against which the adequacy of real-world processes can be measured, and towards which they should be pushed. Eckersley, for example, is critical of presentations of deliberative democracy as model or blueprint, finding it more compelling as 'a counterfactual regulative ideal that provides a critical vantage point from which we might impugn particular institutions and decisions that have fallen short of the ideal' (2000; see also Dobson 1996c: 141).

Turning to postmodern environmentalism, we find a prime difficulty for the project of constructing a postmodern ground for environmental activism is the fact that a great many postmodernists, like Foucault, 'turn their back' on 'nature'. They have no interest in it (except as a common evocation for essentialist positions, and therefore as something to be attacked), may not concede it even exists, and hold to the view that a politics constructed on its behalf cannot be made compatible with key postmodern precepts. 'The Greens declare themselves to be neither right nor left, but ahead', Walter Truett Anderson observes, 'but in many ways they are behind, the very model of an obsolete modern-era political movement weighted down with ideology, self-righteousness, simplistic doctrines, and quasi-totalitarian notions about how everybody ought to think and act' (1990: 248).

From a movement perspective even the postmodern environmentalisms that have been generated to date (with Deleuze and Guattari possible exceptions) so determinedly render nature a sub-set of culture, and so firmly insist on the priority of a subjectivity-creation freed from such contextualising influences as 'society' or 'environment', that the environmentalisms that result will be limited, qualified, and inadequate on that account. Though Van Wyck may have a specific usage in mind when he uses the term 'weak' ecologism, most environmentalists are likely to see 'weak' as an accurate designator in the word's conventional sense. Postmodern theorists like to see themselves as 'radical' — Darier uses this term in specific contrast to ecological thought — but it seems clear that what constitutes 'radical' is very much perspective-dependent.

The point at which ecological thought has most strongly contested postmodernism is the latter's denial of the real in nature (in chapter 1 we have briefly noted the response by Sessions to this). 'Earth is our home in the full, genetic sense, where humanity and its ancestors existed for all the millions of years of their evolution', writes Wilson (1996: 190). It is a given; a context prior to human subjectivity. Similarly, J.B.

Kirkpatrick, noting that postmodernism 'increases the alienation of humans from nature', argues that 'ecologists suspect that there really is a reality', and thus they are 'definitely not postmodern in their outlook' (1998: 36–37). And for Spretnak, postmodernism's prime ecological failing is that it is oblivious to *modernism's* 'repression of the real' (1997: 5). She sets herself the task of recovering the real — body, place, nature — and despairs over postmodernism's 'retreat from engagement with the wildness, the amazing novelty, the vast and relational flux and evanescence of the cosmological *processes* of being, which cannot be captured and pigeonholed by conceptualization' (1991: 126).

An interesting example of this sort of case-making is provided by Paul Shepard. In *Nature and Madness* (1982), Shepard argued that there is an 'ontogenetic' need within humans to bond with nature, and individuals who fail to so bond remain developmentally stunted. From such a perspective collapsing the culture–nature distinction — particularly in the form envisaged by Haraway — would have disastrous consequences. In his contribution to Soule and Lease's *revisionist* 'reinvention of nature', he argues that in the postmodern vision of reality as text 'life is indistinguishable from a video game', and that 'deconstructionist postmodernism rationalizes the final step away from connection' (Shepard 1995: 25). Against the 'aesthetic dandyism' and 'psychopathology of High Culture', Shepard sets this case:

> as for 'truth', 'origins', or 'essentials' beyond the 'metanarratives', the naturalist has a peculiar advantage — by attending to species who have no words and no text other than context and yet among whom there is an unspoken consensus about the contingency of life and real substructures ... To argue that because we interpose talk or pictures between us and this shared immanence, and that it is therefore meaningless, contradicts the testimony of life itself (1995: 27).

It is only a small step from defence of the real to rejection of the claimed human exclusivity in 'constructing' the real. '"Constructs" ... suggests an activity on the part of us humans which is too pretentious', writes Naess. 'We do not construct things. We construct concepts of things — that is remarkable enough for me' (1997: 181). Evernden, too, reaches this conclusion — though he reverses the emphasis. He writes of 'things themselves ... *before* they were "nature", that is, before they were captured and explained, in which transaction they ceased to be themselves and became instead functionaries in the world of social discourse' (1992: 110). Soule also asks whether nature is socially constructed, and he adds another dimension to the question: if so, how much? He argues that the dichotomy between nature as pristine and nature as social construct is a false one. Humans modify landscapes and have always done so, but 'to claim that *Homo sapiens* has produced or invented the forest ignores the basic taxonomic integrity of biogeographic units: species today still have geographic distributions determined largely by economic tolerances and geological history and climate, rather than by human activities' (1995: 157). Some social construction, then — but only up to a point.

Much of Soule's essay rubs against the grain of environmental thought. He stoutly defends the ascendant neo-Darwinist biological paradigm with its emphasis on individualistic competition, and so, Zimmerman observes, 'even though Soule fears that postmodernism will undermine the scientific truth needed to motivate environmental activism, he does not seem to appreciate the extent to which the new, competition- and chaos-oriented ecological theories resonate with the deconstructionist view that there are no stable truths' (1996: 136). But Soule makes one point very much germane to our current purposes. 'Variability is nature's other name', he writes (1995: 155; see also Worster 1995: 81–82). Whilst his insistence that there is no tendency to equilibrium in natural systems (1995: 143) is not universally accepted, there *is* general agreement within environmentalism that natural systems are dynamic, fluid, always in process of becoming. If meaning is found in nature, it is in natural *process* — not in some unchanging monolithic entity. To this extent, postmodernism's criticism that the environment movement 'essentialises' nature is too crude: it simplifies a conceptual richness to the point of distortion and misrepresentation.

We have noted the progressivist criticism that postmodernism so partakes of the key assumptions of liberalism as to be to culture what neo-liberalism is to economics. There are some specifically ecological versions of this argument. 'The very tendency of mature capitalism to fragment traditional social and cultural relations by means of commodification yields reactionary cultural sequelae of its own', writes Bookchin, 'specifically, a consolidating ideology that holds the mind captive to the social order *in the very name of fragmentation and its alleged virtues*' (1995: 175). A variation on the argument that postmodernism is liberalism in disguise is the claim that postmodernism is much closer to the modernism it affects to disdain than it realises. Again, this is a persistent charge within left critiques. In a specifically ecological context it is most powerfully argued by Spretnak: '"postmodern" is really "*most*modern"', and this is so because its proponents 'have been socialized and educated in the scientific-humanist world view, which is dedicated to the denial of the power and presence of nature' (1997: 66). Again, we have a rejection of postmodernism's radical credentials. Postmodernists are wrong to think their attack on nature is radical or revolutionary, because modernism also denies value in nature. Similar doubts are cast upon the status of Derrida's dictum, 'there is nothing outside the text': 'environmentalists assert that there is nothing postmodern about this provocative phrase. Viewing everything as text is consistent with modernity's tendency to disclose everything in *human* terms (as text, image, commodity, object) thereby denying nature its otherness' (Zimmerman 1996: 134; though Zimmerman delineates the position, it is not his own). And Spretnak argues that, having intended demolition of the false boundaries imposed by modernism, in true modernist style postmodernism then establishes an absolute and rigid discursive dualism of its own, that of '*either* fixed essence *or* social construction' (1991: 126), when in fact most conceptualisation is far more complex than this.

Of course, postmodernists can claim in defence that they centrally theorise power, and in a radical way: 'postmodernists have made an important point by insisting that powerful groups have more means for promulgating and defending truth claims', and that decisions 'should be negotiated between local individuals and coalitions, not imposed by centralized political structures' (Zimmerman 1996: 134). The trouble with such an argument is that it does not take a postmodernist perspective to arrive at this 'important point'. Many other 'narratives' — including most forms of environmental thought — come to the same conclusion. Furthermore, in the absence of a theory of agency, in its advanced relativism, and in its a-historicism, postmodern thought has difficulty mounting a coherent defence of even this non-exceptional position (Gare 1995: 97–98).

If postmodernism is a fraudulent radicalism, one whose discourse serves the truth claims and political power of liberal capitalism, it follows that a radical *praxis* cannot be expected from postmodernism. Frodeman argues that the postmodern critique is 'incomplete' in its failure to provide at least a sketch of how things might change: 'postmodernists have seldom offered a program for bettering society. The insistence upon respect for difference embodies only the negative goal of *freedom from* coercion' (1992: 318). Similarly, the 'tenets of deconstructive practice', writes Salleh, lead to 'an impractical nihilism when applied to everyday life', the effect of which is 'to massage the liberal political status quo' (1997: 9). By denying history, reason and agency, postmodernism 'fosters social quietism' and 'neutralizes an activist and interventionist mentality oriented toward the public sphere' (Bookchin 1995: 176). And Gare argues that the extreme relativism promoted within the postmodernist canon 'would fragment all oppositional political movements completely, and make any concerted response to the environmental crisis impossible' (1995: 97).

It can even be argued that, in the case of environmental action, the postmodernist attack on 'nature' and 'wilderness' is *actively contributing* to ongoing exploitation. Opponents of the environment movement are already citing academic pronouncements that there is no objective meaning or value in nature. The message is out that 'if there is no single natural reality which has integrity and evolutionary direction, and the world of Nature is just a social construction, then there is no moral reason why we should not redesign it however we please' (Drengson 1996: 2; see also Snyder 1998; Soule 1995: 154–55;

Worster 1997). Sessions puts this in a still broader context, slamming Cronon and his *Reinventing Nature* collaborators for their 'Disneyland theme park approach to "reinventing Nature", which goes hand-in-hand with the multinational corporate attempt to create a world of universal consumerism' (Sessions 1996: 36).

Some trace postmodernism's ambivalence over oppositional political engagement to a barely acknowledged technophiliac approval of the world as it is and as it is becoming. 'I find it eerie that one rarely encounters an apolitical deconstructive-postmodern analyst who is the least bit wishful over what has been lost', writes Spretnak. 'Instead, the attitude is one of triumph at naming the perceived disempowerment of everyone and everything and a "sophisticated" passivity that mocks any attempt to change the situation' (1991: 127; see also Shepard 1995).

Of course, Donna Haraway is sufficiently 'a-modernist' to *not* adopt a cool cynicism on the subject of human agency; yet, on the crucial question of emergent technology, she calls an exception, counselling acceptance, and advising us to roll with the punches. And it is Haraway who most critics of technological emergence will call to mind when the charge of harbouring a semi-screened technophiliac bliss is made. Several environmentalist critiques have singled her out for especial criticism. For Jhan Hochman, Haraway's collapsing of the technology–nature boundary makes her a 'problematic advocate of nature', for it involves 'a veiled attempt to call upon nature to justify technology, to naturalize technology as intrinsic to humanity' (1998: 12–13). Even Zimmerman observes — though cautiously — that, 'in attempting to shift the informational age in a liberatory direction', Haraway's cyborg may 'concede too much to technological possibilities' (1994: 372). George Sessions (whose views on postmodernism we noted in chapter 1) is much more emphatic. For him the cyborg is an 'abhorrent and dehumanizing vision', and we should resist Haraway's advice to roll over and take the technology that is coming to us because a high-tech society 'will be totalitarian regardless of whatever political structures permit it to develop' (1995c: 154). Sessions puts it bluntly, but this is nevertheless a view with a long pedigree, one that gathers up Lewis Mumford and Jacques Ellul along the way.

One of the discourses within which Haraway is situated is feminism. But one ecofeminist critic sees postmodern thought as a continuance of such patriarchal values as 'autonomy from relationship, separateness, and control through abstraction' (Spretnak 1991: 121), and the charge that postmodernism embodies 'the ancient patriarchal dream: transcendence beyond the body' (1991: 127) would seem particularly applicable to Haraway's animal-machine cyborg. Even the privileging of culture over nature — social construction/mind over essence/body — partakes of a key patriarchal dualism. Thus, 'the patriarchal desire to disempower the body is served by the deconstructive-postmodern assertion that abstraction, or conceptualization, is all' (Spretnak 1991: 124). The focus on individual human subjectivity, on this reading, is a continuance of the heroic 'man alone' patriarchal project.

Whilst conceding that Deleuze and Guattari may provide links that have, to date, been insufficiently recognised and developed, the points of incompatibility between environmmental and postmodern thought would seem to far outweigh the grounds for synergy. As postmodern thought is the current orthodoxy in the humanities and much of the social sciences, it is worth considering whether it will oversee the eclipse of ecological thought. We will visit the question of ecologism's future by way of conclusion.

THOUGHTS BY WAY OF CONCLUSION: THE TENACITY OF ENVIRONMENTALISM

If this book has achieved anything, it will have demonstrated that there is such dynamism in environmental thought, such rapidly ramifying enrichment, that the only credible trajectory is one of burgeoning diversification into the foreseeable future. David Pearce has lampooned the tendency in environmental thought to a proliferation of labels. 'Taking the journal *Environmental Ethics* as a leading example of a forum for philosophical investigations into moral discourse on the environment', he writes, 'the first impression is that environmental ethics suffer from an extreme form of "labilitis"' (1998: 17). Notwithstanding that I have here contributed greatly to this tendency, I had earlier arrived at a similar observation, and worried over this tendency, myself. But under circumstances of rapid intellectual diversification, recourse to the convenience of labels seems unavoidable.

In 1972 the economist Anthony Downs published a paper entitled 'Up and Down with Ecology: The Issue-Attention Cycle'. Downs was moved to theorise the rise and fall of issues on the public agenda, having just witnessed, so he believed, the demise of 'ecology' as an issue of public concern. Though his observation resulted in a useful and widely read paper, his catalytic perception was, in fact, in error. 'Environment' did not disappear from popular concern. Nor has it since, though its demise has been persistently predicted, mainly, one suspects, on the wings of wishful thinking (*The Fading of the Greens* is the bravely ill-considered title of Bramwell's 1994 book).

And where intellectual ferment concerning matters of such real-world import is to be found, so too will that ferment present to the world a robust political face. 'Environment', as spur to thought and action, is here to stay. It manifests because of 'the extent and intensity of ecological issues for people all across the planet, and the poverty that our current discourses and practices exhibit when dealing with them' (Coles 1992: 194).

Ecologism continues to flourish, then, and it does so in defiance of determined attempts on the part of great and powerful people and structures to shut it down. When Paul Sears labelled ecology 'a subversive subject' in 1964, he can have had no idea how prophetic his words would prove to be. With the real-world decline of socialism, environmentalism has become the major vantage point of opposition to business-as-usual. Unwilling to concede *any* ground for behavioural modification, some (not all) powerful components of capital have gone to extraordinary lengths to discredit and defuse environmentalism. And when the apologists for these groups rushed to print in the early 1990s, we were awash in books pointing out the conspiracies, sins and errors of environmentalism. They received a great deal of media boosting and critical promotion — considerably more than most of the theorists I have discussed. I am familiar with almost all of them: so determinedly partial and misleading are they, so determinedly do they engage with obsolete or never-were stereotypes, and so determinedly do they eschew fact and logic for rhetoric, that they are risible. Of them all, only Martin Lewis's *Green Delusions* (1992) presented a case that provoked thoughtful engagement and deserved to be taken seriously. In the 'real world', by contrast, opposition to the environment interest has been rather more formidable, as instanced by the political clout of the 'Wise Use' movement — it would be more accurate to call it the 'Anything Goes Movement' — in its assault on the United States of America's Endangered Species Act (on the 'Wise Use' movement, and on other anti-green corporate-funded front groups, see Beder 1997a; Deal 1993; Ehrlich and Ehrlich 1996: 11–23; Hager and Burton 1999; Helvarg 1994; Kibel 1999: 41–47).

Behind the assumption that environmentalism is 'killable' lies a large, mistakenly founded prejudice: that there is not, objectively there in the biophysical world, anything to explain or justify an environment movement. The whole green edifice is merely an elaborately wrought castle in the air. When a politician (say) confidently asserts that environmentalism is dead, what is left unsaid is the belief that there is no link between an environmental commitment and what is, or is not happening in the wider world. It is an assertion that all is well with the world; environmentalism can then only be assumed to be an internally fuelled phenomenon, possibly created by cant and mischief, but certainly not by the state of the world. It is not even a tacit voicing of the familiar expression of technocratic confidence, 'human history shows that our energy and creativity will surmount whatever difficulties we encounter' (Bailey 1993: xii), because it assumes that no 'difficulties', either in the present or immediate past, have manifested themselves in the first place.

It can be expected, though, that there will be a politics (and a theory) of the environment:

- as long as stocks of ecological capital decline in response to unsustainable growth in industrial production (the latter grew by a factor of 50 between 1900 and 1990, according to Macneill *et al.* 1991: 3);
- as long as it can be held, as do Wackernagel and Rees (1996), that the developed world is consuming the resources of three planets equal to Earth's endowment;
- and, perhaps most crucially — if my attribution of the psychology behind an environmental commitment is correct — as long as the earth continues to lose over 17,000 species per year, in Edward O. Wilson's authoritative, though now outdated, estimate (1988: 13). (Jared Diamond has estimated this figure at a staggering 150,000 per year, or 17 per hour [1992b: 70]; and, as Vandana Shiva points out [1993: 65–66], with half the world's species in the 7 per cent of the Earth's land surface currently under rapidly disappearing tropical moist forests, this rate of disappearance can only increase.)

There seems to be broad agreement that human nature is not fixed but contextual upon institutional and technological conditions. It may be, nevertheless, that the differences between the technological optimists (who see all apparent pathologies as resolvable through human ingenuity) and those who invariably fix upon the loss and opportunities forgone in 'progress' cannot be made amenable to reasoned resolution — that these difficulties stem from some entrenched predisposition of character. Ecologically minded people cannot help but mourn, with Sale, the 'reduction in face-to-face contacts, social discourse, human autonomy, individual choice, and personal skills' (1995: 28) that are seen to be the inevitable consequence for the human condition of technological change, and to ask fearfully 'what will happen to the species and ecosystems destroyed, what will be the consequences if the line of ecological peril is crossed?' (1995: 22).

Ultimately, though, whether or not these irreconcilable perspectives are the product of different mixes in brain chemistry, one or other will prove to have been *objectively* right, and the other — perhaps tragically — wrong. From the environmentalist perspective, the claim that no environmental crisis exists (a confidently recurring claim within anti-green writings) simply cannot be taken seriously. It 'contradicts everything people are experiencing about the state of the world's environment' — to say nothing of the vast weight of scientific evidence. It is best treated as 'an example of the uncanny ability of self-interest to override common sense' (Manes 1990: 152). This observation even applies in the case of global warming, concerning which the popular impression is that the scientific evidence is 'scattered'. But, as Jamieson demonstrates, this impression has been deliberately — and irresponsibly — created by people and institutions unwilling to put the global good above sectional interests, whereas in fact, 'most of the differences of opinion within the scientific community are differences of emphasis rather than differences of kind' (1992: 142).

Technological optimists may well believe, then, with Fukuyama, that history's teleological end has been reached; but ecologically informed people will, with Peter Singer, insist that:

> Fukuyama may have predicted the permanence of the liberal free enterprise system just when it is about to face its gravest crisis ... Throughout human history, we have been able to freely use the oceans and the atmosphere as a vast sink for our wastes. The liberal democratic free enterprise system that Fukuyama proposes as the ultimate outcome of history is built on the idea that we can keep doing this forever. In contrast, responsible scientific opinion now tells us that we are passengers on a runaway train that is heading rapidly towards an abyss (1993b: 14).

I would say one other thing on this matter of tenacious states of mind. If decline in biophysical systems continues, and if the brunt of this is borne by non-human life in the form of runaway-species extinction, environmental politics and ecologically sourced theoretical analysis will continue to attract adherents — and will continue to party-poop upon cornucopian celebration. It will continue because it will be fuelled by that most powerful of emotions — grief. The phenomenon of biophiliac grief is already potently with us. It has spawned yet another offshoot of environmental thought, environmental psychology — Jungian in character (for example, Aizenstat 1995; Hillman 1995; Tacey 1995), and with therapies (Conn 1995; T. O'Connor 1995; Windle 1995) and collective rituals (Macy 1995; Seed, J. with J. Macy *et al.* 1988) to assist mourning and help cope with despair. 'Unlike those who deny the existence of a socioecological crisis, environmentalists directly confront extremely disturbing realities', writes Tim Boston. They can become 'mentally suffocated by the consequences and significance of such problems. Essentially, to be a passionate environmentalist demands formidable sociopolitical and psychological nerve' (1996: 75).

Though it is a comparatively recent branch of ecological thought, Boston (1996) identifies several different perspectives within ecopsychology (cultural, spiritual, political, Gestalt, perceptual, depth, wilderness, and abnormal ecopsychologies). He himself places much emphasis on the concept of 'wellness', which he turns to positive account: 'ultimately, caring for the earth and its communities is analogous to caring for oneself. Conversely, to pollute, degrade and destroy the earth and its communities is like hurting one's physical, psychological and spiritual well-being' (1997: 28). Though activist depression and biophiliac grief go with the green territory, 'wellness is an expression of involvement in issues to achieve socioecological justice', whilst 'a key element of socioecological resistance is the conviction that diversity is necessary, but threatened by monocultures of the mind' (1997: 26).

And here is yet another reason why environmentalism will not be going away for some time yet. If Boston is right, and I think he is, the many people who are distressed at the standardising and blenderising trends in present-day life, distressed at the loss of the particular and the unique in the human realm, will join those similarly distressed at trends towards biological impoverishment and gravitate to a form of ecopolitics. For its part, environmentalism must be prepared to receive them: this is why it is important to resist those who would expunge from environmentalism all orientations that are different from their own. Fortunately, the cause of the 'one way' brigade is hopeless anyway: biodiversity may be in decline, but diversity of ecological thought is not. 'It is fair to say', observes Andrew Ross, 'that no social movement has had such discordant voices within its own ranks, and, one might add, for such good reasons' (1994: 5). Environmentalism's capacity for resilience lies in this plurality: it is to pattern out, and present its multiplicity of ideas that I have written this book.

REFERENCES

Abram, D. (1988), 'The Mechanical and the Organic: Epistemological Consequences of the Gaia Hypothesis', in P. Bunyard and E. Goldsmith (eds), *GAIA, the Thesis, the Mechanisms and the Implications: Proceedings of the First Annual Camelford Conference on the Implications of the Gaia Hypothesis*, Wadebridge Ecological Centre, Camelford (Eng.), 119–32.

———. (1995), 'Merleau-Ponty and the Voice of the Earth', in M. Oelschlaeger (ed.), *Postmodern Environmental Ethics*, State University of New York Press, Albany (N.Y.), 57–77.

———. (1996), *The Spell of the Sensuous: Perception and Language in a More-than-Human World*, Pantheon, New York.

Achterberg, W. (1993), 'Can Liberal Democracy Survive the Environmental Crisis? Sustainability, Liberal Neutrality and Overlapping Consensus', in A. Dobson and P. Lucardie (eds), *The Politics of Nature: Explorations in Green Political Theory*, Routledge, London, 81–101.

Adams, C.J. (1992), *The Sexual Politics of Meat: A Feminist-Vegetarian Critical Theory*, Continuum, New York.

———. (1993), 'Introduction', in C.J. Adams (ed.), *Ecofeminism and the Sacred*, Continuum, New York, 1–9.

———. (1994), *Neither Man Nor Beast: Feminism and the Defense of Animals*, Continuum, New York.

Adams, C.J., and Donovan, J. (eds) (1995), *Animals and Women: Feminist Theoretical Approaches*, Duke University Press, Durham (N.C.).

Adler, M. (1989), 'The Juice and the Mystery', in J. Plant (ed.), *Healing the Wounds: The Promise of Ecofeminism*, New Society, Philadelphia (Pa.), 151–54.

Aizenstat, S. (1995), 'Jungian Psychology and the World Unconscious', in T. Roszak, M.E. Gomes, and A.D. Kanner (eds), *Ecopsychology: Restoring the Earth, Healing the Mind*, Sierra Club, San Francisco (Calif.), 92–100.

Allaby, M. (1986), 'Decentralisation: An Alternative to Ecocide', *New Scientist*, 6 March, 59–60.

———. (1989), *Guide to Gaia*, Optima, London.

Allen, P.G. (1990), 'The Woman I Love is a Planet; the Planet I Love is a Tree', in I. Diamond and G.F. Orenstein (eds), *Reweaving the World: The Emergence of Ecofeminism*, Sierra Club, San Francisco (Calif.), 52–57.

Alston, W.P. (1967), 'Religion', in *The Encyclopedia of Philosophy*, Macmillan & The Free Press, New York, 140–45.

Ames, R.T. (1986), 'Taoism and the Nature of Nature', *Environmental Ethics*, 8, 317–50.

Anderson, T., and Leal, D.R. (1991), *Free Market Environmentalism*, Pacific Research Institute for Public Policy, San Francisco (Calif.).

Anderson, W.T. (1990), *Reality Isn't What It Used to Be: Theatrical Politics, Ready-to-Wear Religion, Global Myths, Primitive Chic, and Other Wonders of the Postmodern World*, HarperCollins, San Francisco (Calif.).

Anker, P. (1996), 'From Scepticism to Dogmatism and Back: Remarks on the History of Deep Ecology', in N. Witoszek (ed.), *Rethinking Deep Ecology: Proceedings from a Seminar at SUM, University of Oslo, 5 September 1995*, Centre for Development and the Environment, University of Oslo, Oslo, 40–60.

Arnot, R.P. (1934), *William Morris: A Vindication*, Martin Lawrence, London.

Athanasiou, T. (1996), 'The Age of Greenwashing', *Capitalism, Nature, Socialism*, 7(1), 1–36.

Attfield, R. (1983), 'Western Traditions and Environmental Ethics', in R. Elliot and A. Gare (eds), *Environmental Philosophy*, University of Queensland Press, St Lucia (Qld), 201–30.

———. (1991), *The Ethics of Environmental Concern*, 2nd edn, The University of Georgia Press, Athens (Ga.).

———. (1994), 'The Precautionary Principle and Moral Values', in T. O'Riordan and J. Cameron (eds), *Interpreting the Precautionary Principle*, Earthscan, London, 152–64.

Bachelard, G. (1969), *The Poetics of Space*, Beacon Press, Boston (Mass.).

Bacon, F. (1878; first published 1620), *Novum Organum*, Clarendon, Oxford.

———. (1990; first published 1626), *The New Atlantis*, Cambridge University Press, Cambridge.

Badiner, A.H. (ed.) (1990), *Dharma Gaia: A Harvest of Essays in Buddhism and Ecology*, Parallax, Berkeley (Calif.).

Bahro, R. (1978), *The Alternative in Eastern Europe*, New Left Books, London.

———. (1982), *Socialism and Survival*, Heretic Books, London.

———. (1984), *From Red to Green*, Verso/New Left Books, London.

———. (1986), *Building the Green Movement*, New Society, Philadelphia (Pa.).

———. (1994), *Avoiding Social and Ecological Disaster*, Gateway Books, Bath (Eng.).

Bailey, R. (1993), *Ecoscam: The False Prophets of Ecological Apocalypse*, St Martin's Press, New York.

Baker, J.A. (1990), 'Biblical Views of Nature', in C. Birch, W. Eakin and J.B. McDaniel (eds), *Liberating Life: Contemporary Approaches to Ecological Theology*, Orbis, Mayknoll (N.Y.), 9–26.

Barry, J. (1994), 'The Limits of the Shallow and the Deep: Green Politics, Philosophy and Praxis', *Environmental Politics*, 3, 369–94.

———. (1996), 'Sustainability, Political Judgement and Citizenship: Connecting Green Politics and Democracy', in B. Doherty and M. de Geus (eds), *Democracy and Green Political Thought: Sustainability, Rights and Citizenship*, Routledge, London, 115–31.

Barth, K. (1959), *Dogmatics in Outline*, Harper & Row, New York.

Bartlett, R.V. (1990), Ecological Reason in Administration: Environmental Impact Assessment and Administrative Theory', in R. Paehlke and D. Torgerson (eds), *Managing Leviathan: Environmental Politics and the Administrative State*, Broadview, Peterborough (Ont.), 81–96.

Bate, J. (1991), *Romantic Ecology: Wordsworth and the Environmental Tradition*, Routledge, London.

Beauvoir, S. de (1982), *The Second Sex*, Penguin, Harmondsworth (Eng.).

Beck, U. (1992), *Risk Society: Towards a New Modernity*, Sage, London.

———. (1995a), *Ecological Politics in an Age of Risk*, Polity, Cambridge.

———. (1995b), *Ecological Enlightenment: Essays on the Politics of the Risk Society*, Humanities Press, Atlantic Highlands (N.J.).

———. (1997), *The Reinvention of Politics: Rethinking Modernity in the Global Social Order*, Polity, Cambridge.

———. (1998), 'From Industrial Society to the Risk Society: Questions of Survival, Social Structure, and Ecological Enlightenment', in J.S. Dryzek and D. Schlosberg (eds), *Debating the Earth: The Environmental Politics Reader*, Oxford University Press, Oxford, 327–46.

Beckerman, W. (1974), *In Defense of Economic Growth*, Jonathan Cape, London.

———. (1975), *Two Cheers for the Affluent Society: A Spirited Defense of Economic Growth*, St Martin's Press, New York.

———. (1994), 'Sustainable Development: Is It a Useful Concept?', *Environmental Values*, 3, 191–209.

———. (1995a), *Small Is Stupid: Blowing the Whistle on the Greens*, Duckworth, London.

———. (1995b), 'How Would You Like Your "Sustainability", Sir? Weak or Strong?', *Environmental Values*, 4, 169–79.

Beder, S. (1993), *The Nature of Sustainable Development*, Scribe, Newnham (Vic.).

———. (1997a), *Global Spin: The Corporate Assault on Environmentalism*, Scribe, Carlton North (Vic.).

———. (1997b), 'The Environment Goes to Market', *Democracy & Nature*, 3(3), 90–106.

Bell, A. (1994), 'Cost-Benefit Analysis: A Conservation Caveat', *The Trumpeter: Journal of Ecosophy*, 11, 93–94.

Bell, S. (1992), 'Socialism and Ecology: Will Ever the Twain Meet?', in P.R. Hay and R. Eckersley (eds), *Ecopolitical Theory: Essays from Australia*, Occasional Paper 24, Centre for Environmental Studies, University of Tasmania, Hobart (Tas.), 207–22.

Benello, C.G., Swann, R., and Turnbull, S. (1989), *Building Sustainable Communities: Tools and Concepts for Self-Reliant Economic Change*, Bootstrap, New York.

Bennett, D.H. (1985), *Interspecies Ethics: A Brief Aboriginal and Non-Aboriginal Comparison*, Discussion Paper in Environmental Philosophy No. 7, Department of Philosophy, Australian National University, Canberra.

Bennett, J. (1993), 'Primate Visions and Alter-Tales', in J. Bennett and W. Chaloupka (eds), *In the Nature of Things: Language, Politics and the Environment*, University of Minnesota Press, Minneapolis (Minn.), 250–65.

Bentham, J. (1970; first published 1823), *An Introduction to the Principles of Morals and Legislation*, Athlone Press, London.

Benton, T. (1988), 'Humanism=Speciesism: Marx on Humans and Animals', *Radical Philosophy*, Autumn, 4–18.

———. (1989), 'Marxism and Natural Limits: An Ecological Critique and

Reconstruction', *New Left Review*, 178, 51–87.
———. (1993a), 'Review of Janet Biehl, *Rethinking Ecofeminist Politics*', *Capitalism, Nature, Socialism*, 4(2), 138–41.
———. (1993b), *Natural Relations: Ecology, Animal Rights and Social Justice*, Verso, London.
———. (1994), 'Biology and Social Theory in the Environmental Debate', in M. Redclift and T. Benton (eds), *Social Theory and the Global Environment*, Routledge, London, 28–50.
———. (1997), 'Beyond Left and Right?: Ecology, Politics, Capitalism and Modernity', in M. Jacobs (ed.), *Greening the Millenium: The New Politics of the Environment*, Blackwell, Oxford, 34–46.
Benton, T., and Redfearn, S. (1996), 'The Politics of Animal Rights - Where is the Left?', *New Left Review*, 215, 43–58.
Beresford, M. (1977), 'Doomsayers and Eco-nuts: A Critique of the Ecology Movement', *Politics* (Aust.), 12, 98–106.
Berg, P. (1990), 'Bioregional and Wild! A New Cultural Image ...', in C. Plant and J. Plant (eds), *Turtle Talk: Voices for a Sustainable Future*, New Society, Philadelphia (Pa.), 22–31.
Berg, P., and Dasmann, R.F. (1978), ''Reinhabiting California', in P. Berg (ed.), *Reinhabiting a Separate Country: A Bioregional Anthology of Northern California*, Planet Drum Foundation, San Francisco (Calif.), 217–20.
Berlin, I. (1958), *Two Concepts of Liberty*, Oxford University Press, Oxford.
Berman, M. (1981), *The Reenchantment of the World*, Cornell University Press, Ithaca (N.Y.).
Berman, T. (1994), 'The Rape of Mother Nature?: Women in the Language of Environmental Discourse', *The Trumpeter: Journal of Ecosophy*, 11, 173–79.
Berry, J. (1993), 'The Universe Story as Told by Brian Swimme and Thomas Berry', *The Trumpeter: Journal of Ecosophy*, 10, 79–80.
Berry, T. (1988), *The Dream of the Earth*, Sierra Club, San Francisco (Calif.).
———. (1990), 'The Spirituality of the Earth', in C. Birch, W. Eakin and J.B. McDaniel (eds), *Liberating Life: Contemporary Approaches to Ecological Theology*, Orbis, Mayknoll (N.Y.), 151–58.
Berry, T., with T. Clarke (1991), *Befriending the Earth: A Theology of Reconcilation between Humans and Earth*, Twenty-Third Publications, Mystic (Conn.).
Biehl, J. (1987), 'It's Deep, But Is It Broad?: An Eco-feminist Looks at Deep Ecology', *Kick It Over*, Winter (special supplement), 2A–4A.
———. (1991), *Rethinking Ecofeminist Politics*, South End Press, Boston (Mass.).
———. (1993), 'Problems in Ecofeminism', *Society and Nature*, 2(1), 52–71.
———. (1994), '"Ecology" and the Modernization of Fascism in the German Ultra-right', *Society and Nature*, 2(2), 130–70.
Biggins, D. (1978), 'The Social Context of Ecology', *Ecologist Quarterly*, Autumn, 218–26.
———. (1979), 'Scientific Knowledge and Values: Imperatives in Ecology', *Ethics in Science and Medicine*, 6, 49–57.
Bilimoria, P. (1998), 'Indian Religious Traditions', in D.E. Cooper and J.A. Palmer (eds), *Spirit of the Environment: Religion, Value and Environmental Concern*, Routledge, London, 1–14.
Birch, C. (1995), *Feelings*, University of New South Wales Press, Sydney.
Birch, C., and Cobb Jr, J. (1981), *The Liberation of Life: From the Cell to the Community*, Cambridge University Press, Cambridge.
Birkeland, J. (1992), 'Bionomics', in I. Thomas (ed.), *Interactions and Actions:*

Ecopolitics VI Proceedings, Faculty of Environmental Design and Construction, Royal Melbourne Institute of Technology, Melbourne (Vic.), E9–E15.
———. (1993), 'Ecofeminism: Linking Theory and Practice', in G. Gaard (ed.), *Ecofeminism: Women, Animals, Nature*, Temple University Press, Philadelphia (Pa.), 13–59.
Birkeland, J., Dodds, S., and Hamilton, C. (1997), 'Values and Ethics', in M. Diesendorf and C. Hamilton (eds), *Human Ecology, Human Economy: Ideas for an Ecologically Sustainable Future*, Allen & Unwin, St Leonards (NSW), 125–47.
Black, J.N. (1970), *The Dominion of Man: The Search for Ecological Responsibility*, Edinburgh University Press, Edinburgh.
Black Elk (1932), *Black Elk Speaks*, J.G. Neihardt (ed.), University of Nebraska Press, Lincoln (Neb.).
Blackwell, T., and Seabrook, J. (1993), *The Revolt Against Change: Towards a Conserving Radicalism*, Vintage, London.
Blake, T. (1992), 'Ecological Contradiction: The Grounding of the Green Critique of Industrialism', in P.R. Hay and R. Eckersley (eds), *Ecopolitical Theory: Essays from Australia*, Occasional Paper 24, Centre for Environmental Studies, University of Tasmania, Hobart (Tas.), 189–205.
Bohm, D. (1983), *Wholeness and the Implicate Order*, Ark, London.
———. (1994), 'Postmodern Science and a Postmodern World', in C. Merchant (ed.), *Ecology: Key Concepts in Critical Theory*, Humanities Press, Atlantic Highlands (N.J.), 342–50.
Bohm, D., and Peat, F.D. (1989), *Science, Order, and Creativity*, Routledge, London.
Boland, J. (1994), 'Ecological Modernization', *Capitalism, Nature, Socialism*, 5(3), 135–41.
Bookchin, M., as L. Herber (1962), *Our Synthetic Environment*, Knopf, New York.
Bookchin, M. (1971), *Post-Scarcity Anarchism*, Ramparts, Berkeley (Calif.).
———. (1980), *Toward an Ecological Society*, Black Rose, Montreal.
———. (1982), *The Ecology of Freedom: The Emergence and Dissolution of Hierarchy*, Cheshire, Palo Alto (Calif.).
———. (1985), 'Toward a Philosophy of Nature - the Bases for an Ecological Ethics', in M. Tobias (ed.), *Deep Ecology*, Avant Books, San Diego (Calif.), 213–39.
———. (1986a), 'A Green Course' (interviewed by Satish Kumar), *Resurgence*, 115, 10–13.
———. (1986b), *The Modern Crisis*, New Society, Philadelphia (Pa.).
———. (1987a), 'Social Ecology Versus "Deep" Ecology', *Green Perspectives: Newsletter of the Green Program Project*, 4/5 Double Issue, Summer, 1–23.
———. (1987b), 'Thinking Ecologically: A Dialectical Approach', *Our Generation*, 18, 3–40.
———. (1988), 'A Reply to My Critics', *Green Synthesis*, December, 5–7.
———. (1990), *Remaking Society: Pathways to a Green Future*, South End Press, Boston (Mass.).
———. (1992), 'Libertarian Municipalism: An Overview', *Society and Nature*, 1(1), 93–104.
———. (1995), *Re-Enchanting Humanity: A Defense of the Human Spirit against Antihumanism, Misanthropy, and Primitivism*, Cassell, London.
———. (1997), 'Comments on the International Social Ecology Network Gathering and the "Deep Social Ecology" of John Clark', *Democracy & Nature*, 3(3), 154–97.
Booth, A.L. (1997), 'Critical Questions in Environmental Philosophy', in A. Light and J.M. Smith (eds), *Philosophy and Geography I: Space, Place and Environmental*

Ethics, Rowman and Littlefield, Lanham (Md.), 255–73.
———. (1999), 'Does the Spirit Move You? Environmental Spirituality', *Environmental Values*, 8, 89–105.
Booth, A.L., and Jacobs, H.M. (1990), 'Ties that Bind: Native American Beliefs as a Foundation for Environmental Consciousness', *Environmental Ethics*, 12, 27–43.
Booth, W.J. (1989), 'Gone Fishing: Making Sense of Marx's Concept of Communism', *Political Theory*, 17, 205–22.
Bortoft, H. (1996), *The Wholeness of Nature: Goethe's Way towards a Science of Conscious Participation in Nature*, Lindisfarne Press, Hudson (N.Y.).
Boston, T. (1996), 'Ecopsychology: An Earth-Psyche Bond', *The Trumpeter: Journal of Ecosophy*, 13, 72–79.
———. (1997), 'Situating a Subjective Project: A Radical Ecologist's Interpretation of Wellness', *The Trumpeter: Journal of Ecosophy*, 14, 23–29.
Bowling, S., and Martin, B. (1985), 'Science: A Masculine Disorder', *Science and Public Policy*, 12, 308–16.
Bramwell, A. (1985), *Blood and Soil: Richard Walther Darre and Hitler's 'Green Party'*, Kensal, Bourne End (Eng.).
———. (1989), *Ecology in the 20th Century: A History*, Yale University Press, New Haven (Conn.).
———. (1994), *The Fading of the Greens: The Decline of Environmental Politics in the West*, Yale University Press, New Haven (Conn.).
Bratton, S.P. (1993), *Christianity, Wilderness, and Wildlife: The Original Desert Solitaire*, University of Scranton Press, Scranton (Pa.).
Brennan, A. (1988), *Thinking About Nature: An Investigation of Nature, Value and Ecology*, Routledge, London.
———. (1996), 'Reconsidering Deep Ecology', in N. Witoszek (ed.), *Rethinking Deep Ecology: Proceedings from a Seminar at SUM, University of Oslo, 5 September 1995*, Centre for Development and the Environment, University of Oslo, Oslo, 4–23.
Brook, I. (1998), 'Goethean Science as a Way to Read Landscape', *Landscape Research*, 23, 51–69.
Bryant, P. (1995, 10 February). 'Constructing Nature Again', *ASLE Network* [online]. Available e-mail: <ASLE@UNR.EDU>.
Burkett, P. (1996), 'On Some Common Misconceptions about Nature and Marx's Critique of Political Economy', *Capitalism, Nature, Socialism*, 7(3), 57–80.
———. (1997), 'Nature in Marx Reconsidered: A Silver Anniversary Assessment of Alfred Schmidt's Concept of Nature in Marx', *Organization & Society*, 10, 164–83.
Byers, B. (1992), 'Deep Ecology and its Critics: A Buddhist Perspective', *The Trumpeter: Journal of Ecosophy*, 9, 33–35.

Callicott, J.B. (1979), 'Elements of an Environmental Ethic: Moral Considerability and the Biotic Community', *Environmental Ethics*, 1, 71–81.
———. (1980), 'Animal Liberation: A Triangular Affair', *Environmental Ethics*, 2, 311–38.
———. (1982), 'Hume's Is/Ought Dichotomy and the Relation of Ecology to Leopold's Land Ethic', *Environmental Ethics*, 4, 163–74.
———. (1983), 'Traditional American Indian and Traditional Western European Attitudes towards Nature: An Overview', in R. Elliot and A. Gare (eds), *Environmental Philosophy*, University of Queensland Press, St Lucia (Qld), 231–59.

———. (1986a), 'On the Intrinsic Value of Nonhuman Species', in B.G. Norton (ed.), *The Preservation of Species: The Value of Biological Diversity*, Princeton University Press, Princeton (N.J.), 138–72.
———. (1986b), 'The Search for an Environmental Ethic', in T. Regan (ed.), *Matters of Life and Death*, 2nd edn, Random House, New York, 381–423.
———. (1987), 'The Conceptual Foundations of the Land Ethic', in J.B. Callicott (ed.), *Companion to a Sand County Almanac: Interpretive and Critical Essays*, University of Wisconsin Press, Madison (Wis.), 186–217.
———. (1992), 'Animal Liberation and Environmental Ethics: Back Together Again', in E.C. Hargrove (ed.), *The Animal Rights/Environmental Ethics Debate: The Environmental Perspective*, State University of New York Press, Albany (N.Y.), 249–55.
———. (1993), 'Introduction', in M.E. Zimmerman, J.B. Callicott, G. Sessions, K.J. Warren and J. Clark (eds), *Environmental Philosophy: From Animal Rights to Radical Ecology*, Prentice-Hall, Englewood Cliffs (N.J.), 3–11.
———. (1994), *Earth's Insights: A Multicultural Survey of Ecological Ethics from the Mediterranean Basin to the Australian Outback*, University of California Press, Berkeley (Calif.).
Cameron, A. (1989), 'First Mother and the Rainbow Children', in J. Plant (ed.), *Healing the Wounds: The Promise of Ecofeminism*, New Society, Philadelphia (Pa.), 54–66.
Cameron, J., and Abouchar, J. (1991), 'The Precautionary Principle: A Fundamental Principle of Law and Policy for the Protection of the Global Environment', *Boston College International and Comparative Law Review*, 14, 1–27.
Campolo, T. (1992), *How to Rescue the Earth without Worshipping Nature: A Christian's Call to Save Creation*, Word Publishing, Milton Keynes (Eng.).
Capra, F. (1983), *The Turning Point: Science, Society and the Rising Culture*, Bantam, New York.
———. (1988), 'Physics and the Current Change of Paradigms', in R.F. Kitchener (ed.), *The World View of Contemporary Physics: Does It Need a New Metaphysics?*, State University of New York Press, Albany (N.Y.), 144–52.
———. (1992), *The Tao of Physics: An Exploration of the Parallels between Modern Physics and Eastern Mysticism*, Flamingo, London.
———. (1993), 'Turning of the Tide', *ReVision*, 16, 59–72.
———. (1997), *The Web of Life*, Flamingo, London.
Card, C. (1990), 'Gender and Moral Luck', in O. Flanagan and A.O. Rorty (eds), *Identity, Character and Morality: Essays in Moral Psychology*, MIT Press, Cambridge (Mass.), 199–218.
Carpenter, G.P. (1997), 'Redefining Scarcity: Marxism and Ecology Reconciled', *Democracy & Nature*, 9(3), 129–53.
Carroll, J. (1993), *Humanism: The Wreck of Western Culture*, Fontana, London.
Carson, R. (1965), *Silent Spring*, Penguin, Harmondsworth (Eng.).
Carter, A. (1993), 'Towards a Green Political Theory', in A. Dobson and P. Lucardie (eds), *The Politics of Nature: Explorations in Green Political Theory*, Routledge, London, 63–80.
Casey, E.S. (1998), *The Fate of Place: A Philosophical History*, University of California Press, Berkeley (Calif.).
Catton Jr, W.R., and Dunlap, R.E. (1980), 'A New Ecological Paradigm for Post-Exuberant Sociology', *American Behavioral Scientist*, 24, 15–47.
Chaloupka, W., and Cawley, R.M. (1993), 'The Great Wild Hope: Nature, Environmentalism, and the Open Secret', in J. Bennett and W. Chaloupka (eds),

In the Nature of Things: Language, Politics and the Environment, University of Minnesota Press, Minneapolis (Minn.), 3–23.
Cheney, J. (1987), 'Ecofeminism and Deep Ecology', *Environmental Ethics*, 9, 115–45.
———. (1989), 'Postmodern Environmental Ethics: Ethics as Bioregional Narrative', *Environmental Ethics*, 11, 117–35.
Cheng, C-Y. (1986), 'On the Environmental Ethics of the Tao and the Ch'i', *Environmental Ethics*, 8, 351–70.
Chiras, D. (1992), *Lessons from Nature: Learning to Live Sustainably on the Earth*, Island Press, Washington (D.C.).
Chodorow, N. (1978), *The Reproduction of Mothering: Psychoanalysis and the Sociology of Gender*, University of California Press, Berkeley (Calif.).
Christoff, P. (1996a), 'Ecological Modernisation, Ecological Modernities', *Environmental Politics*, 5, 476–500.
———. (1996b), 'Ecological Citizens and Ecologically Guided Democracy', in B. Doherty and M. de Geus (eds), *Democracy and Green Political Thought: Sustainability, Rights and Citizenship*, Routledge, London, 151–69.
Clark, J. (1982), 'Review of M. Bookchin, *Toward an Ecological Society*', *Telos*, 52, 224–29.
———. (1984), *The Anarchist Moment: Reflections on Culture, Nature and Power*, Black Rose, Montreal.
———. (1989), 'Marx's Inorganic Body', *Environmental Ethics*, 11, 243–58.
———. (1992), 'What Is Social Ecology?', *Society and Nature*, 1(1), 85–92.
Clark, S.L.R. (1975), *The Moral Status of Animals*, Clarendon Press, Oxford.
———. (1997), *Animals and their Moral Standing*, Routledge, London.
Clarke, A.C. (1962), *Profiles of the Future*, Harper & Row, New York.
Clifford, S. (1994), 'Common Ground, Cultural Landscapes and Conservation', in J. de Gryse and A. Sant (eds), *Our Common Ground: A Celebration of Art, Place and Environment*, Australian Institute of Landscape Architects (Tasmania)/Centre for Environmental Studies, University of Tasmania, Hobart (Tas.), 16–30.
Clifford, S. and King, A. (1993), *Local Distinctiveness: Place, Particularity and Identity*, Common Ground, London.
Cobb Jr, J.B. (1982), *Beyond Dialogue: Toward a Mutual Transformation of Christianity and Buddhism*, Fortress Press, Philadelphia (Pa.).
Coddington, W. (1993), *Environmental Marketing: Positive Strategies for Reaching the Green Consumer*, McGraw Hill, New York.
Coles, R. (1992), *Self/Power/Other: Political Theory and Dialogical Ethics*, Cornell University Press, Ithaca (N.Y.).
———. (1993), 'Ecotones and Environmental Ethics: Adorno and Lopez', in J. Bennett and W. Chaloupka (eds), *In the Nature of Things: Language, Politics and the Environment*, University of Minnesota Press, Minneapolis (Minn.), 226–49.
Collard, A., with J. Contrucci (1988), *Rape of the Wild: Man's Violence against Animals and the Earth*, Indiana University Press, Indianapolis (Ind.).
Common, M., and Perrings, C. (1992), 'Towards an Ecological Economics and Sustainability', *Ecological Economics*, 6, 7–34.
Commoner, B. (1972), *The Closing Circle: Confronting the Environmental Crisis*, Jonathan Cape, London.
Conn, S.A. (1995), 'When the Earth Hurts, Who Responds?', in T. Roszak, M.E. Gomes, and A.D. Kanner (eds), *Ecopsychology: Restoring the Earth, Healing the Mind*, Sierra Club, San Francisco (Calif.), 156–71.
Coombs, H.C. (1990), *The Return of Scarcity: Strategies for an Economic Future*, Cambridge University Press, Cambridge.

Coren S. (1994), *The Intelligence of Dogs: Canine Consciousness and Capabilities*, Free Press, New York.

Costanza, R. (1989), 'What Is Ecological Economics?', *Ecological Economics*, 1, 1–7.

Costanza, R., and Daly, H.E. (1987), 'Toward an Ecological Economics', *Ecological Modelling*, 38, 1–7.

Costanza, R., Daly, H.E., and Bartholomew, J.A. (1991), 'Goals, Agenda and Policy Recommendations for Ecological Economics', in R. Costanza (ed.), *Ecological Economics: The Science and Management of Sustainability*, Columbia University Press, New York, 1–20.

Cotgrove, S. (1982), *Catastrophe or Cornucopia: The Environment, Politics and the Future*, Wiley & Sons, Chichester (Eng.).

Cotgrove, S., and Duff, A. (1980), 'Environmentalism, Middle-Class Radicalism and Politics', *Sociological Review*, 28, 333–51.

Crawford, M. (1991), 'Green Futures on Wall Street', *New Scientist*, 5 January.

Cristaudo, W. (1990), 'Heidegger's Political Judgement: Nazism and After', *Australian Journal of Political Science*, 25, 289–308.

Cronon, W. (1995a), 'Introduction: In Search of Nature', in W. Cronon (ed.), *Uncommon Ground: Toward Reinventing Nature*, W.W. Norton, New York, 23–67.

———. (1995b), 'The Trouble with Wilderness; or, Getting Back to the Wrong Nature', in W. Cronon (ed.), *Uncommon Ground: Toward Reinventing Nature*, W.W. Norton, New York, 69–90.

Cuomo, C.J. (1992), 'Unravelling the Problems in Ecofeminism', *Environmental Ethics*, 14, 351–63.

———. (1994), 'Ecofeminism, Deep Ecology, and Human Population', in K.J. Warren (ed.), *Ecological Feminism*, Routledge, London, 88–105.

———. (1998), *Feminism and Ecological Communities: An Ethic of Flourishing*, Routledge, New York.

Curtin, D. (1994), 'Dogen, Deep Ecology, and the Ecological Self', *Environmental Ethics*, 16, 195–213.

Daly, H.E. (1973a), 'Introduction', in H.E. Daly (ed.), *Towards a Steady State Economy*, W.H. Freeman, San Francisco (Calif.), 1–36.

———. (1973b), 'The Steady State Economy: Towards a Political Economy of Biophysical Equilibrium and Moral Growth', in H.E. Daly (ed.), *Towards a Steady State Economy*, W.H. Freeman, San Francisco (Calif.), 149–74.

———. (1977), *Steady-State Economics: The Economics of Biophysical Equilibrium and Moral Growth*, W.H. Freeman, San Francisco (Calif.).

———. (1991), 'Elements of Environmental Macroeconomics', in R. Costanza (ed.), *Ecological Economics: The Science and Management of Sustainability*, Columbia University Press, New York, 32–46.

———. (1992), 'Free Market Environmentalism: Turning a Good Servant into a Bad Master', *Critical Review*, 6, 171–83.

———. (1993), 'Sustainable Growth: An Impossible Theorem', in H.E. Daly and K.N. Townsend (eds), *Valuing the Earth: Economics, Ecology, Ethics*, MIT Press, Cambridge (Mass.), 267–73.

———. (1994), 'Operationalizing Sustainable Development by Investing in Natural Capital', in R. Goodland and V. Edmundson (eds), *Environmental Assessment and Development*, World Bank, Washington (D.C.), 152–59.

———. (1995), 'On Wilfred Beckerman's Critique of Sustainable Development',

Environmental Values, 4, 49–55.
———. (1996), *Beyond Growth: The Economics of Sustainable Development*, Beacon Press, Boston (Mass.).
Daly, H.E., and Cobb Jr, J.B., with C.W. Cobb (1989), *For the Common Good: Redirecting the Economy Towards Community, the Environment and a Sustainable Future*, Beacon Press, Boston (Mass.).
Daly, H.E., and Townsend, K.N. (1993), 'Introduction to Part II', in H.E. Daly and K.N. Townsend (eds), *Valuing the Earth: Economics, Ecology, Ethics*, MIT Press, Cambridge (Mass.), 155–57.
Daly, M. (1987), *Gyn/Ecology: The Metaethics of Radical Feminism*, Women's Press, London.
Darier, E. (1999a), 'Foucault and the Environment: An Introduction', in E. Darier (ed.), *Discourses of the Environment*, Blackwell, Oxford, 1–33.
———. (1999b), 'Foucault Against Environmental Ethics', in E. Darier (ed.), *Discourses of the Environment*, Blackwell, Oxford, 217–40.
Dauncey, G. (1989), 'A New Local Economic Order', in P. Ekins (ed.), *The Living Economy: A New Economics in the Making*, Routledge, London, 264–71.
Davidson, J. (1998), 'Sustainable Development: Business as Usual or a New Way of Living?', in C. Starr (ed.), *Green Politics in Grey Times: Ecopolitics XI Conference Proceedings*, Department of Political Science, University of Melbourne, Melbourne (Vic.), 236–48.
Davies, P. (1992a), 'Fascinating, Wonderful and Weird: An Interview with Paul Davies', *Island*, 51, 22–30.
———. (1992b), *The Mind Of God: Science and the Search for Ultimate Meaning*, Simon & Schuster, London.
Davies, P., and Gribbin, J. (1991), *The Matter Myth: Towards 21st-Century Science*, Viking/Penguin, Harmondsworth (Eng.).
Davion, V. (1994), 'Is Ecofeminism Feminist?', in K.J. Warren (ed.), *Ecological Feminism*, Routledge, London, 8–28.
Dawkins, M.S. (1993), *Through Our Eyes Only?: The Search for Animal Consciousness*, W.H. Freeman, Oxford.
Dawkins, R. (1976), *The Selfish Gene*, Oxford University Press, Oxford.
———. (1986), *The Blind Watchmaker*, Longmans, London.
d'Eaubonne, F. (1974), *La Féminisme ou la Mort*, Pierre Horay, Paris.
de Geus, M. (1996), 'The Ecological Restructuring of the State', in B. Doherty and M. de Geus (eds), *Democracy and Green Political Thought: Sustainability, Rights and Citizenship*, Routledge, London, 188–211.
de Silva, P. (1998), *Environmental Philosophy and Ethics in Buddhism*, Macmillan, London.
de Waal, F. (1996), *Good Natured: The Origins of Right and Wrong in Humans and Other Animals*, Harvard University Press, Cambridge (Mass.).
Deal, C. (1993), *The Greenpeace Guide to Anti-Environmental Organizations*, Odonian Press, Berkeley (Calif.).
DeGrazia, D. (1996), *Taking Animals Seriously: Mental Life and Moral Status*, Cambridge University Press, Cambridge.
Deleage, J-P. (1994), 'Eco-Marxist Critique of Political Economy', in M. O'Connor (ed.), *Is Capitalism Sustainable? Political Economy and the Politics of Ecology*, Guildford, New York, 37–52.
Deleuze, G. (1990), *The Logic of Sense*, C.V. Boundas (ed.), Columbia University Press, New York.
Deleuze, G., and Guattari, F. (1988), *A Thousand Plateaus: Capitalism and*

Schizophrenia, B. Massumi (trans.), Athlone, London.
———. (1994), *What Is Philosophy?*, H. Tomlinson and G. Burchell (trans.), Columbia University Press, New York.
Dennett, D.C. (1995), *Darwin's Dangerous Idea: Evolution and the Meanings of Life*, Allen Lane/Penguin, London.
Derrida, J. (1976), *Of Grammatology*, John Hopkins University Press, G. Chakravorty (trans.), Baltimore (Md.).
Deutsch, E. (1986), 'A Metaphysical Grounding for Nature Reverence: East-West', *Environmental Ethics*, 8, 293–99.
Devall, B. (1977), 'Currents in the River of Environmentalism', *Econews*, April, 9.
———. (1980), 'The Deep Ecology Movement', *Natural Resources Journal*, 20, 299–322.
———. (1988), *Simple in Means, Rich in Ends: Practicing Deep Ecology*, Gibbs M. Smith, Salt Lake City (Utah).
———. (1991), 'Deep Ecology and Radical Environmentalism', *Society and Natural Resources*, 4, 247–58.
———. (1992), 'Compassionate Activism for Hard Realities: Reflections on Arne Naess' Commitment to Nonviolence', *The Trumpeter: Journal of Ecosophy*, 9, 77–79.
Devall, B., and Sessions, G. (1985), *Deep Ecology: Living as if Nature Mattered*, Gibbs M. Smith, Salt Lake City (Utah).
Diamond, I., and Kuppler, L. (1990), 'Frontiers of the Imagination: Women, History, and Nature', *Journal of Women's History*, 1, 160–80.
Diamond, J. (1992a), *The Rise and Fall of the Third Chimpanzee*, Vintage, London.
———. (1992b), 'Playing Dice with Megadeath', in S. Morgan and D. Okerstrom (eds), *The Endangered Earth: Readings for Writers*, Allyn & Bacon, Boston (Mass.), 63–71.
Diesendorf, M. (1997), 'Principles of Ecological Sustainability', in M. Diesendorf and C. Hamilton (eds), *Human Ecology, Human Economy: Ideas for an Ecologically Sustainable Future*, Allen & Unwin, St Leonards (NSW), 64–97.
Dillard, A. (1976), *Pilgrim at Tinker Creek*, Pan, London.
diZerega, G. (1995), 'Individuality, Human and Natural Communities, and the Foundations of Ethics', *Environmental Ethics*, 17, 23–37.
———. (1996a), 'Deep Ecology and Liberalism: The Greener Implications of Evolutionary Liberal Theory', *Review of Politics*, 58, 699–734.
———. (1996b), 'Deep Ecology and Liberalism', *The Trumpeter: Journal of Ecosophy*, 13, 173–80.
Dobson, A. (1990), *Green Political Thought*, Unwin Hyman, London.
———. (1993), 'Critical Theory and Green Politics', in A. Dobson and P. Lucardie (eds), *The Politics of Nature: Explorations in Green Political Theory*, Routledge, London, 190–209.
———. (1995), *Green Political Thought: Second Edition*, Routledge, London.
———. (1996a), 'Environmental Sustainabilities: An Analysis and a Typology', *Environmental Politics*, 5, 401–28.
———. (1996b), 'Representative Democracy and the Environment', in W.M. Lafferty and J. Meadowcroft (eds), *Democracy and the Environment: Problems and Prospects*, Edward Elgar, Cheltenham (Eng.), 124–39.
———. (1996c), 'Democratising Green Theory: Preconditions and Principles', in B. Doherty and M. de Geus (eds), *Democracy and Green Thought: Sustainability, Rights and Citizenship*, Routledge, London, 132–48.
Dodge, J. (1981), 'Living By Life: Some Bioregional Theory and Practice', *The*

CoEvolution Quarterly, Winter, 6–12.

Doherty, B., and de Geus, M. (1996), 'Introduction', in B. Doherty and M. de Geus (eds), *Democracy and Green Thought: Sustainability, Rights and Citizenship*, Routledge, London, 1–15.

Doubiago, S. (1989), 'Mama Coyote Talks to the Boys', in J. Plant (ed.), *Healing the Wounds: The Promise of Ecofeminism*, New Society, Philadelphia (Pa.), 40–44.

Douthwaite, R. (1996), *Short Circuit: Strengthening Local Economies for Security in an Unstable World*, Green Books, Totnes (Eng.).

Downs, A. (1972), 'Up and Down with Ecology: The Issue-Attention Cycle', *Public Interest*, 28, 38–50.

Doyle, T. (1998), 'Sustainable Development and Agenda 21: The Secular Bible of Global Free Markets and Pluralist Democracy', *Third World Quarterly*, 19, 771–85.

Drengson, A. (1980), 'Shifting Paradigms: From the Technocratic to the Person-Planetary', *Environmental Ethics*, 2, 221–40.

———. (1989), *Beyond Environmental Crisis: From Technocrat to Planetary Person*, Peter Lang, New York.

———. (1992), 'Tacit Knowledge, the Ecological Unconscious, and Narrative', *The Trumpeter: Journal of Ecosophy*, 9, 1–2.

———. (1995), 'The Deep Ecology Movement', *The Trumpeter: Journal of Ecosophy*, 12, 143–45.

———. (1996), 'How Many Realities?', *The Trumpeter: Journal of Ecosophy*, 13, 2–3.

Drengson, A., and Inoue, Y. (eds) (1995), *The Deep Ecology Movement: An Introductory Anthology*, North Atlantic Books, Berkeley (Calif.).

Dryzek, J.S. (1987), *Rational Ecology: Environment and Political Economy*, Blackwell, Oxford.

———. (1990a), 'Green Reason: Communicative Ethics for the Biosphere', *Environmental Ethics*, 12, 195–210.

———. (1990b), *Discursive Democracy: Politics, Policy, and Political Science*, Cambridge University Press, Cambridge.

———. (1992), 'Ecology and Discursive Democracy: Beyond Liberal Capitalism and the Administrative State', *Capitalism, Nature, Socialism*, 3(2), 18–42.

———. (1996a), *Democracy in Capitalist Times: Ideals, Limits and Struggles*, Oxford University Press, New York.

———. (1996b), 'Political and Ecological Communication', in F. Mathews (ed.), *Ecology and Democracy*, Frank Cass, London, 13–30.

———. (1996c), 'Foundations for Environmental Political Economy: The Search for Homo Ecologicus', *New Political Economy*, 1, 27–40.

———. (1996d), 'Strategies of Ecological Democratization', in W.M. Lafferty and J. Meadowcroft (eds), *Democracy and the Environment: Problems and Prospects*, Edward Elgar, Cheltenham (Eng.), 108–23.

———. (1997), *The Politics of the Earth: Environmental Discourses*, Oxford University Press, Oxford.

Dunkley, G. (1992), *The Greening of the Red: Sustainability, Socialism and the Environmental Crisis*, Pluto Press/Australian Fabian Society/Socialist Forum, Sydney, 1992.

Dunlap, R.E., and Van Liere, K.D. (1978), 'The New Environmental Paradigm', *Environmental Education*, 9, 10–19.

Dutney, A. (1992), 'Creation and the Church', in P.R. Hay and R. Eckersley (eds), *Ecopolitical Theory: Essays from Australia*, Occasional Paper 24, Centre for Environmental Studies, University of Tasmania, Hobart (Tas.), 77–92.

Dwivedi, O.P. (1990), '*Satyagraha* for Conservation: Awakening the Spirit of Hinduism', in J.R. Engel and J.G. Engel (eds), *Ethics of Environment and Development: Global Challenge, International Response*, Belhaven, London, 201–12.

Easlea, B. (1973), *Liberation and the Aims of Science: An Essay on Obstacles to the Building of a Beautiful World*, Chatto & Windus/Sussex University Press, London.

———. (1980), *Witch-Hunting, Magic and the New Philosophy: An Introduction to the Debates of the Scientific Revolution 1450–1750*, Harvester Press, Brighton (Eng.).

———. (1981), *Science and Sexual Oppression: Patriarchy's Confrontation with Women and Nature*, Weidenfeld and Nicolson, London.

———. (1983), *Fathering the Unthinkable: Masculinity, Science and the Nuclear Arms Race*, Pluto Press, London.

Eckersley, R. (1988), 'The Road to Ecotopia? Socialism Versus Environmentalism', *The Ecologist*, 18, 142–47.

———. (1989), 'Divining Evolution: The Ecological Ethics of Murray Bookchin', *Environmental Ethics*, 11, 99–116.

———. (1990), 'Habermas and Green Political Theory: Two Roads Diverging', *Theory and Society*, 19, 739–76.

———. (1992), *Environmentalism and Political Theory: Toward an Ecocentric Approach*, State University of New York Press, New York.

———. (1993a), 'Rationalising the Environment: How Much Am I Bid?', in S. Rees, G. Rodley and F. Stilwell (eds), *Beyond the Market: Alternatives to Economic Rationalism*, Pluto Press, Leichhardt (NSW), 237–50.

———. (1993b), 'Free Market Environmentalism: Friend or Foe?', *Environmental Politics*, 2, 1–19.

———. (1993c), 'Just Natural Relations? Recent Developments in Environmental Political Theory', *Political Theory Newsletter*, 5, 110–25.

———. (1996a), 'Liberal Democracy and the Rights of Nature: The Struggle for Inclusion', in F. Mathews (ed.), *Ecology and Democracy*, Frank Cass, London, 169–98.

———. (1996b), 'Greening Liberal Democracy: The Rights Discourse Revisited', in B. Doherty and M. de Geus (eds), *Democracy and Green Political Thought: Sustainability, Rights and Citizenship*, Routledge, London, 212–36.

———. (1999), 'The Discourse Ethic and the Problem of Representing Nature', *Environmental Politics*, 8, 24–49.

———. (2000), 'Deliberative Democracy, Ecological Representation and Risk: Towards a Democracy of the Affected', in M. Saward (ed.), *Democratic Innovation: Deliberation, Representation and Association*, Routledge, London, 117–32.

Ecologist, The (1993), *Whose Common Future?: Reclaiming the Commons*, Earthscan, London.

Ehrenfeld, D. (1988), 'Why Put a Value on Biodiversity?, in E.O. Wilson (ed.), *Biodiversity*, National Academy Press, Washington (D.C.), 212–16.

———. (1993), *Beginning Again*, Oxford University Press, New York.

Ehrlich, P.[R.] (1972), *The Population Bomb*, Pan/Ballantyne, London, 1972.

Ehrlich, P.R., and Ehrlich, A.H. (1996), *The Betrayal of Science and Reason*, Island Press, Washington (D.C.).

Eisler, R. (1987), *The Chalice and the Blade: Our History, Our Future*, Harper & Row, San Francisco (Calif.).

Ekins, P. (1992), 'Sustainability First', in P. Ekins and M. Max-Neef (eds), *Real-Life*

Economics: Understanding Wealth Creation, Routledge, London, 412–22.
———. (1996), *Environmental Taxes and Charges: National Experiences and Plans*, European Foundation for the Improvement of Living and Working Conditions, Dublin.
Ekins, P. (ed.) (1989), *The Living Economy: A New Economics in the Making*, Routledge, London.
Ekins, P., Hillman, M., and Hutchison, R. (1992), *Wealth Beyond Measure: An Atlas of the New Economics*, Gaia Books, London.
Elgin, D. (1981), *Voluntary Simplicity*, William Morrow, New York.
Elkington, J., and Burke, T. (1987), *The Green Capitalists*, Victor Gollancz, London.
Elsdon, R. (1981), *Bent World: Science, the Bible and the Environment*, Inter-Varsity Press, Leicester (Eng.).
Emerson, R.W. (nd), *The Complete Prose Works*, Ward, Lock & Co., London.
Enzensberger, H.M. (1974), 'A Critique of Political Ecology', *New Left Review*, 84, 3–31.
Evans, Jeremy (1973), 'The Tao and the Eco-System', in J. Nurser (ed.), *Living with Nature: Proceedings of the 4th National Summer School on Religion*, Centre for Continuing Education, Australian National University, Canberra, 37–46.
Evans, Judy (1993), 'Ecofeminism and the Politics of the Gendered Self', in A. Dobson and P. Lucardie (eds), *The Politics of Nature: Explorations in Green Political Theory*, Routledge, London, 177–89.
Evernden, N. (1978), 'Beyond Ecology: Self, Place, and the Pathetic Fallacy', *North American Review*, 263(4), 16–20.
———. (1985a), *The Natural Alien: Humankind and Environment*, University of Toronto Press, Toronto.
———. (1985b), 'Constructing the Natural: The Darker Side of the Environmental Movement', *North American Review*, 270(1), 15–19.
———. (1992), *The Social Creation of Nature*, John Hopkins University Press, Baltimore (Md.).

Faber, D., and O'Connor, J. (1989a), 'The Struggle for Nature: Environmental Crisis and the Crisis of Environmentalism in the United States', *Capitalism, Nature, Socialism*, 1(2), 12–39.
———. (1989b), 'The Struggle for Nature: Rejoinders', *Capitalism, Nature, Socialism*, 1(3), 174–78.
———. (1991), 'Eco-Socialism/Eco-Feminism: Reply', *Capitalism, Nature, Socialism*, 2(1), 137–40.
Farias, V. (1989), *Heidegger and Nazism*, Temple University Press, Philadelphia (Pa.).
Feeny, D., Berkes, F., McCay, B.J., and Acheson, J.M. (1990), 'The Tragedy of the Commons: Twenty-Two Years Later', *Human Ecology*, 18, 1–19.
Ferre, F. (1994), 'Personalistic Organicism: Paradox or Paradigm?', in R. Attfield and A. Belsey (eds), *Philosophy and the Natural Environment*, Cambridge University Press, Cambridge, 59–73.
Finger, M. (1988), 'Gaia: Implications for the Social Sciences', in P. Bunyard and E. Goldsmith (eds), *GAIA, the Thesis, the Mechanisms and the Implications: Proceedings of the First Annual Camelford Conference on the Implications of the Gaia Hypothesis*, Wadebridge Ecological Centre, Camelford (Eng.), 201–18.
Firestone, S. (1970), *The Dialectic of Sex: The Case for Feminist Revolution*, Morrow, New York.
Flynn, A., and Lowe, P. (1992), 'The Conservative Party and the Environment', in W.

Rudig (ed.), *Green Politics Two*, Edinburgh University Press, Edinburgh, 9–36.

Foster, J. (1997), 'Introduction: Environmental Value and the Scope of Economics', in J. Foster (ed.), *Valuing Nature? Ethics, Economics and the Environment*, Routledge, London, 1–17.

Foster, J.B. (1992), 'The Absolute General Law of Environmental Degradation under Capitalism', *Capitalism, Nature, Socialism*, 3(3), 77–82.

———. (1997), 'The Crisis of the Earth: Marx's Theory of Ecological Sustainability as a Nature-Imposed Necessity for Human Production', *Organization & Environment*, 10, 278–95.

Fotopolous, T. (1994), 'The End of Socialist Statism', *Society and Nature*, 2(3), 11–68.

Foucault, M. (1980), *Power/Knowledge: Selected Interviews and Other Writings, 1972–1977*, C. Gordon (ed.), Pantheon, New York.

———. (1991), 'Governmentality', in G. Burchell, C. Gordon, and P. Miller (eds), *The Foucault Effect: Studies in Governmentality*, Routledge, London, 78–82.

Fox, M. (1983), 'Preface', in G. Uhlein, *Meditations with Hildegard of Bingen*, Bear & Co., Santa Fé (N.Mex.), 7–11.

———. (1984), 'Creation-Centred Spirituality from Hildegard of Bingen to Julian of Norwich: 300 Years of an Ecological Spirituality in the West', in P.N. Joranson and K. Butigan (eds), *Cry of the Environment: Building the Christian Creation Tradition*, Bear & Company, Santa Fé (N.Mex.), 84–106.

———. (1989), *The Coming of the Cosmic Christ: The Healing of Mother Earth and the Birth of a Global Renaissance*, Collins Dove, Melbourne (Vic.).

Fox, M.A. (1997), 'Humans and Other Animals: Ethical Reflections on their Future Relationship', *Island*, 70, 31–47.

Fox, S. (1981), *John Muir and His Legacy*, Little, Brown & Co., Boston (Mass.).

Fox, W. (1986), *Approaching Deep Ecology: A Response to Richard Sylvan's Critique of Deep Ecology*, Occasional Paper No. 20, Centre for Environmental Studies, University of Tasmania, Hobart (Tas.).

———. (1989), 'The Deep Ecology-Ecofeminism Debate and its Parallels', *Environmental Ethics*, 11, 5–25.

———. (1990), *Toward a Transpersonal Ecology: Developing New Foundations for Environmentalism*, Shambhala, Boston (Mass.).

———. (1992), 'New Philosophical Directions in Environmental Decision-Making', in P.R. Hay and R. Eckersley (eds), *Ecopolitical Theory: Essays from Australia*, Occasional Paper 24, Centre for Environmental Studies, University of Tasmania, Hobart (Tas.), 1–20.

———. (1993), 'From Anthropocentrism to Deep Ecology', *ReVision*, 16, 75–76.

———. (1994), 'Ecophilosophy and Science', *The Environmentalist*, 14, 207–13.

———. (1995a), 'Deep Ecology: Meanings', in R. Paehlke (ed.), *Conservation and Environmentalism: An Encylopedia*, Garland, New York, 165–67.

———. (1995b), 'Education, the Interpretive Agenda of Science, and the Obligation of Scientists to Promote this Agenda', *Environmental Values*, 4, 109–14.

———. (1996), 'A Critical Overview of Environmental Ethics', *World Futures*, 46, 1–21.

Frankel, B. (1987), *The Post-Industrial Utopians*, Polity/Basil Blackwell, Cambridge.

Frankena, W.K. (1979), 'Ethics and the Environment', in K.E. Goodpaster and K.M. Sayre (eds), *Ethics and Problems of the 21st Century*, University of Notre Dame Press, Notre Dame (Ind.), 3–20.

Freeman, R.E., Pierce, J., and Dodd, R.H. (2000), *Environmentalism and the New Logic of Business*, Oxford University Press, Oxford.

Frodeman, R. (1992), 'Radical Environmentalism and the Political Roots of Postmodernism: Differences that Make a Difference' *Environmental Ethics*, 14, 307–19.
Fuchs, S. (1970), 'Ecology Movement Exposed', *Progressive Labor*, 7, 50–63.
Fukuyama, F. (1992), *The End of History and the Last Man*, Hamilton, London.

Gaard, G. (1993), 'Ecofeminism and Native American Cultures: Pushing the Limits of Cultural Imperialism?', in G. Gaard (ed.), *Ecofeminism: Women, Animals, Nature*, Temple University Press, Philadelphia (Pa.), 295–314.
Galtung, J. (1979), *Development, Environment and Technology: Towards a Technology for Self-Reliance*, United Nations, New York.
———. (1980a), 'The Basic Needs Approach', in K. Lederer (ed.), *Human Needs: A Contribution to the Current Debate*, Oelgeschlager, Gunn & Hain, Cambridge (Mass.), 55–125.
———. (1980b), *Basic Needs and the Green Movement*, United Nations University, Tokyo.
———. (1989), 'Towards a New Economics: On The Theory and Practice of Self-reliance', in P. Ekins (ed.), *The Living Economy: A New Economics in the Making*, Routledge, London, 97–109.
Garb, Y.J. (1990), 'Perspective or Escape?: Ecofeminist Musings on Contemporary Earth Imagery', in I. Diamond and G.F. Orenstein (eds), *Reweaving the World: The Emergence of Ecofeminism*, Sierra Club, San Francisco (Calif.), 264–78.
Gare, A.E. (1995), *Postmodernism and the Environmental Crisis*, Routledge, London.
Giddens, A. (1991), *Modernity and Self-Identity: Self and Society in the Late Modern Age*, Polity Press, Cambridge.
———. (1994), *Beyond Left and Right: The Future of Radical Politics*, Polity Press, Cambridge.
Gilligan, C. (1982), *In a Different Voice: Psychological Theory and Women's Development*, Harvard University Press, Cambridge (Mass.).
Glaser, H. (1978), *The Cultural Roots of National Socialism*, Croom Helm, London.
Godfrey-Smith, W. (1979), 'The Value of Wilderness', *Environmental Ethics*, 1, 309–19.
Goethe, J.W. von (1987), *Scientific Studies*, D. Miller (ed.), Princeton University Press, Princeton (N.J.).
Goldblatt, D. (1996), *Social Theory and the Environment*, Polity, Cambridge.
Goldsmith, E. (1972), 'Conclusion: What of the Future?', in E. Goldsmith (ed.), *Can Britain Survive?*, Sphere, London, 277–90.
———. (1988a), 'Gaia: Some Implications for Theoretical Ecology', *The Ecologist*, 18, 64–74.
———. (1988b), 'The Way: An Ecological World-View', *The Ecologist*, 18, 160–85.
———. (1988c), *The Great U-Turn: De-Industrializing Society*, Green Books, Bideford (Eng.).
Goldsmith, E., Allen, R., Allaby, M., Davoll, T., and Lawrence, S. (1972), *Blueprint for Survival*, Penguin, Harmondsworth (Eng.).
Goodin, R.E. (1992), *Green Political Theory*, Polity, Cambridge.
———. (1996), 'Enfranchising the Earth, and its Alternatives', *Political Studies*, 44, 835–49.
———. (1998), 'Selling Environmental Indulgences', in J.S. Dryzek and D. Schlosberg (eds), *Debating the Earth: The Environmental Politics Reader*, Oxford University Press, Oxford, 237–54.

Goodman, R. (1980), 'Taoism and Ecology', *Environmental Ethics*, 2, 73–80.
Goodpaster, K. (1978), 'On Being Morally Considerable', *Journal of Philosophy*, 75, 308–25.
———. (1979), 'From Egoism to Environmentalism', in K.E. Goodpaster and K.M. Sayre (eds), *Ethics and Problems of the 21st Century*, University of Notre Dame Press, Notre Dame (Ind.), 21–35.
Gorz, A. (1980), *Ecology as Politics*, South End Press, Boston (Mass.).
———. (1982), *Farewell to the Working Class: An Essay on Post-Industrial Socialism*, Pluto Press, London.
Gould, N. (1974), 'William Morris', *The Ecologist*, 4, 210–12.
Gouldson, A., and Murphy, J. (1997), 'Ecological Modernisation: Restructuring Industrial Economics', in M. Jacobs (ed.), *Greening the Millenium: The New Politics of the Environment*, Blackwell, Oxford, 74–86.
Gowdy, J.M., and Olsen, P.R. (1995), 'Further Problems with Neo-Classical Environmental Economics', *Environmental Ethics*, 161–71.
Granberg-Michaelson, W. (1992), *Redeeming the Creation: The Rio Earth Summit: Challenges for the Churches*, World Council of Churches Publications, Geneva.
Gray, J. (1993), *Beyond the New Right: Markets, Government and the Common Environment*, Routledge, London.
Griffin, D.R. (1992), *Animal Minds*, University of Chicago Press, Chicago (Ill.).
Griffin, S. (1978), *Woman and Nature: The Roaring Inside Her*, Harper & Row, New York.
———. (1995), *The Eros of Everyday Life: Essays on Ecology, Gender and Society*, Doubleday, New York.
Griscom, J.L. (1981), 'On Healing the Nature/History Split in Feminist Thought', *Heresies*, 13(4), 4–9.
Grove-White, R. (1997), 'The Environmental "Valuation" Controversy: Observations on its Recent History and Significance', in J. Foster (ed.), *Valuing Nature: Economics, Ethics and the Environment*, Routledge, London, 21–31.
Guattari, F. (1990), *Les Trois Ecologies*, Editions Galilee, Paris.
Guha, R. (1989), 'Radical American Environmentalism and Wilderness Preservation: A Third World Critique', *Environmental Ethics*, 11, 71–83.
Gullett, W. (1997), 'Environmental Protection and the "Precautionary Principle": A Response to Scientific Uncertainty in Environmental Management', *Environmental and Planning Law Journal*, 14, 52–69.
Gundersen, A. (1995), *The Environmental Promise of Democratic Deliberation*, University of Wisconsin Press, Madison (Wis.).
Gupta, L. (1993), 'Purity, Pollution and Hinduism', in C.J. Adams (ed.), *Ecofeminism and the Sacred*, Continuum, New York, 99–116.

Habermas, J. (1981), 'New Social Movements', *Telos*, 49, 33–37.
———. (1984), *The Theory of Communicative Action I: Reason and the Rationalization of Society*, Beacon Press, Boston (Mass.).
———. (1987), *The Theory of Communicative Action II: Lifeworld and System*, Beacon Press, Boston (Mass.).
———. (1989), 'Work and *Weltanschauung*: The Heidegger Controversy from a German Perspective', *Critical Inquiry*, 15, 431–56.
Haenke, D. (1984), 'A Short History of the Bioregional Movement', in *North American Bioregional Congress Proceedings*, New Life Farm, Drury (Md.), 3–6.
Hager, N., and Burton, B. (1999), *Secrets and Lies: The Anatomy of an Anti-*

Environmental Campaign, Craig Potton, Nelson (NZ).

Hajer, M.A. (1995), *The Politics of Environmental Discourse: Ecological Modernization and the Policy Process*, Oxford University Press, Oxford.

———. (1996), 'Ecological Modernisation as Cultural Politics', in S. Lash, B. Szerszynski and B. Wynne (eds), *Risk, Environment and Modernity*, Sage, London, 246–68.

Hall, D.L. (1987) 'On Seeking a Change of Environment: A Quasi-Taoist Proposal', *Philosophy East and West*, 37, 160–71.

Hallen, P. (1992), 'Some Ecofeminist Dimensions of Ecopolitics', in P.R. Hay and R. Eckersley (eds), *Ecopolitical Theory: Essays from Australia*, Occasional Paper 24, Centre for Environmental Studies, University of Tasmania, Hobart (Tas.), 49–76.

———. (1994), 'Reawakening the Erotic: Why the Conservation Movement Needs Ecofeminism', *Habitat Australia*, February, 18–21.

Hamilton, C. (1994), *The Mystic Economist*, Willow Park, Fyshwick (ACT).

———. (1997), 'Foundations of Ecological Economics', in M. Diesendorf and C. Hamilton (eds), *Human Ecology, Human Economy: Ideas for an Ecologically Sustainable Future*, Allen & Unwin, St Leonards (NSW), 35–66.

———. (1998), 'Economic Growth and Social Decline', *Australian Quarterly*, 70(3), 22–30.

Haraway, D. (1985), 'A Manifesto for Cyborgs: Science, Technology and Socialist Feminism in the 1980s', *Socialist Review*, 15(2), 64–107.

———. (1991), *Simians, Cyborgs, and Women: The Reinvention of Nature*, Routledge, New York.

———. (1992), 'The Promises of Monsters: A Regenerative Politics for Inappropriate/d Others', in L. Grossberg, C. Nelson, and P. Treichler (eds), *Cultural Studies*, Routledge, New York, 295–337.

Hardin, G. (1968), 'The Tragedy of the Commons', *Science*, 162, 1243–248.

———. (1973), *Exploring New Ethics for Survival: The Voyage of the Spaceship Beagle*, Penguin, Baltimore (Md.).

———. (1974a), 'Living on a Lifeboat', *Bio Science*, 24, 561–68.

———. (1974b), 'The Economics of Wilderness', *The Ecologist*, 4, 44–46.

———. (1985), *Filters Against Folly: How to Survive Despite Economists, Ecologists, and the Merely Eloquent*, Penguin, New York.

Hare, W.L. (1990), *Ecologically Sustainable Development*, Australian Conservation Foundation, Fitzroy (Vic.).

Hargrove, E.C. (1992), 'Weak Anthropocentric Intrinsic Value', *The Monist*, 75, 183–207.

Harries, K. (1993), 'Thoughts on a Non-Arbitrary Architecture', in D. Seamon (ed.), *Dwelling, Seeing, and Designing: Toward a Phenomenological Ecology*, State University of New York Press, Albany (N.Y.), 41–60.

Harris, J. (1989), *Wilderness and Garden: Biblical Principles for a Christian View of the Environment*, Theology of Everyday Life Study Guide No. 1, Zadok Institute for Christianity and Society, Canberra.

Hartley, R.F. (1993), *Business Ethics: Violations of the Public Trust*, Wiley, New York.

Harvey, D. (1989), *The Condition of Postmodernity: An Enquiry into the Origins of Cultural Change*, Blackwell, Cambridge (Mass.).

———. (1996), *Justice, Nature and the Geography of Difference*, Blackwell, Malden (Mass.).

Hawken, P. (1995), *The Ecology of Commerce: A Declaration of Sustainability*, Phoenix, London.

Hay, J. (1985), 'The Resplendent Quetzel', in M. Tobias (ed.), *Deep Ecology*, Avant

Books, San Diego (Calif.), 161–68.
Hay, P.R. (1988), 'The Contemporary Environment Movement as Neo-Romanticism: A Re-appraisal from Tasmania', *Environmental Review*, 12(4), 39–59.
———. (1990), 'The Triumph of Market Liberalism: A Threat to the Environmentalist Agenda', *New Zealand Environment*, 65, 17–23.
———. (1992), 'Getting from Here to There: Reflections on a Green Praxis', in P.R. Hay and R. Eckersley (eds), *Ecopolitical Theory: Essays from Australia*, Occasional Paper 24, Centre for Environmental Studies, University of Tasmania, Hobart (Tas.), 223–38.
———. (1994a), 'Introduction', in J. de Gryse and A. Sant (eds), *Our Common Ground: A Celebration of Art, Place and Environment*, Australian Institute of Landscape Architects (Tasmania)/Centre for Environmental Studies, University of Tasmania, Hobart (Tas.), 11–15.
———. (1994b), 'The Politics of Tasmania's World Heritage Area: Contesting the Democratic Subject', *Environmental Politics*, 3, 1–21.
———. (1998), 'Green Politics "In the System": Assessing the Obstacles to Labor/Green Power Sharing', in S. Crook, J. Pakulski and E. Papadakis (eds), *Ebbing of the Green Tide? Environmentalism, Public Opinion and the Media in Australia*, Occasional Paper Series, No. 5, School of Sociology and Social Work, University of Tasmania, Hobart (Tas.), 103–25.
Hay, P.R., and Haward, M.G. (1988), 'Comparative Green Politics: Beyond the European Context?', *Political Studies*, 36, 433–48.
Hay, R.B. (1988), 'Toward a Theory of Sense of Place', *The Trumpeter: Journal of Ecosophy*, 5, 159–64.
———. (1992), 'An Appraisal of Our Meaningful Relationships in Place', *The Trumpeter: Journal of Ecosophy*, 9, 98–105.
Hayden, P. (1997), 'Gilles Deleuze and Naturalism: A Convergence with Ecological Theory and Politics', *Environmental Ethics*, 19, 185–204.
Hayward, B.M. (1996), 'The Greening of Participatory Democracy: Reconsideration of Theory', in F. Mathews (ed.), *Ecology and Democracy*, Frank Cass, London, 215–36.
Hayward, T. (1995), *Ecological Thought: An Introduction*, Polity, Cambridge.
Heidegger, M. (1978), *Basic Writings: From* Being and Time *(1927) to* The Task of Thinking *(1964)*, Routledge & Kegan Paul, London.
Heilbroner, R. (1974), *An Inquiry into the Human Prospect*, W.W. Norton, New York.
Heller, C. (1993), 'For the Love of Nature: Ecology and the Cult of the Romantic', in G. Gaard (ed.), *Ecofeminism: Women, Animals, Nature*, Temple University Press, Philadelphia (Pa.), 219–42.
Helvarg, D. (1994), *The War Against the Greens: The 'Wise Use' Movement, the New Right, and Anti-Environmental Violence*, Sierra Club, San Francisco (Calif.).
Henderson, H. (1988), *The Politics of the Solar Age: Alternatives to Economics*, Knowledge Systems, Indianapolis (Ind.).
Heyd, T. (1999), 'Art and Foucauldian Heterotopias', in E. Darier (ed.), *Discourses of the Environment*, Blackwell, Oxford, 152–62.
Hill, S.B. (1999), 'Social Ecology as Future Story', *A Social Ecology Journal*, 1, 197–208.
Hillman, J. (1995), 'A Psyche the Size of the Earth', in T. Roszak, M.E. Gomes, and A.D. Kanner (eds), *Ecopsychology: Restoring the Earth, Healing the Mind*, Sierra Club, San Francisco (Calif.), xvii–xxiii.
Hindess, B. (1992), 'Heidegger and the Nazis: Cautionary Tales of the Relations between Theory and Practice', *Thesis Eleven*, 31, 115–30.

Hirsch, F. (1976), *Social Limits to Growth*, Harvard University Press, Cambridge (Mass.).
Hochman, J. (1998), *Green Cultural Studies: Nature in Film, Novel and Theory*, University of Idaho Press, Moscow (Id.).
Hodgson, G.M. (1993), *Economics and Evolution: Bringing Life Back into Economics*, Polity, London.
Hoffman, N. (1994), 'Beyond Constructivism: A Goethean Approach to Environmental Education', *Australian Journal of Environmental Education*, 10 (Sept.), 71–90.
———. (1996), 'Beyond the Division of Art and Science: Goethe and the Organic Tradition', *Social Alternatives*, 15(1), 46–49.
Holland, A. (1994), 'Natural Capital', in R. Attfield and A. Belsey (eds), *Philosophy and the Natural Environment*, Cambridge University Press, Cambridge, 1994, 169–82.
———. (1997), 'Substitutability: Or, Why Strong Sustainability is Weak and Absurdly Strong Sustainability is Not Absurd', in J. Foster (ed.), *Valuing Nature: Ethics, Economics and the Environment*, Routledge, London, 119–34.
Holsworth, R. (1979), 'Recycling Hobbes: The Limits to Political Ecology', *Massachusetts Review*, 20, 9–40.
Hooper, P., and Palmer, M. (1992), 'St. Francis and Ecology', in E. Breuilly and M. Palmer (eds), *Christianity and Ecology*, Cassell, London, 76–85.
Hughes, J.D. (1983), *American Indian Ecology*, Texas Western Press, El Paso (Tex.).
———. (1996), 'Francis of Assisi and the Diversity of Creation', *Environmental Ethics*, 18, 311–20.

Illich, I. (1973), *Tools for Conviviality*, Harper & Row, New York.
Inada, K.K. (1987), 'Environmental Problematics in the Buddhist Context', *Philosophy East and West*, 37, 135–49.
Innes, K. (1987), *Caring for the Earth - the Environment, Christians and the Church*, Grove Books, Bramcote (Eng.).
Inoue, Y. (1992), 'Arne Naess and the Deep Ecology Movement: Reflections from Japan', *The Trumpeter: Journal of Ecosophy*, 9, 82.
Irigaray, L. (1993), *An Ethics of Sexual Difference*, Cornell University Press, Ithaca (N.Y.).
Izzi Dien, M. (1990), 'Islamic Environmental Ethics: Law and Society', in J.R. Engel and J.G. Engel (eds), *Ethics of Environment and Development: Global Challenge, International Response*, Belhaven, London, 189–98.
———. (1997), 'Islam and the Environment: Theory and Practice', *Journal of Beliefs and Values*, 18, 47–58.

Jackson, M. (1992), 'Revitalising Local Communities through Local Employment and Trading Schemes', in I. Thomas (ed.), *Interactions and Actions: Ecopolitics VI Proceedings*, Faculty of Environmental Design and Construction, Royal Melbourne Institute of Technology, Melbourne (Vic.), E34–E40.
Jacobs, M. (1991), *The Green Economy*, Pluto Press, London.
———. (1994), 'The Limits of Neoclassicism: Towards an Institutional Economics', in M. Redclift and T. Benton (eds), *Social Theory and the Global Environment*, Routledge, London, 67–91.
———. (1995a), *Sustainability and Socialism*, Socialist Environment and Resources

Association, London.

———. (1995b), 'Sustainable Development, Capital Substitution and Economic Humility: A Response to Beckerman', *Environmental Values*, 4, 57–68.

———. (1995c), 'Sustainability and "the Market": A Typology of Environmental Economics', in R. Eckersley (ed.), *Markets, the State and the Environment: Towards Integration*, Macmillan Education, South Melbourne (Vic.), 46–70.

———. (1997), 'Environmental Valuation, Deliberative Democracy and Public Decision-Making Institutions', in J. Foster (ed.), *Valuing Nature: Ethics, Economics and the Environment*, Routledge, London, 211–31.

———. (1999), 'Sustainable Development as a Contested Concept', in A. Dobson (ed.), *Fairness and Futurity: Essays on Environmental Sustainability and Social Justice*, Oxford University Press, Oxford, 21–45.

Jacobs, M./Real World Coalition (1996), *The Politics of the Real World: Meeting the New Century*, Earthscan, London.

Jacobsen, K.A. (1994), 'The Institutionalization of the Ethics of "Non-injury" toward all "Beings" in Ancient India', *Environmental Ethics*, 16, 287–302.

Jamieson, D. (1992), 'Ethics, Public Policy, and Global Warming', *Science, Technology, & Human Values*, 17, 139–53.

———. (1998), 'Animal Liberation Is an Environmental Ethic?', *Environmental Values*, 7, 41–57.

Jeffers, R. (1977, first published 1948), *The Double Axe, and Other Poems*, Random House, New York.

Johnson, E. (1984), 'Treating the Dirt: Environmental Ethics and Moral Theory', in T. Regan (ed.), *Earthbound: New Introductory Essays in Environmental Ethics*, Temple University Press, Philadelphia (Pa.), 336–65.

Johnson, L.E. (1991), *A Morally Deep World: An Essay on Moral Significance and Environmental Ethics*, Cambridge University Press, Cambridge.

Jones, A. (1987), 'From Fragmentation to Wholeness: A Green Approach to Science and Society (Part 1)', *The Ecologist*, 17, 236–40.

Jones, K.H. (1993), *Beyond Optimism: A Buddhist Political Ecology*, Jon Carpenter, Oxford.

Joranson, P.N., and Butigan, K. (1984), 'Transforming the Creation Tradition: The Present Challenge', in P.N. Joranson and K. Butigan (eds), *Cry of the Environment: Rebuilding the Christian Creation Tradition*, Bear & Co., Santa Fé (N.Mex.), 1–18.

Jung, H.Y. (1983), 'Marxism, Ecology and Technology', *Environmental Ethics*, 5, 169–71.

Kalupahana, D.J. (1986), 'Man and Nature: Toward a Middle Path of Survival', *Environmental Ethics*, 8, 371–80.

Kassiola, J.K. (1990), *The Death of Industrial Civilization: The Limits to Economic Growth and the Repoliticization of Advanced Industrial Society*, State University of New York Press, Albany (N.Y.).

Katz, E. (1992), 'Organism, Community and the Substitution Problem', in R.E. Hart (ed.), *Ethics and the Environment*, University Press of America, Lanham (Md.), 55–64.

———. (1999), 'A Pragmatic Reconsideration of Anthropocentrism', *Environmental Ethics*, 21, 377–90.

Kaza, S. (1993), 'Buddhism, Feminism, and the Environmental Crisis', in C.J. Adams (ed.), *Ecofeminism and the Sacred*, Continuum, New York, 50–69.

Keat, R. (1997), 'Values and Preferences in Neo-Classical Environmental Economics', in J. Foster (ed.), *Valuing Nature: Ethics, Economics and the Environment*, Routledge, London, 32–47.

Keller, E.F. (1983), *A Feeling for the Organism: The Life and Work of Barbara McClintock*, W.H. Freeman and Company, New York.

———. (1985), *Reflections on Gender and Science*, Yale University Press, New Haven (Conn.).

———. (1992), *Secrets of Life, Secrets of Death: Essays on Language, Gender and Science*, Routledge, New York.

Kellert, S.R., and Wilson, E.O. (eds) (1993), *The Biophilia Hypothesis*, Island Press, Washington (D.C.).

Kenny, M. (1996), 'Paradoxes of Community', in B. Doherty and M. de Geus (eds), *Democracy and Green Political Thought: Sustainability, Rights and Citizenship*, Routledge, London, 19–35.

Khalid, F., and O'Brien, J. (eds) (1992), *Islam and Ecology*, Cassell, London.

Khavari, F.A. (1993), *Environomics - The Economics of Environmentally Safe Prosperity*, Praeger, Westport (Conn.).

Kheel, M. (1990), 'Ecofeminism and Deep Ecology: Reflections on Identity and Difference', in I. Diamond and G.F. Orenstein (eds), *Reweaving the World: The Emergence of Ecofeminism*, Sierra Club, San Francisco (Calif.), 128–37.

Kibel, P.S. (1999), *The Earth on Trial: Environmental Law on the International Stage*, Routledge, New York.

King, A., and Clifford, S. (1987), *Holding Your Ground: An Action Guide to Local Conservation*, Wildwood House, Aldershot (Eng.).

King, Y. (1981), 'Feminism and the Revolt of Nature', *Heresies*, 13(4), 12–16.

———. (1989), 'The Ecology of Feminism and the Feminism of Ecology', in J. Plant (ed.), *Healing the Wounds: The Promise of Ecofeminism*, New Society, Philadelphia (Pa.), 1–28.

———. (1990), 'Healing the Wounds: Feminism, Ecology, and the Nature/Culture Dualism', in I. Diamond and G.F. Orenstein (eds), *Reweaving the World: The Emergence of Ecofeminism*, Sierra Club, San Francisco (Calif.), 106–21.

Kinrade, P. (1995), 'Towards Ecologically Sustainable Development: The Role and Shortcomings of Markets', in R. Eckersley (ed.), *Markets, the State and the Environment: Towards Integration*, Macmillan Education, South Melbourne (Vic.), 86–109.

Kirkpatrick, J.B. (1998), 'The Politics of the Media and Ecological Ethics', in R. Wills and R. Hobbs (eds), *Ecology for Everyone: Communicating Ecology to Scientists, the Public and the Politicians*, Surrey, Beatty & Sons, Sydney, 36–41.

Kneese, A.V. (1980), 'Environmental Policy', in P. Duignan and A. Rabushka (eds), *The United States in the 1980s*, Hoover Institute Press, Stanford (Calif.), 253–83.

Knudtson, P., and Suzuki, D. (1992), *Wisdom of the Elders*, Stoddart, Toronto.

Kornfield, J. (1993), 'The Buddhist Path and Social Responsibility', *ReVision*, 16, 77–82.

Kovel, J. (1995), 'Ecological Marxism and Dialectic', *Capitalism, Nature, Socialism*, 6(4), 31–50.

Krebs, A. (1997), 'Discourse Ethics and Nature', *Environmental Values*, 6, 267–79.

Kroeber, K. (1994), *Ecological Literary Criticism: Romantic Imagining and Biology of Mind*, Columbia University Press, New York.

Kropotkin, P. (1972, first published 1902), *Mutual Aid*, Allen Lane/The Penguin Press, London.

———. (1974, first published 1899), *Fields, Factories and Workshops Tomorrow*, C. Ward (ed.), George Allen and Unwin, London.

Kuhn, T. (1970), *The Structure of Scientific Revolutions*, 2nd edn, University of Chicago Press, Chicago (Ill.).

La Chapelle, D. (1978), *Earth Wisdom*, Guild of Tutors Press, Los Angeles.

———. (1985), 'Ritual is Essential', in B. Devall and G. Sessions, *Deep Ecology: Living as if Nature Mattered*, Gibbs M. Smith, Salt Lake City (Utah), 247–50.

———. (1988), *Sacred Land, Sacred Sex, Rapture of the Deep: Concerning Deep Ecology and Celebrating Life*, Fine Hill Arts, Silverton (Colo.).

———. (1989), 'No, I'm not an Ecofeminist: A Few Words in Defense of Men', *Earth First!*, 9(4), 6–7.

Labour Party (UK) (1987a), *Charter for the Environment*, Labour Party, London.

———. (1987b), *Industrial Strength for Britain*, Labour Party, London.

Lafferty, W.M. (1996), 'The Politics of Sustainable Development: Global Norms for National Implementation', *Environmental Politics*, 5, 185–208.

Lafferty, W.M., and Meadowcroft, J. (1996), 'Democracy and the Environment: Congruence and Conflict – Preliminary Reflections', in W.M. Lafferty and J. Meadowcroft (eds), *Democracy and the Environment: Problems and Prospects*, Edward Elgar, Cheltenham (Eng.), 1–17.

Lame Deer, J.F. (1972), *Lame Deer: Seeker of Visions*, R. Erdoes (ed.), Pocket Books, New York.

Langlais, R. (1992), 'Living in the World: Mountain Humility, Great Humility', *The Trumpeter: Journal of Ecosophy*, 9, 63–66.

Lappe, F. (1999), 'The Art of Democracy: An Interview with Frances Moore Lappe' *Orion Afield*, 3(3), 24–26.

Lash, S., and Wynne, B. (1992), 'Introduction', in U. Beck, *Risk Society: Towards a New Modernity*, Sage, London, 1–8.

Lawson, T. (1992), 'Abstraction, Tendencies and Stylized Facts', in P. Ekins and M. Max-Neef (eds), *Real-Life Economics: Understanding Wealth Creation*, Routledge, London, 21–38.

Le Guin, U. (1974), *The Dispossessed*, Victor Gollancz, London.

Lee, D.C. (1980), 'On the Marxian View of the Relationship between Man and Nature', *Environmental Ethics*, 2, 3–16.

———. (1982), 'Toward a Marxian Ecological Ethic: A Response to Two Critics', *Environmental Ethics*, 4, 339–43.

Leff, E. (1993), 'Marxism and the Environmental Question', *Capitalism, Nature, Socialism*, 4(1), 44–66.

Leipert, C. (1989), 'From Gross to Adjusted National Product', in P. Ekins (ed.), *The Living Economy: A New Economics in the Making*, Routledge, London, 132–40.

Leopold, A. (1968, first published 1949), *A Sand County Almanac, and Sketches Here and There*, Oxford University Press, New York.

Lewis, M.W. (1992), *Green Delusions: An Environmentalist Critique of Radical Environmentalism*, Duke University Press, Durham (N.C.).

Light, A. (1996), 'Compatibilism in Political Ecology', in A. Light and E. Katz (eds), *Environmental Pragmatism*, Routledge, New York, 161–84.

Light, A., and Katz, E. (1996), 'Introduction: Environmental Pragmatism and Environmental Ethics as Contested Terrain', in A. Light and E. Katz (eds), *Environmental Pragmatism*, Routledge, New York, 1–18.

Lindblom, C.E. (1977), *Politics and Markets*, Basic Books, New York.

———. (1982), 'The Market as Prison', *Journal of Politics*, 44, 324–36.

Linton, M., Turnbull, S., and Weston, D. (1989), 'Financial Futures', in P. Ekins (ed.), *The Living Economy: A New Economics in the Making*, Routledge, London, 194–203.

Linzey, A. (1994), *Animal Theology*, SCM Press, London.

Livingston, J.A. (1981), *The Fallacy of Wildlife Conservation*, McClelland & Stewart, Toronto.

———. (1984), 'The Dilemma of the Deep Ecologist', in N. Evernden (ed.), *The Paradox of Environmentalism*, Faculty of Environmental Studies, York University, Downsview (Ont.), 61–72.

Lopez, B. (1986), 'The Stone Horse', in D. Halpern (ed.), *On Nature*, North Point Press, San Francisco (Calif.), 220–29.

———. (1987), *Arctic Dreams: Imagination and Desire in a Northern Landscape*, Pan, London.

———. (1997), 'A Literature of Place', *Portland: The University of Portland Magazine*, Summer, 22–25.

Lovelock, J. (1979), *Gaia: A New Look at Life on Earth*, Oxford University Press, Oxford.

———. (1988a), *The Ages of Gaia: A Biography of Our Living Earth*, Oxford University Press, Oxford.

———. (1988b), 'The Gaia Hypothesis', in P. Bunyard and E. Goldsmith (eds), *GAIA, the Thesis, the Mechanisms and the Implications: Proceedings of the First Annual Camelford Conference on the Implications of the Gaia Hypothesis*, Wadebridge Ecological Centre, Camelford (Eng.), 35–49.

Lowenthal, D. (1985), *The Past is a Foreign Country*, Cambridge University Press, Cambridge.

Lowry, R.P. (1971), 'Toward a Radical View of the Ecological Crisis', *Environmental Affairs*, 1, 350–59.

Luke, T.W. (1983), 'Notes for a Deconstructionist Ecology', *New Political Science*, 11, 21–32.

———. (1993), 'Green Consumerism: Ecology and the Ruse of Recycling', in J. Bennett and W. Chaloupka (eds), *In the Nature of Things: Language, Politics and the Environment*, University of Minnesota Press, Minneapolis (Minn.), 154–72.

———. (1997), *Ecocritique: Contesting the Politics of Nature, Economy and Culture*, University of Minnesota Press, Minneapolis (Minn.).

Lutzenberger, J. (1985), 'Gaian Economics', in M. Inglis and S. Kramer (eds), *The New Economic Agenda*, Findhorn, Forres (Scot.), 154–66.

———. (1996), 'Re-thinking Progress', *New Internationalist*, 278, 20–23.

Luzon, M.S. (1992), 'Political Ecological Considerations in Marx', *Capitalism, Nature, Socialism*, 3(1), 37–48.

Lyotard, J-F. (1990), *Heidegger and The Jews*, University of Minnesota Press, Minneapolis (Minn.).

Maby, R. (1980), *The Common Ground: A Place for Nature in Britain's Future?*, Hutchinson, London.

McBurney, S. (1990), *Ecology into Economics Won't Go, or, Life is not a Concept*, Green Books, Totnes (Eng.).

———. (1998), *Ecology into Economics Won't Go, or, Life is not a Concept*, revised edn, Green Books, Totnes (Eng.).

McDaniel, J.B. (1986), 'Christian Spirituality as Openness to Fellow Creatures', *Environmental Ethics*, 8, 33–44.

———. (1990a), *Earth, Sky, Gods & Mortals: Developing an Ecological Spirituality*, Twenty-Third Publications, Mystic (Conn.).

———. (1990b), 'Revisioning God and the Self: Lessons from Buddhism', in C. Birch, W. Eakin and J.B. McDaniel (eds), *Liberating Life: Contemporary Approaches to Ecological Theology*, Orbis, Mayknoll (N.Y.), 228–58.

McDonagh, S. (1986), *To Care for the Earth: A Call to a New Theology*, Geoffrey Chapman, London.

McFague, S. (1990), 'Imaging a Theology of Nature: The World as God's Body', in C. Birch, W. Eakin and J.B. McDaniel (eds), *Liberating Life: Contemporary Approaches to Ecological Theology*, Orbis, Mayknoll (N.Y.), 201–27.

McHallam, A. (1991), *The New Authoritarians: Reflections on the Greens*, Institute for European and Defence Studies, London.

McKibben, B. (1989), *The End of Nature*, Random House, New York.

McLaughlin, A. (1993), *Regarding Nature: Industrialism and Deep Ecology*, State University of New York Press, Albany (N.Y.).

———. (1995), 'The Heart of Deep Ecology', in G. Sessions (ed.), *Deep Ecology for the Twenty-First Century*, Shambhala, Boston (Mass.), 85–93.

MacNeil, J., Winsemius, P., and Yakushiji, T. (1991), *Beyond Interdependence: The Meshing of the World's Economy and the Earth's Ecology*, Oxford University Press, New York.

Macy, J. (1991), *Mutual Causality in Buddhism and General Systems Theory: The Dharma of Natural Systems*, State University of New York Press, New York.

———. (1993), *World as Lover, World as Self*, Random House, London.

———. (1995), 'Working Through Environmental Despair', in T. Roszak, M.E. Gomes, and A.D. Kanner (eds), *Ecopsychology: Restoring the Earth, Healing the Mind*, Sierra Club, San Francisco (Calif.), 241–59.

Malpas, J. (1999), *Place and Experience: A Philosophical Topography*, Cambridge University Press, Cambridge.

Malthus, T.R. (1969; first published 1798), 'An Essay on the Principle of Population', in G. Hardin (ed.), *Population, Evolution and Birth Control*, W.H. Freeman, San Francisco (Calif.), 4–17.

Mander, J. (1991), *In the Absence of the Sacred: The Failure of Technology and the Survival of the Indian Nations*, Sierra Club, San Francisco (Calif.).

Manes, C. (1990), *Green Rage: Radical Environmentalism and the Unmaking of Civilization*, Little, Brown, and Co., Boston (Mass.).

Margulis, L. (1988), 'Jim Lovelock's Gaia', in P. Bunyard and E. Goldsmith (eds), *GAIA, the Thesis, the Mechanisms and the Implications: Proceedings of the First Annual Camelford Conference on the Theoretical Implications of the Gaia Hypothesis*, Wadebridge Ecological Centre, Camelford (Eng.), 50–65.

Marietta Jr, D.E. (1979), 'The Interrelationship of Ecological Science and Environmental Ethics', *Environmental Ethics*, 1, 195–207.

———. (1982), 'Knowledge and Obligation in Environmental Ethics: A Phenomenological Analysis', *Environmental Ethics*, 4, 153–62.

Markandya, A., Pearce, D., and Barbier, E.B. (1990), 'Environmental Sustainability and Cost-Benefit Analysis', *Environment and Planning: Series A*, 22, 1259–266.

Marshall, P. (1992), *Nature's Web: An Exploration of Ecological Thinking*, Simon & Schuster, London.

Martell, L. (1994), *Ecology and Society: An Introduction*, Polity, Cambridge.

Martin, B. (1981), 'The Scientific Straightjacket: The Power Structure of Science and the Suppression of Environmental Scholarship', *The Ecologist*, 11, 33–43.

———. (1993), 'Is the "New Paradigm" of Physics Inherently Ecological?', *Chain*

Reaction, 68, 38–39.
Martin, C. (1978), *Keepers of the Game: Indian Animal Relationship and the Fur Trade*, University of California Press, Berkeley (Calif.).
———. (1979), 'The Metaphysics of Writing Indian-White History', *Ethnohistory*, 26, 153–59.
Martinez-Alier, J., and Schlupmann, K. (1987), *Ecological Economics: Energy, Environment and Society*, Basil Blackwell, Oxford.
Marx, K. (1964; ms 1844, first published in full 1932), *The Economic and Philosophic Manuscripts of 1844*, International Publishers, New York.
Marx, K., and Engels, E. (1965; ms 1846, first published in full 1932), *The German Ideology*, Lawrence & Wishart, London.
Masson, J., and McCarthy, S. (1995), *When Elephants Weep: The Emotional Lives of Animals*, Delacorte Press, New York.
Mathews, F. (1991), *The Ecological Self*, Routledge, London.
———. (1994a), 'Deep Ecology or Ecofeminism?', *The Trumpeter: Journal of Ecosophy*, 11, 159–66.
———. (1994b), '*Terra Incognita*: Carnal Legacies', in L. Cosgrove, D.G. Evans and D. Yencken (eds), *Restoring the Land: Environmental Knowledge and Action*, Melbourne University Press, Melbourne (Vic.), 37–46.
———. (1994c), 'Deep Ecology and Ecofeminism', in C. Merchant (ed.), *Ecology: Key Concepts in Critical Theory*, Humanities Press, Atlantic Highlands (N.J.), 235–45.
———. (1996), 'Community and the Ecological Self', in F. Mathews (ed.), *Ecology and Democracy*, Frank Cass, London, 66–100.
Max-Neef, M. (1985), 'Reflections on a Paradigm Shift in Economics', in M. Inglis and S. Kramer (eds), *The New Economic Agenda*, Findhorn, Forres (Scot.), 143–53.
———. (1989), 'Human-Scale Economics: The Challenges Ahead', in P. Ekins (ed.), *The Living Economy: A New Economics in the Making*, Routledge, London, 45–54.
———. (1992a), *From the Outside Looking In: Experiences in Barefoot Economics*, Zed, London.
———. (1992b), 'Development and Human Needs', in P. Ekins and M. Max-Neef (eds), *Real-Life Economics: Understanding Wealth Creation*, Routledge, London, 197–214.
Meadows, D.H., Meadows, D.L., Randers, J., and Behrens III, W.W. (1972), *The Limits to Growth: A Report for the Club of Rome's Project on the Predicament of Mankind*, Universe, New York.
Mellor, M. (1992a), *Breaking the Boundaries: Towards a Feminist Green Socialism*, Virago, London.
———. (1992b), 'Eco-Feminism and Eco-Socialism: Dilemmas of Essentialism and Materialism', *Capitalism, Nature, Socialism*, 3(2), 43–62.
Melucci, A. (1988), 'Social Movements and the Democratization of Everyday Life', in J. Keane (ed.), *Civil Society and the State: New European Perspectives*, Verso, London, 245–60.
Merchant, C. (1980), *The Death of Nature: Women, Ecology and the Scientific Revolution*, Harper & Row, San Francisco (Calif.).
———. (1990), 'Ecofeminism and Feminist Theory', in I. Diamond and G.F. Orenstein (eds), *Reweaving the World: The Emergence of Ecofeminism*, Sierra Club, San Francisco (Calif.), 100–05.
———. (1992), *Radical Ecology: The Search for a Livable World*, Routledge, New York.
———. (1996), *Earthcare: Women and the Environment*, Routledge, New York.
Merleau-Ponty, M. (1962), *Phenomenology of Perception*, Routledge & Kegan Paul,

London.

Metzger, D. (1989), 'Invoking the Grove', in J. Plant (ed.), *Healing the Wounds: The Promise of Ecofeminism*, New Society, Philadelphia (Pa.), 115–17.

Midgley, M. (1983a), *Animals and Why They Matter*, Penguin, Harmondsworth (Eng.).

———. (1983b), 'Duties Concerning Islands', in R. Elliot and A. Gare (eds), *Environmental Philosophy*, University of Queensland Press, St Lucia (Qld), 166–81.

———. (1992a), 'The Significance of Species', in E.C. Hargrove (ed.), *The Animal Rights/Environmental Ethics Debate: The Environmental Perspective*, State University of New York Press, Albany (N.Y.), 121–35.

———. (1992b), 'The Mixed Community', in E.C. Hargrove (ed.), *The Animal Rights/Environmental Ethics Debate: The Environmental Perspective*, State University of New York Press, Albany (N.Y.), 211–25.

———. (1994), 'The End of Anthropocentrism?', in R. Attfield and A. Belsey (eds), *Philosophy and the Natural Environment*, Cambridge University Press, Cambridge, 103–12.

Mies, M. (1986), *Patriarchy and Accumulation on a World Scale*, Zed, London.

Mies, M., and Shiva, V. (1993), *Ecofeminism*, Fernwood, Halifax (N.S.).

Milbrath, L.W. (1984), *Environmentalists: Vanguard for a New Society*, State University of New York Press, Albany (N.Y.).

———. (1989), *Envisioning a Sustainable Society: Learning Our Way Out*, State University of New York Press, Albany (N.Y.).

Mills, M. (1996), 'Green Democracy: The Search for an Ethical Solution', in B. Doherty and M. de Geus (eds), *Democracy and Green Political Thought: Sustainability, Rights and Citizenship*, Routledge, London, 97–114.

Mishan, E.J. (1969), *The Costs of Economic Growth*, Penguin, Harmondsworth (Eng.).

'Miss Ann Thropy' (1987), 'Population and AIDS', *Earth First!*, 1 May, 32.

Mol, A.P.J. (1996), 'Ecological Modernisation and Institutional Reflexivity: Environmental Reform in the Late Modern Age', *Environmental Politics*, 5, 302–23.

Momaday, N.S. (1970), 'An American Land Ethic', in J.G. Mitchell and C.L. Stallings (eds), *Ecotactics: The Sierra Club Handbook for Environment Activists*, Pocket Books, New York, 97–105.

Moncrief, L.W. (1970), 'The Cultural Basis for our Environmental Crisis', *Science*, 170, 508–12.

Montaigne, M.E. de (1991; first published 1580), *The Complete Essays*, M.A. Screech (trans. and ed.), Penguin, London.

Moran, A., Chisholm, A. and Porter, M. (1991) (eds), *Markets, Resources and the Environment*, Allen & Unwin, Sydney.

Muir, J. (1970; first published in 1901), *Our National Parks*, AMS Press, New York.

Murdy, W.H. (1983), 'Anthropocentrism: A Modern Version', in D. Scherer and T. Attig (eds), *Ethics and the Environment*, Prentice-Hall, Englewood Cliffs (N.J.), 12–21.

Naess, A. (1973), 'The Shallow and the Deep, Long-Range Ecology Movement: A Summary', *Inquiry*, 16, 95–100.

———. (1979), 'Self-realization in Mixed Communities of Humans, Bears, Sheep and Wolves', *Inquiry*, 22, 231–41.

———. (1984), 'Intuition, Intrinsic Value and Deep Ecology', *The Ecologist*, 14, 201–03.

———. (1985), 'Identification as a Source of Deep Ecological Attitudes', in M. Tobias (ed.), *Deep Ecology*, Avant Books, San Diego (Calif.), 256–70.

———. (1988), 'Deep Ecology and Ultimate Premises', *The Ecologist*, 18, 128–31.

———. (1989), *Ecology, Community and Lifestyle*, D. Rothenberg (trans.), Cambridge University Press, New York.

———. (1990), 'Sustainable Development and Deep Ecology', in J.R. Engel and J.G. Engel (eds), *Ethics of Environment and Development: Global Challenge, International Response*, Belhaven, London, 88–96.

———. (1993), 'The Deep Ecological Movement: Some Philosophical Aspects', in M.E. Zimmerman, J.B. Callicott, G. Sessions, K.J. Warren, and J. Clark (eds), *Environmental Philosophy: From Animal Rights to Radical Ecology*, Prentice-Hall, Englewood Cliffs (N.J.), 193–212.

———. (1995a), 'Deepness of Questions and the Deep Ecology Movement', in G. Sessions (ed.), *Deep Ecology for the Twenty-First Century*, Shambhala, Boston (Mass.), 204–12.

———. (1995b), 'The Deep Ecology "Eight Points" Revisited', in G. Sessions (ed.), *Deep Ecology for the Twenty-First Century*, Shambhala, Boston (Mass.), 213–21.

———. (1995c), 'The Third World, Wilderness, and Deep Ecology', in G. Sessions (ed.), *Deep Ecology for the Twenty-First Century*, Shambhala, Boston (Mass.), 398–407.

———. (1997), 'Heidegger, Postmodern Theory and Deep Ecology', *The Trumpeter: Journal of Ecosophy*, 14, 181–83.

Naess, A., and Sessions, G. (1984), 'Basic Principles of Deep Ecology', *Ecophilosophy*, 6, 3–7.

Nash, R. (1982), *Wilderness and the American Mind*, 3rd edn, Yale University Press, New Haven (Conn.).

———. (1990), *The Rights of Nature: A History of Environmental Ethics*, Primavera/The Wilderness Society, Leichhardt (NSW).

Neuhaus, R. (1971), *In Defence of People: Ecology and the Seduction of Radicalism*, Macmillan, New York.

Newby, H. (1985), *Green and Pleasant Land?: Social Change in Rural England*, Wildwood House, London.

Norberg-Hodge, H. (1991), *Ancient Futures: Learning from Ladakh*, Sierra Club, San Francisco (Calif.).

———. (1997), 'Buddhism in the Global Economy', *Resurgence*, 181, 18–22.

Norberg-Schulz, C. (1980), *Genius Loci: Towards a Phenomenology of Architecture*, Rizzoli, New York.

Norgaard, R. (1985), 'Environmental Economics: An Evolutionary Critique and a Plea for Pluralism', *Journal of Environmental Economics and Management*, 12, 382–93.

———. (1987), 'Economics as Mechanics and the Demise of Biological Diversity', *Ecological Modelling*, 38, 107–21.

———. (1988), 'Sustainable Development: A Coevolutionary View', *Futures*, 20, 606–20.

———. (1992), 'Coevolution of Economy, Society and Environment', in P. Ekins and M. Max-Neef (eds), *Real-Life Economics: Understanding Wealth Creation*, Routledge, London, 76–86.

———. (1994), *Development Betrayed: The End of Progress and a Coevolutionary Revisioning of the Future*, Routledge, London.

Norris, C. (1990), *What's Wrong with Postmodernism: Critical Theory and the Ends of Philosophy*, Harvester Wheatsheaf, New York.

North, R.D. (1995a), 'End of the Green Crusade', *New Scientist*, 4 March, 38–41.

———. (1995b), *Life on a Modern Planet*, Manchester University Press, Manchester.
Norton, B.G. (1984), 'Environmental Ethics and Weak Anthropocentrism', *Environmental Ethics*, 6, 131–48.
———. (1986), 'Conservation and Preservation: A Conceptual Rehabilitation', *Environmental Ethics*, 8, 195–220.
———. (1987), 'Review of Paul W. Taylor, *Respect for Nature*', *Environmental Ethics*, 9, 261–67.
———. (1991), *Toward Unity Among Environmentalists*, Oxford University Press, New York.
———. (1992a), 'Epistemology and Environmental Values', *The Monist*, 75, 208–26.
———. (1992b), 'Environmental Ethics and Nonhuman Rights', in E.C. Hargrove (ed.), *The Animal Rights/Environmental Ethics Debate: The Environmental Perspective*, State University of New York Press, Albany (N.Y.), 71–93.
———. (1995), 'Why I Am Not a Non-anthropocentrist: Callicott and the Failure of Monistic Inherentism', *Environmental Ethics*, 17, 341–58.
Norton, B.G., and Hannon, B. (1997), 'Environmental Values: A Place-Based Approach', *Environmental Ethics*, 19, 227–46.
Noske, B. (1989), *Humans and Other Animals*, Pluto Press, London.
Novak, P. (1993), 'Tao How?: Asian Religions and the Problem of Environmental Degradation', *ReVision*, 16, 77–82.

O'Connor, J. (1987), *The Meaning of Crisis: A Theoretical Introduction*, Basil Blackwell, Oxford.
———. (1989), 'Political Economy of Ecology of Socialism and Capitalism', *Capitalism, Nature, Socialism*, No. 3, 93–108.
———. (1991a), 'Socialism and Ecology', *Capitalism, Nature, Socialism*, 2(3), 1–11.
———. (1991b), 'On the First and Second Contradiction of Capitalism', *Capitalism, Nature, Socialism*, 2(3), 107–10.
———. (1992a), 'A Political Strategy for Ecology Movements', *Capitalism, Nature, Socialism*, 3(1), 1–5.
———. (1992b), 'Think Globally, Act Locally?', *Capitalism, Nature, Socialism*, 3(4), 1–7.
———. (1993), 'A Political Strategy for Ecology Movements', *Society and Nature*, 1(3), 86–91.
———. (1994), 'Is Sustainable Capitalism Possible?' in M. O'Connor (ed.), *Is Capitalism Sustainable? Political Economy and the Politics of Ecology*, Guildford, New York, 152–75.
———. (1998), *Natural Causes: Essays in Ecological Marxism*, Guildford, New York, 1998.
O'Connor, J., and Orton, D. (1991), 'Discussion: Socialist Biocentrism', *Capitalism, Nature, Socialism*, 2(3), 93–99.
O'Connor, M. (1991), 'Discussion: Eco-Socialism/Eco-Feminism', *Capitalism, Nature, Socialism*, 2(1), 135–37.
———. (1994a), 'Codependency and Indeterminacy: A Critique of the Theory of Production', in M. O'Connor (ed.), *Is Capitalism Sustainable? Political Economy and the Politics of Ecology*, Guildford, New York, 53–75.
———. (1994b), 'Introduction: Liberate, Accumulate - and Bust?', in M. O'Connor (ed.), *Is Capitalism Sustainable? Political Economy and the Politics of Ecology*, Guildford, New York, 1–22.
———. (1994c), 'On the Misadventures of Capitalist Nature', in M. O'Connor (ed.),

Is Capitalism Sustainable? Political Economy and the Politics of Ecology, Guildford, New York, 125–51.
O'Connor, T. (1995), 'Therapy for a Dying Planet', in T. Roszak, M.E. Gomes, and A.D. Kanner (eds), *Ecopsychology: Restoring the Earth, Healing the Mind*, Sierra Club, San Francisco (Calif.), 149–55.
Oelschlaeger, M. (1991), *The Idea of Wilderness: From Prehistory to the Age of Ecology*, Yale University Press, New Haven (Conn.).
———. (1994), *Caring for Creation: An Ecumenical Approach to the Environmental Crisis*, Yale University Press, New Haven (Conn.).
———. (1995), 'Introduction', in M. Oelschlaeger (ed.), *Postmodern Environmental Ethics*, State University of New York Press, Albany (N.Y.), 1–20.
Okruhlik, K. (1992), 'Birth of a New Physics or Death of Nature?', in E.D. Harvey and K. Okruhlik (eds), *Women and Reason*, University of Michigan Press, Ann Arbor (Mich.), 63–76.
O'Neill, J. (1993), *Ecology, Policy and Politics: Human Well-Being and the Natural World*, Routledge, London.
———. (1996a), 'Cost-Benefit Analysis, Rationality and the Plurality of Values', *The Ecologist*, 26, 98–103.
———. (1996b), 'Contingent Valuation and Qualitative Democracy', *Environmental Politics*, 5, 752–59.
———. (1997), 'Value Pluralism, Incommensurability and Institutions', in J. Foster (ed.), *Valuing Nature: Ethics, Economics and the Environment*, Routledge, London, 75–88.
———. (1998), *The Market: Ethics, Knowledge and Politics*, Routledge, London.
Ophuls, W. (1973), 'Leviathan or Oblivion?', in H. Daly (ed.), *Toward a Steady State Economy*, W.H. Freeman, San Francisco (Calif.), 215–30.
———. (1977), *Ecology and the Politics of Scarcity: A Prologue to a Political Theory of the Steady State*, W.H. Freeman, San Francisco (Calif.).
O'Riordan, T. (1981), *Environmentalism*, 2nd edn, Pion, London.
———. (1992), *The Precaution Principle in Environmental Management*, CSERGE Working Paper GEC 92–03, University of East Anglia, Norwich (Eng.).
———. (1993), 'The Politics of Sustainability', in R.K. Turner (ed.), *Sustainable Environmental Economics and Management: Principles and Practices*, Belhaven, London, 37–69.
———. (1996), *Ecotaxation*, Earthscan, London.
O'Riordan, T., and Cameron, J. (1994a), 'The History and Contemporary Significance of the Precautionary Principle', in T. O'Riordan and J. Cameron (eds), *Interpreting the Precautionary Principle*, Earthscan, London, 12–30.
———. (1994b), 'Seeping through the Pores', in T. O'Riordan and J. Cameron (eds), *Interpreting the Precautionary Principle*, Earthscan, London, 292–98.
O'Riordan, T., and Voisey, H. (1997), 'Beyond the Early Stages of the Sustainability Transition', *Environmental Politics*, 6, 174–77.
Orr, D.W. (1992), *Ecological Literacy: Education and the Transition to a Postmodern World*, State University of New York Press, Albany (N.Y.).
Orr, D.W., and Hill, S. (1978), 'Leviathan, the Open Society, and the Crisis of Ecology', *Western Political Quarterly*, 31, 457–69.

Paden, R. (1994), 'Against Grand Theory in Environmental Ethics', *Environmental Values*, 3, 61–70.
Paehlke, R. (1988), 'Democracy, Bureaucracy, and Environmentalism', *Environmental*

Ethics, 10, 291–308.

———. (1989), *Environmentalism and the Future of Progressive Politics*, Yale University Press, New Haven (Conn.).

———. (1990), 'Democracy and Environmentalism: Opening a Door to the Administrative State', in R. Paehlke and D. Torgerson (eds), *Managing Leviathan: Environmental Politics and the Administrative State*, Broadview, Peterborough (Ont.), 35–55.

———. (1996), 'Environmental Challenges to Democratic Practice', in W.M. Lafferty and J. Meadowcroft (eds), *Democracy and the Environment: Problems and Prospects*, Edward Elgar, Cheltenham (Eng.), 18–38.

———. (1998), 'Environmental Values for a Sustainable Society: The Democratic Challenge', in J.S. Dryzek and D. Schlosberg (eds), *Debating the Earth: The Environmental Politics Reader*, Oxford University Press, Oxford, 147–60.

Pakulski, J. (1991), *Social Movements: The Politics of Moral Protest*, Longman Cheshire, Sydney.

Palmer, M. (1990), 'The Encounter of Religion and Conservation', in J.R. Engel and J.G. Engel (eds), *Ethics of Environment and Development: Global Challenge, International Response*, Belhaven, London, 50–62.

———. (1998), 'Chinese Religion and Ecology', in D.E. Cooper and J.A. Palmer (eds), *Spirit of the Environment: Religion, Value and Environmental Concern*, Routledge, London, 15–29.

Parsons, H. (1978), *Marx and Engels on Ecology*, Greenwood, Westview (Conn.).

Partridge, M. (1987), 'Building a Sustainable Green Economy: Ethical Investment, Ethical Work', in A. Hutton (ed.), *Green Politics in Australia*, Angus & Robertson, Sydney, 123–71.

Passmore, J. (1974), *Man's Responsibility for Nature: Ecological Problems and Western Traditions*, Charles Scribner's Sons, New York.

———. (1993), 'Environmentalism', in R. Goodin and P. Pettit (eds), *A Companion to Contemporary Political Theory*, Cambridge University Press, Cambridge, 471–88.

Patterson, J. (1994), 'Maori Environmental Virtues', *Environmental Ethics*, 16, 397–408.

Pearce, D. (1983), *Cost-Benefit Analysis*, 2nd edn, Macmillan, London.

———. (1992a), 'Economics, Equity and Sustainable Development', in P. Ekins and M. Max-Neef (eds), *Real-Life Economics: Understanding Wealth Creation*, Routledge, London, 69–76.

———. (1992b), 'The Practical Implications of Sustainable Development', in P. Ekins and M. Max-Neef (eds), *Real-Life Economics: Understanding Wealth Creation*, Routledge, London, 403–11.

———. (1993), *Economic Values and the Natural World*, Earthscan, London.

———. (1994), 'The Precautionary Principle and Economic Analysis', in T. O'Riordan and J. Cameron (eds), *Interpreting the Precautionary Principle*, Earthscan, London, 132–51.

———. (1998), *Economics and Environment: Essays on Ecological Economics and Sustainable Development*, Edward Elgar, Cheltenham (Eng.).

Pearce *et al.* (1989) [Pearce, D., Markandya, A., and Barbier, E.B.], *Blueprint for a Green Economy*, Earthscan, London.

Pearce *et al.* (1990) [Pearce, D., Barbier, E.B., and Markandya, A.], *Sustainable Development: Economics and Environment in the Third World*, Edward Elgar, Aldershot (Eng.).

Pearce *et al.* (1991) [Pearce, D., Barbier, E., Markandya, A., Barrett, S., Turner, R.K.,

and Swanson, T.], *Blueprint 2: Greening the World Economy*, Earthscan, London.
Pearce *et al.* (1993) [Pearce, D., Turner, R.K., O'Riordan, T., Adger, N., Atkinson, G., Brisson, I., Brown, K., Dubourg, R., Fankhauser, S., Jordan, A., Maddison, D., Moran, D., and Powell, J.], *Blueprint 3: Measuring Sustainable Development*, Earthscan, London.
Peerenboom, R.P. (1991), 'Beyond Naturalism: A Reconstruction of Daoist Environmental Ethics', *Environmental Ethics*, 13, 3–22.
Pepper, D. (1984), *The Roots of Modern Environmentalism*, Routledge, London.
———. (1985), 'Determinism, Idealism and the Politics of Environmentalism - A Viewpoint', *International Journal of Environmental Studies*, 26, 11–19.
———. (1993a), *Ecosocialism: From Deep Ecology to Social Justice*, Routledge, London.
———. (1993b), 'Anthropocentrism, Humanism and Eco-Socialism: A Blueprint for the Survival of Ecological Politics', *Environmental Politics*, 2, 428–52.
Perelman, M. (1993), Marx and Resource Scarcity', *Capitalism, Nature, Socialism*, 4(2), 65–84.
Perrings, C. (1987), *Economy and the Environment: A Theoretical Essay on the Interdependence of Economic and Environmental Systems*, Cambridge University Press, Cambridge.
Peterson, A. (1999), 'Environmental Ethics and the Social Construction of Nature', *Environmental Ethics*, 21, 339–57.
Phelan, S. (1993), 'Intimate Distance: The Dislocation of Nature in Modernity', in J. Bennett and W. Chaloupka (eds), *In the Nature of Things: Language, Politics and the Environment*, University of Minnesota Press, Minneapolis (Minn.), 44–62.
Pigou, A.C. (1920), *The Economics of Welfare*, Macmillan, London.
Plumwood, V., as V. Routley (1981), 'On Karl Marx as an Environmental Hero', *Environmental Ethics*, 3, 237–44.
Plumwood, V. (1986), 'Ecofeminism: An Overview and Discussion of the Arguments', *Australian Journal of Philosophy*, 64 (Supplement), 120–38.
———. (1988), 'Women, Humanity and Nature', *Radical Philosophy*, Spring, 16–24.
———. (1993a), *Feminism and the Mastery of Nature*, Routledge, London.
———. (1993b), 'Feminism and Ecofeminism', *Society and Nature*, 1(2), 36–51.
———. (1996), 'Has Democracy Failed Ecology? An Ecofeminist Perspective', in F. Mathews (ed.), *Ecology and Democracy*, Frank Cass, London, 134–68.
———. (1998), 'Inequality, Ecojustice, and Ecological Rationality', in J. Dryzek and D. Schlosberg (eds), *Debating the Earth: The Environmental Politics Reader*, Oxford University Press, Oxford, 559–83.
Porritt, J. (1984), *Seeing Green: The Politics of Ecology Explained*, Basil Blackwell, Oxford.
Porritt, J., and Winner, D. (1988), *The Coming of the Greens*, Fontana/Collins, London.
Prigogine, I. (1994), 'Science in a World of Limited Predictability', in C. Merchant (ed.), *Ecology: Key Concepts in Critical Theory*, Humanities Press, Atlantic Highlands (N.J.), 363–69.

Quigley, P. (1995), 'Rethinking Resistance: Environmentalism, Literature, and Poststructural Theory', in M. Oelschlaeger (ed.), *Postmodern Environmental Ethics*, State University of New York Press, Albany (N.Y.), 173–91.
———. (1999), 'Nature as Dangerous Space', in E. Darier (ed.), *Discourses of the Environment*, Blackwell, Oxford, 181–202.

Raberg, P. (ed.) (1997), *The Life Region: The Social and Cultural Ecology of Sustainable Development*, Routledge, London.
Rainbow, S. (1993), *Green Politics*, Oxford University Press, Auckland.
Ray, D.L., with L. Guzzo (1993), *Environmental Overkill: Whatever Happened to Common Sense?*, HarperCollins, New York.
Razak, A. (1990), 'Toward a Womanist Analysis of Birth', in I. Diamond and G.F. Orenstein (eds), *Reweaving the Earth: The Emergence of Ecofeminism*, Sierra Club, San Francisco (Calif.), 165–72.
Redclift, M. (1984), *Development and the Environmental Crisis: Red or Green Alternatives*, Methuen, London.
———. (1987), *Sustainable Development: Exploring the Contradictions*, Methuen, London.
———. (1989), 'The Green Movement in Eastern Europe', *The Ecologist*, 19, 177–83.
———. (1993a), 'Sustainable Development: Needs, Values, Rights', *Environmental Values*, 2, 3–20.
———. (1993b), 'Environmental Economics, Policy Consensus and Political Empowerment', in R.K. Turner (ed.), *Sustainable Environmental Economics and Management: Principles and Practices*, Belhaven, London, 106–19.
Reed. P., and Rothenberg, D. (eds) (1987), *Wisdom and the Open Air: Selections from Norwegian Ecophilosophy*, Council for Environmental Studies, University of Oslo, Oslo.
Regan T. (1976), 'Do Animals Have a Right to Life?', in T. Regan and P. Singer (eds), *Animal Rights and Human Obligations*, Prentice-Hall, Englewood Cliffs (N.J.), 197–204.
———. (1980), 'Animal Rights, Animal Wrongs', *Environmental Ethics*, 2, 99–120.
———. (1982), *All That Dwell Therein: Essays on Animal Rights and Environmental Ethics*, University of California Press, Berkeley (Calif.).
———. (1983), *The Case for Animal Rights*, University of California Press, Berkeley (Calif.).
Relph, E. (1976), *Place and Placelessness*, Pion, London.
———. (1981), *Rational Landscapes and Humanistic Geography*, Croom Helm, London.
———. (1984), 'Seeing, Thinking, and Describing Landscapes', in T.F. Saarinen, D. Seamon, and J.L. Sell (eds), *Environmental Perception and Behavior: An Inventory and Prospect*, Research Paper No. 209, Department of Geography, University of Chicago, Chicago (Ill.), 209–24.
Robertson, J. (1983), *The Sane Alternative: A Choice of Futures*, self-published, Wallingford (Eng.).
———. (1985), *Future Work: Jobs, Self-Employment, and Leisure after the Industrial Age*, Gower, Aldershot (Eng.).
———. (1989a), 'The Mismatch between Health and Economics', in P. Ekins (ed.), *The Living Economy: A New Economics in the Making*, Routledge, London, 113–21.
———. (1989b), 'What Comes After Full Employment?', in P. Ekins (ed.), *The Living Economy: A New Economics in the Making*, Routledge, London, 85–96.
———. (1997), 'Institutional Restructuring and Liberatory Transition', *Democracy & Nature*, 3(3), 1–20.
———. (1998), *Transforming Economic Life: A Millennial Challenge*, Green Books, Foxhole (Eng.).
Rodd, R. (1992), *Biology, Ethics and Animals*, Clarendon Press, Oxford.
Rodman, J. (1973), 'What Is Living and What Is Dead in the Political Philosophy of

T.H. Green?', *Western Political Quarterly*, 26, 566–86.
———. (1974), 'The Dolphin Papers', *North American Review*, 259(1), 12–26.
———. (1975), 'On the Human Question: Being the Report of the Erewhonian High Commission to Evaluate Technological Society', *Inquiry*, 18, 127–66.
———. (1976a), 'Analysis and History: Or, How the Invisible Hand Works through Robert Nozick', *Western Political Quarterly*, 29, 197–201.
———. (1976b), Four Stages of Ecological Consciousness. Part One: Resource Conservation - Economics and After, unpublished paper presented at the Annual Meeting of the American Political Science Association, Washington (D.C.), September.
———. (1977a), Ecological Resistance: John Stuart Mill and the Case of the Kentish Orchids, unpublished paper presented at the Annual Meeting of the American Political Science Association, Washington (D.C.), September.
———. (1977b), 'The Liberation of Nature?', *Inquiry*, 20, 83–145.
———. (1978), 'Theory and Practice in the Environmental Movement: Notes Towards an Ecology of Experience', in *The Search for Absolute Values in a Changing World: Proceedings of the Sixth International Conference on the Unity of the Sciences*, Vol. 1, The International Cultural Foundation, San Francisco (Calif.), 45–56.
———. (1980), 'Paradigm Change in Political Science: An Ecological Perspective', *American Behavioral Scientist*, 24, 49–78.
———. (1983), 'Four Forms of Ecological Consciousness Reconsidered', in D. Scherer and T. Attig (eds), *Ethics and the Environment*, Prentice-Hall, Englewood Cliffs (N.J.), 82–92.
Rogers, L.J. (1997), *Minds of their Own: Thinking and Awareness in Animals*, Allen & Unwin, St Leonards (NSW).
Rogers, R.A. (1994), *Nature and the Crisis of Modernity: A Critique of Contemporary Discourse on Managing the Earth*, Black Rose, Montreal.
Rolston III, H. (1975), 'Is There an Environmental Ethic?', *Ethics*, 85, 93–109.
———. (1982), 'Are Values in Nature Subjective or Objective?', *Environmental Ethics*, 4, 125–51.
———. (1985), 'Duties to Endangered Species', *BioScience*, 35, 718–26.
———. (1986), *Philosophy Gone Wild: Essays in Environmental Ethics*, Prometheus, Buffalo (N.Y.).
———. (1987), 'Duties to Ecosystems', in J.B. Callicott (ed.), *Companion to A Sand County Almanac: Interpretive and Critical Essays*, University of Wisconsin Press, Madison (Wis.), 246–74.
———. (1988), *Environmental Ethics: Duties and Values in the Natural World*, Temple University Press, Philadelphia (Pa.).
———. (1990), 'Science-based Versus Traditional Ethics', in J.R. Engel and J.G. Engel (eds), *Ethics of Environment and Development: Global Challenge, International Response*, Belhaven, London, 63–72.
———. (1994), 'Value in Nature and the Nature of Value', in R. Attfield and A. Belsey (eds), *Philosophy and the Natural Environment*, Cambridge University Press, Cambridge, 13–30.
Rose, D., Saunders, P., Newby, H., and Bell, C. (1976), 'Ideologies of Property: A Case Study', *Sociological Review*, 24, 697–730.
Rose, H. (1986), 'Beyond Masculinist Realities', in R. Bleier (ed.), *Feminist Approaches to Science*, Pergamon Press, New York, 57–76.
Rosewarne, S. (1993), 'Selling the Environment: A Critique of Market Ideology', in S. Rees, G. Rodley and F. Stilwell (eds), *Beyond the Market: Alternatives to*

Economic Rationalism, Pluto Press, Leichhardt (NSW), 53–71.
———. (1995), 'On Ecological Economics', *Capitalism, Nature, Socialism*, 6(3), 105–15.
Ross, A. (1994), *The Chicago Gangster Theory of Life: Nature's Debt to Society*, Verso, London.
Roszak, T. (1971), *The Making of a Counter Culture*, Faber and Faber, London.
———. (1972), *Where the Wasteland Ends: Politics and Transcendence in Postindustrial Society*, Doubleday, New York.
———. (1981), *Person/Planet: The Creative Disintegration of Industrial Society*, Granada, London.
Rothchild, M.L. (1992), *Bionomics: The Inevitability of Capitalism*, Futura, London.
Rothenberg, D. (1989), 'Introduction: Ecosophy T - from Intuition to System', in A. Naess, *Ecology, Community and Lifestyle*, D. Rothenberg (trans.), Cambridge University Press, New York, 1–22.
———. (1992), 'Out to Nowhere: Travels with Arne Naess', *The Trumpeter: Journal of Ecosophy*, 9, 49–53.
———. (1995), 'A Platform of Deep Ecology', in A. Drengson and Y. Inoue (eds), *The Deep Ecology Movement: An Introductory Anthology*, North Atlantic Books, Berkeley (Calif.), 155–66.
Rudy, A. (1994), 'Dialectics of Capital and Nature', *Capitalism, Nature, Socialism*, 5(2), 95–106.
Ruether, R.R. (1975), *New Woman, New Earth: Sexist Ideologies and Human Liberation*, Seabury, New York.
———. (1992), *Gaia and God: An Ecofeminist Theology of Earth Healing*, HarperCollins, San Francisco.
———. (1993), 'Ecofeminism: Symbolic and Social Connections of the Oppression of Women and the Domination of Nature', in C.J. Adams (ed.), *Ecofeminism and the Sacred*, Continuum, New York, 13–23.
Rutherford, P. (1994), 'The Administration of Life: Ecological Discourse as "Intellectual Machinery of Government"', *Australian Journal of Communication*, 21(3), 40–55.
———. (1999a), '"The Entry of Life into History"', in E. Darier (ed.), *Discourses of the Environment*, Blackwell, Oxford, 37–62.
———. (1999b), 'Ecological Modernization and Environmental Risk', in E. Darier (ed.), *Discourses of the Environment*, Blackwell, Oxford, 95–118.
Ryle, M. (1986), 'From Green to Red', *New Socialist*, December, 36–38.
———. (1988), *Ecology and Socialism*, Century Hutchinson, London.

Sachs, W. (1989), 'Delinking from the World Market', in P. Ekins (ed.), *The Living Economy: A New Economics in the Making*, Routledge, London, 333–43.
Sagoff, M. (1984), 'Animal Liberation and Environmental Ethics: Bad Marriage, Quick Divorce', *Osgood Hall Law Journal*, 22, 297–307.
———. (1988), *The Economy of the Earth: Philosophy, Law and the Environment*, Cambridge University Press, Cambridge.
———. (1992), 'Settling America: The Concept of Place in Environmental Ethics', *Journal of Energy, Natural Resources and Environmental Law*, 12, 351–418.
———. (1993), 'Environmental Ethics: An Epitaph', *Resources*, Spring, 2–7.
Sale, K. (1984a), 'Bioregionalism - A New Way to Treat the Land', *The Ecologist*, 14, 167–73.
———. (1984b), 'Mother of All: An Introduction to Bioregionalism', in S. Kumar

(ed.), *The Schumacher Lectures: Second Volume*, Blond & Briggs, London, 219–50.
———. (1985a), 'Anarchy and Ecology - A Review Essay', *Social Anarchism*, 10, 14–23.
———. (1985b), *Dwellers in the Land: The Bioregional Vision*, Sierra Club, San Francisco (Calif.).
———. (1988), 'Deep Ecology and Its Critics', *The Nation*, 14 May, 670–75.
———. (1995), *Rebels Against the Future: The Luddites and their War on the Industrial Revolution*, Addison Wesley, Reading (Mass.).
Salleh, A.K. (1984), 'Deeper than Deep Ecology: The Eco-Feminist Connection', *Environmental Ethics*, 6, 339–45.
———. (1989), 'Stirrings of a New Renaissance', *Island*, 38, 26–31.
———. (1990), 'Living with Nature: Reciprocity or Control?', in J.R. Engel and J.G. Engel (eds), *Ethics of Environment and Development: Global Challenge, International Response*, Belhaven, London, 245–53.
———. (1991), 'Discussion: Eco-Socialism/Eco-Feminism', *Capitalism, Nature, Socialism*, 2(1), 129–34.
———. (1992), 'The Ecofeminist/Deep Ecology Debate: A Reply to Patriarchal Reason', *Environmental Ethics*, 14, 195–216.
———. (1993a), 'Class, Race and Gender Discourse in the Ecofeminist/Deep Ecology Debate', *Environmental Ethics*, 15, 225–44.
———. (1993b), 'Second Thoughts on Rethinking Ecofeminist Politics: A Dialectical Critique', *Interdisciplinary Studies in Literature and Environment*, 1(2), 93–106.
———. (1994), 'Nature, Woman, Labor, Capital: Living the Deepest Contradiction', in M. O'Connor (ed.), *Is Capitalism Sustainable?: Political Economy and the Politics of Ecology*, Guildford, New York, 106–24.
———. (1997), *Ecofeminism as Politics: Nature, Marx and the Postmodern*, Zed, London.
Santmire, H.P. (1977), 'Historical Dimensions of the American Crisis', in I.G. Barbour (ed.), *Western Man and Environmental Ethics: Attitudes Toward Nature and Technology*, Addison Wesley, Reading (Mass.), 1977, 66–93.
Sapontzis, S.F. (1987), *Morals, Reason and Animals*, Temple University Press, Philadelphia (Pa.).
Sardar, Z. (1989), *The Future of Knowledge and Environment in Islam*, Mansell, London.
———. (1990), 'Islam and Intellect', *The Listener*, 123, 25 January, 12–15.
Saward, M. (1993), 'Green Democracy', in A. Dobson and P. Lucardie (eds), *The Politics of Nature: Explorations in Green Political Theory*, Routledge, London, 63–80.
———. (1996), 'Must Democrats Be Environmentalists?', in B. Doherty and M. de Geus (eds), *Democracy and Green Political Thought: Sustainability, Rights and Citizenship*, Routledge, London, 79–96.
Sayer, A. (1995), *Radical Political Economy: A Critique*, Blackwell, Oxford.
Schama, S. (1995), *Landscape and Memory*, HarperCollins, London.
Schmidheiny, D. (1992), *Changing Course: A Global Business Perspective on Development and the Environment*, MIT Press, Cambridge (Mass.).
Schmidt, A. (1971), *The Concept of Nature in Marx*, New Left Books, London.
Schumacher, E.F. (1974), *Small Is Beautiful: A Study of Economics as if People Mattered*, Sphere, London.
———. (1980), *Good Work*, Abacus, London.
Schwarz, W., and Schwarz, D. (1987), *Breaking Through: Theory and Practice of Wholistic Living*, Green Books, Bideford (Eng.).
Schwarzer, A. (1984), *After the Second Sex: Conversations with Simone de Beauvoir*,

Pantheon, New York.

Schweitzer, A. (1923), *The Philosophy of Civilization. Part II: Civilization and Ethics*, trans. J. Naish, A. and C. Black, London.

Scullion, J. (1973), 'Man, Nature & the Old Testament', in J. Nurser (ed.), *Living with Nature: Proceedings of the 4th National Summer School on Religion*, Centre for Continuing Education, Australian National University, Canberra, 20–28.

Seager, J. (1993), *Earth Follies: Feminism, Politics and the Environment*, Earthscan, London.

Seamon, D. (1976), 'Goethe's Approach to the Natural World: Implications for Environmental Theory and Education', in D. Ley and M. Samuels (eds), *Humanistic Geography: Prospects and Problems*, Maaroufa Press, Chicago (Ill.), 238–50.

———. (1982), 'The Phenomenological Contribution to Environmental Psychology', *Journal of Environmental Psychology*, 2, 119–40.

———. (1984a), 'Emotional Experience of the Environment', *American Behavioral Scientist*, 27, 757–70.

———. (1984b), 'Heidegger's Notion of Dwelling and One Concrete Interpretation as Indicated by Hassan Fathy's *Architecture for the Poor*', *Geoscience & Man*, 24, 43–53.

———. (1993), 'Dwelling, Seeing, and Designing: An Introduction', in D. Seamon (ed.), *Dwelling, Seeing, and Designing: Toward a Phenomenological Ecology*, State University of New York Press, Albany (N.Y.), 1–21.

Seamon, D., and Zajonc, A. (eds) (1998), *Goethe's Way of Science: A Phenomenology of Nature*, State University of New York Press, Albany (N.Y.).

Sears, P.B. (1964), 'Ecology - a Subversive Subject', *BioScience*, 14(7), 11–13.

Seed, J. (1990), 'Deep Ecology Down Under', in C. Plant and J. Plant (eds), *Turtle Talk: Voices for a Sustainable Future*, New Society, Philadelphia (Pa.), 68–75.

Seed, J., with J. Macy, P. Fleming, and A. Naess (1988), *Thinking Like a Mountain: Towards a Council of All Beings*, New Society, Philadelphia (Pa.).

Selman, P. (1996), *Local Sustainability: Managing and Planning Ecologically Sound Places*, Paul Chapman, London.

Sessions, G. (1974), 'Anthropocentrism and the Environmental Crisis' *Humboldt Journal of Social Relations*, 2, 71–81.

———. (1977), 'Spinoza and Jeffers on Man in Nature', *Inquiry*, 20, 481–528.

———. (1985a), 'Appendix D: Western Process Metaphysics (Heraclitus, Whitehead, and Spinoza)', in B. Devall and G. Sessions, *Deep Ecology: Living as if Nature Mattered*, Gibbs M. Smith, Salt Lake City (Utah), 236–42.

———. (1985b), 'Ecological Consciousness and Paradigm Change', in M. Tobias (ed.), *Deep Ecology*, Avant Books, San Diego (Calif.), 28–44.

———. (1992a), 'Arne Naess & the Union of Theory & Practice', *The Trumpeter: Journal of Ecosophy*, 9, 73–76.

———. (1992b), 'Radical Environmentalism in the 90s', *Wild Earth*, Fall, 64–67.

———. (1995a), 'Introduction: Arne Naess on Deep Ecology', in G. Sessions (ed.), *Deep Ecology for the Twenty-First Century*, Shambhala, Boston (Mass.), 187–94.

———. (1995b), 'Deep Ecology and the New Age Movement', in G. Sessions (ed.), *Deep Ecology for the Twenty-First Century*, Shambhala, Boston (Mass.), 290–310.

———. (1995c), 'Postmodernism and Environmental Justice', *The Trumpeter: Journal of Ecosophy*, 12, 150–54.

———. (1996), 'Reinventing Nature, ...? A Response to Cronon's *Uncommon Ground*', *The Trumpeter: Journal of Ecosophy*, 13 (1), 33–38.

Sheldrake, R. (1990), *The Rebirth of Nature: The Greening of Science and God*, Century, London.
Shepard, P. (1969), 'Introduction: Ecology and Man - A Viewpoint', in P. Shepard and D. McKinley (eds), *The Subversive Science: Essays Toward an Ecology of Man*, Houghton Mifflin, Boston (Mass.), 1–10.
———. (1982), *Nature and Madness*, Sierra Club, San Francisco (Calif.).
———. (1995), 'Virtual Hunting in the Forests of Simulacra', in M.E. Soule and G. Lease (eds), *Reinventing Nature? Responses to Postmodern Deconstruction*, Island Press, Washington (D.C.), 17–29.
Shiva, V. (1989), *Staying Alive: Women, Ecology, and Development*, Zed, Atlantic Highlands (N.J.).
———. (1993), *Monocultures of the Mind: Perspectives on Biodiversity and Biotechnology*, Zed, London.
Sikorski, W. (1993), 'Building Wilderness', in J. Bennett and W. Chaloupka (eds), *In the Nature of Things: Language, Politics and the Environment*, University of Minnesota Press, Minneapolis (Minn.), 24–43.
Simmons, I.G. (1969), 'Evidence for Vegetation Changes Associated with Mesolithic Man', in P.J. Ucko and G.W. Dimbleby (eds), *The Domestication and Exploitation of Plants and Animals*, Duckworth, London, 111–19.
Singer, B. (1988), 'An Extension of Rawls' Theory of Justice to Environmental Ethics', *Environmental Ethics*, 10, 233–50.
Singer, P. (1975), *Animal Liberation: A New Ethics for Our Treatment of Animals*, New York Review/Random House, New York.
———. (1979a), *Practical Ethics*, Cambridge University Press, Cambridge.
———. (1979b), 'Killing Humans and Killing Animals', *Inquiry*, 22, 145–56.
———. (1985), 'Not for Humans Only: The Place of Non-Humans in Environmental Issues', in M. Velazquez and C. Rostakowski (eds), *Ethics: Theory and Practice*, Prentice-Hall, Englewood Cliffs (N.J.), 476–90.
———. (1990), *Animal Liberation*, 2nd edn, Jonathan Cape, London.
———. (1993a), 'Animal Liberation', *Island*, 54, 62–66.
———. (1993b), *How Are We to Live? Ethics in an Age of Self Interest*, Text, Melbourne (Vic.).
Sjoo, M., and Mor, B. (1987), *The Great Cosmic Mother: Rediscovering the Religion of the Earth*, Harper & Row, San Francisco (Calif.).
Skirbekk, G. (1994), 'Marxism and Ecology', *Capitalism, Nature, Socialism*, 5(4), 95–104.
Skolimowski, H. (1981), *Eco-Philosophy: Designing New Tactics for Living*, Marion Boyars, New York.
———. (1990), 'Reverence for Life', in J.R. Engel and J.G. Engel (eds), *Ethics of Environment and Development: Global Challenge, International Response*, Belhaven, London, 97–103.
———. (1992), *Living Philosophy: Eco-Philosophy as a Tree of Life*, Penguin, London.
———. (1993), 'Early Ecophilosophers among the Tribal People: Letter from India', *The Trumpeter: Journal of Ecosophy*, 10, 136–37.
Slicer, D. (1994), 'Wrongs of Passage: Three Challenges to the Maturing of Ecofeminism', in K.J. Warren (ed.), *Ecological Feminism*, Routledge, London, 29–41.
———. (1995), 'Is There an Ecofeminism-Deep Ecology "Debate"?', *Environmental Ethics*, 17, 151–69.
Smith, A. (1994), 'For All Those Who Were Indian in a Former Life', *Cultural Survival Quarterly*, 17(4), 70.
Smith, F.L. (1991), 'Free-market Eco-management', *Chemtech*, 21, 598–605.

Smith, H. (1972), 'Tao Now: An Ecological Testament', in I.G. Barbour (ed.), *Earth Might Be Fair: Reflections on Ethics, Religion and Ecology*, Prentice-Hall, Englewood Cliffs (N.J.), 62–81.

Smith, M. (1997), 'Against the Enclosure of the Ethical Commons: Radical Environmentalism as an "Ethics of Place"', *Environmental Ethics*, 19, 339–54.

———. (1999), 'To Speak of Trees: Social Constructivism, Environmental Values, and the Future of Deep Ecology', *Environmental Ethics*, 21, 359–76.

Smith, N. (1990), *Uneven Development: Nature, Capital and the Production of Space*, Basil Blackwell, Oxford.

Snyder, G. (1980), *The Real Work: Interviews and Talks, 1964–1979*, W.S. McLean (ed.), New Directions, New York.

———. (1985), 'Buddhism and the Possibilities of a Planetary Culture', in B. Devall and G. Sessions, *Deep Ecology: Living as if Nature Mattered*, Gibbs M. Smith, Salt Lake City (Utah), 251–53.

———. (1990), *The Practice of the Wild*, North Point Press, San Francisco (Calif.).

———. (1998), 'Is Nature Real?', *Resurgence*, 190, 32–33.

Sobosan, J. (1991), *Bless the Beasts: A Spirituality of Animal Care*, Crossroad, New York.

Somervell, D. (1965), *English Thought in the Nineteenth Century*, David McKay, New York.

Soper, K. (1995), *What Is Nature?: Culture, Politics and the Non-Human*, Blackwell, Oxford.

———. (1996), 'Nature/"nature"', in G. Robertson, M. Marsh, L. Tickner, J. Bird. B. Curtis and T. Putnam (eds), *FutureNatural: Nature, Science, Culture*, Routledge, London, 22–34.

Soule, M.E. (1995), 'The Social Siege of Nature', in M.E. Soule and G. Lease (eds), *Reinventing Nature? Responses to Postmodern Deconstruction*, Island Press, Washington (D.C.), 137–70.

Spretnak, C. (1986), *The Spiritual Dimension of Green Politics*, Bear & Co., Santa Fé (N.Mex.).

———. (1988), 'Ecofeminism: Our Roots and Flowering', *The Elmswood Newsletter*, 4, 1.

———. (1989), 'Toward an Ecofeminist Spirituality', in J. Plant (ed.), *Healing the Wounds: The Promise of Ecofeminism*, New Society, Philadelphia (Pa.), 127–32.

———. (1991), *States of Grace: The Recovery of Meaning in the Post-Modern Era*, HarperCollins, New York.

———. (1997), *The Resurgence of the Real: Body, Nature, and Place in a Hypermodern World*, Addison-Wesley, Reading (Mass.).

Standing Bear, L. (1933), *Land of the Spotted Eagle*, Houghton Mifflin, Boston (Mass.).

Stanfield, J.R. (1982), 'Toward an Ecological Economics', *International Journal of Social Economics*, 10(5), 27–37.

Starhawk (1979), 'Witchcraft and Women's Spirit', in C.P. Christ and J. Plaskow (eds), *Womanspirit Rising: A Feminine Reader in Religion*, Harper & Row, New York, 1979.

———. (1989), 'Feminist Earth-based Spirituality and Ecofeminism', in J. Plant (ed.), *Healing the Wounds: The Promise of Ecofeminism*, New Society, Philadelphia (Pa.), 174–85.

———. (1990), 'Bending the Energy: Spirituality, Politics and Culture', in C. Plant and J. Plant (eds), *Turtle Talk: Voices for a Sustainable Future*, New Society, Philadelphia (Pa.), 32–39.

Stavins, R.N. (1989), 'Clean Profits: Using Economic Incentives to Protect the Environment', *Policy Review*, Spring, 58–63.

Steiner, R. (1988), *Goethean Science*, Mercury Spring Press, Spring Valley (N.Y.).

Sterba, J.P. (1995), 'From Biocentric Individualism to Biocentric Pluralism', *Environmental Ethics*, 17, 191–207.

Sterling, S.R. (1990), 'Towards an Ecological World View', in J.R. Engel and J.G. Engel (eds), *Ethics of Environment and Development: Global Challenge, International Response*, Belhaven, London, 77–86.

Stevenson, B.K. (1995), 'Contextualism and Norton's Convergence Hypothesis', *Environmental Ethics*, 17, 135–50.

Stone, C. (1987), *Earth and Other Ethics: The Case for Moral Pluralism*, Harper & Row, New York.

Storm, H. (1972), *Seven Arrows*, Ballantine, New York.

Stretton, H. (1976), *Capitalism, Socialism and the Environment*, Cambridge University Press, Cambridge.

Sturgeon, N. (1997), *Ecofeminist Natures: Race, Gender, Feminist Theory and Political Action*, Routledge, London.

Swanson, B. (1991), 'History, Culture and Nature', *The Trumpeter: Journal of Ecosophy*, 8, 164–69.

Swift, A. (1993), *Global Political Ecology: The Crisis in Economy and Government*, Pluto Press, London.

Swimme, B. (1988), 'The Cosmic Creation Story', in D.R. Griffin (ed.), *The Reenchantment of Science: Postmodern Proposals*, State University of New York Press, Albany (N.Y.), 47–56.

Swimme, B. and Berry, T. (1992), *The Universe Story: From the Primordial Flaring to the Ecozoic Era - A Celebration of the Unfolding of the Cosmos*, HarperCollins, San Francisco (Calif.).

Sylvan, R., as R. Routley (1976), 'Is there a Need for a New, an Environmental Ethic?', *Philosophy Today*, Winter, 306–16.

Sylvan, R. (1985), *A Critique of Deep Ecology*, Discussion Papers in Environmental Philosophy 12, Department of Philosophy, Australian National University, Canberra.

Sylvan, R., and Bennett, D. (1992), 'Tao and Deep (Ecological) Theory: A Preliminary Investigation', in P.R. Hay and R. Eckersley (eds), *Ecopolitical Theory: Essays from Australia*, Occasional Paper 24, Centre for Environmental Studies, University of Tasmania, Hobart (Tas.), 21–48.

———. (1994), *The Greening of Ethics*, White Horse Press/University of Arizona Press, Cambridge and Tucson (Ariz.).

Tacey, D. (1995), *The Edge of the Sacred*, HarperCollins, North Blackburn (Vic.).

Taylor, A. (1997), 'Inhaling All the Forces of Nature: William Morris's Socialist Biophilia', *The Trumpeter: Journal of Ecosophy*, 14, 207–12.

Taylor, C. (1992), 'Heidegger, Language, and Ecology', in H.L. Dreyfus and H. Hall (eds), *Heidegger: A Critical Reader*, Blackwell, Oxford.

Taylor, P.W. (1987), *Respect for Nature: A Theory of Environmental Ethics*, Princeton University Press, Princeton (N.J.).

Taylor, R. (1993), 'The Environmental Implications of Liberalism', *Critical Review*, 6, 265–82.

Teilhard de Chardin, P., (1959), *The Phenomenon of Man*, Collins, London.

Thompson, E.P. (1976), *William Morris: Romantic to Revolutionary*, Pantheon, New York.

Thompson, P. (1977), *The Work of William Morris*, Quartet, London.

Thoreau, H.D. (1965, first published 1864), *Walden - Essay on Civil Disobedience*, Airmont, New York.

Thrupp, L. (1989), 'The Struggle for Nature: Replies', *Capitalism, Nature, Socialism*, 1(3), 169–74.

Tindale, S. (1997), 'The Political Economy of Environmental Tax Reform', in M. Jacobs (ed.), *Greening the Millenium: The New Politics of the Environment*, Blackwell, Oxford, 98–108.

Torgerson, D. (1990), 'Limits of the Administrative Mind: The Problem of Defining Environmental Problems', in R. Paehlke and D. Torgerson (eds), *Managing Leviathan: Environmental Politics and the Administrative State*, Broadview, Peterborough (Ont.), 115–61.

———. (1994), 'Strategy and Ideology in Environmentalism: A Decentered Approach to Sustainability', *Industrial and Environmental Crisis Quarterly*, 8, 295–321.

———. (1998), 'The Political Theory of Ecological Resistance: A Critique', in C. Starr (ed.), *Green Politics in Grey Times: Ecopolitics XI Conference Proceedings*, Department of Political Science, University of Melbourne, Melbourne (Vic.), 318–29.

———. (1999), *The Promise of Green Politics: Environmentalism and the Public Sphere*, Duke University Press, Durham (N.C.).

Trainer, T. (1985), *Abandon Affluence!*, Zed, London.

———. (1989), *Developed to Death*, Merlin, London.

———. (1995a), *The Conserver Society: Alternatives for Sustainability*, Zed, London.

———. (1995b), 'What Is Development?', *Society and Nature*, 7, 26–56.

———. (1998), *Saving the Environment: What It Will Take*, University of New South Wales Press, Sydney.

Tuan, Y-F. (1974a), *Topophilia: A Study of Environmental Perception, Attitudes, and Values*, Prentice-Hall, Englewood Cliffs (N.J.).

———. (1974b), 'Space and Place: Humanistic Perspective', *Progress in Geography*, 6, 211–52.

———. (1977), *Space and Place: The Perspective of Experience*, Edward Arnold, London.

Turner, R.K. (1993), 'Sustainability: Principles and Practice', in R.K. Turner (ed.), *Sustainable Environmental Economics and Management: Principles and Practice*, Belhaven, London, 3–36.

Turner, R.K., Pearce, D., and Bateman, I. (1993), *Environmental Economics: An Elementary Introduction*, John Hopkins University Press, Baltimore (Md.).

Uhlein, G. (1983), *Meditations with Hildegard of Bingen*, Bear & Co., Santa Fé (N.Mex.).

UNCED (United Nations Conference on Environment and Development) (1992), *Agenda 21*, UNCED, New York.

Urry, J. (1995), *Consuming Places*, Routledge, London.

Van Wyck, P.C. (1997), *Primitives in the Wilderness: Deep Ecology and the Missing Human Subject*, State University of New York Press, Albany (N.Y.).

Varner, G.E. (1998), *In Nature's Interests?: Interests, Animal Rights, and*

Environmental Ethics, Oxford University Press, New York.
Vincent, A. (1993), 'The Character of Ecology', *Environmental Politics*, 2, 248–76.
———. (1998), 'Liberalism and the Environment', *Environmental Values*, 7, 443–59.
Vycinas, V. (1961), *Earth and Gods: An Introduction to the Philosophy of Martin Heidegger*, Martinus Nijhoff, The Hague.

Wackernagel, M., and Rees, W.E. (1996), *Our Ecological Footprint: Reducing Human Impact on the Earth*, New Society, Gabriola Island (B.C.).
Walker, K.J. (1985), Economic Growth and Environmental Management: The Dilemma of the State, unpublished paper presented to the 27th Conference of the Australasian Political Studies Association, Adelaide (S.A.), 28–30 August.
Warren, K.J. (1987), 'Feminism and Ecology: Making Connections', *Environmental Ethics*, 9, 3–20.
———. (1990), 'The Power and the Promise of Ecological Feminism', *Environmental Ethics*, 12, 125–46.
———. (1993), 'A Feminist Philosophical Perspective on Ecofeminist Spiritualities', in C.J. Adams (ed.), *Ecofeminism and the Sacred*, Continuum, New York, 119–32.
———. (1994), 'Toward an Ecofeminist Peace Politics', in K.J. Warren (ed.), *Ecological Feminism*, Routledge, London, 179–98.
Warren, M.A. (1992), 'The Rights of the Nonhuman World', in E.C. Hargrove (ed.), *The Animal Rights/Environmental Ethics Debate: The Environmental Perspective*, State University of New York Press, Albany (N.Y.), 185–210.
Watson, M., and Sharpe, D. (1993), 'Green Beliefs and Religion', in A. Dobson and P. Lucardie (eds), *The Politics of Nature: Explorations in Green Political Theory*, Routledge, London, 210–28.
WCED (World Commission on Environment and Development) (1987), *Our Common Future*, Oxford University Press, Oxford.
WCU/UNEP/WWFN (World Conservation Union/United Nations Environment Program/World Wide Fund for Nature) (1991), *Caring for the Earth*, WCU/UNEP/WWFN, Gland.
Weale, A. (1992), *The New Politics of Pollution*, Manchester University Press, Manchester.
———. (1993), 'The Limits of Ecocentrism', *Environmental Politics*, 2, 340–43.
Weil, S. (1978, first published 1952), *The Need for Roots: Prelude to a Declaration of Duties towards Mankind*, Routledge and Kegan Paul, London.
Weizsacker, E.U. von, and Jesinghaus, J. (1992), *Ecological Tax Reform*, Zed, London.
Welbourn, F.B. (1975), 'Man's Dominion', *Theology*, 78, 561–68.
Welford, R. (1995), *Environmental Strategy and Sustainable Development: The Corporate Challenge for the Twenty-first Century*, Routledge, London.
Welford, R., and Gouldson, A. (1993), *Environmental Management and Business Strategy*, Pitman, London.
Wells, D. (1978), 'Radicalism, Conservatism and Environmentalism', *Politics* (Aust.), 13, 299–306.
———. (1982), 'Resurrecting the Dismal Parson: Malthus, Ecology, and Political Thought', *Political Studies*, 30, 1–15.
———. (1993), 'Green Politics and Environmental Ethics: A Defence of Human Welfare Ecology', *Australian Journal of Political Science*, 28, 515–27.
Wenz, P. (1988), *Environmental Justice*, State University of New York Press, Albany (N.Y.).
Wessels, M.A. (1989), 'A Christian Responsibility for Animal Liberty', in J.E. Dixon,

N.J. Ericksen and A.S. Gunn (eds), *Ecopolitics III: Conference Proceedings*, Environmental Studies Unit, University of Waikato, Hamilton (NZ), 177–80.

Weston, J. (1986a), 'Introduction', in J. Weston (ed.), *Red and Green: The New Politics of the Environment*, Pluto Press, London, 1–8.

———. (1986b), 'The Greens, "Nature", and the Social Environment', in J. Weston (ed.), *Red and Green: The New Politics of the Environment*, Pluto Press, London, 11–29.

White Jr, L. (1967), 'The Historical Roots of Our Ecologic Crisis', *Science*, 155, 1203–207.

———. (1977), 'Continuing the Conversation', in I.G. Barbour (ed.), *Western Man and Environmental Ethics: Attitudes Toward Nature and Technology*, Addison Wesley, Reading (Mass.), 55–64.

Williams, R. (nd), *Socialism and Ecology*, Socialist Environment and Resources Association, London.

Willums, J-O. (1998), *The Sustainable Business Challenge: A Briefing for Tomorrow's Business Leaders*, Greenleaf, Sheffield (Eng.).

Wilson, E.O. (1984), *Biophilia*, Cambridge University Press, Cambridge (Mass.).

———. (1988), 'The Current State of Biological Diversity', in E.O. Wilson (ed.), *Biodiversity*, National Academy Press, Washington (D.C.), 3–18.

———. (1993), 'Biophilia and the Environmental Ethic', in S.R. Kellert and E.O. Wilson (eds), *The Biophilia Hypothesis*, Island Press, Washington (D.C.), 31–41.

———. (1996), *In Search of Nature*, Penguin, London.

Windle, P. (1995), 'The Ecology of Grief', in T. Roszak, M.E. Gomes, and A.D. Kanner (eds), *Ecopsychology: Restoring the Earth, Healing the Mind*, Sierra Club, San Francisco (Calif.), 136–55.

Wissenburg, M. (1993), 'The Idea of Nature and the Nature of Distributive Justice', in A. Dobson and P. Lucardie (eds), *The Politics of Nature: Explorations in Green Political Theory*, Routledge, London, 3–20.

———. (1998), *Green Liberalism: The Free and the Green Society*, UCL Press, London.

Witoszek, N., and Brennan, A. (eds) (1999), *Philosophical Dialogues: Arne Naess and the Progress of Ecophilosophy*, Rowman & Littlefield, Lanham (Md.).

Wittbecker, A.E. (1986), 'Deep Anthropology: Ecology and Human Order', *Environmental Ethics*, 8, 261–70.

Wittfogel. K.A. (1957), *Oriental Despotism: A Comparative Study of Total Power*, Yale University Press, New Haven (Conn.).

Wolin, R. (1990), *The Politics of Being: the Political Thought of Martin Heidegger*, Columbia University Press, New York.

Woodcock, G. (1974), 'Anarchism and Ecology', *The Ecologist*, 4, 84–89.

Worster, D. (1983), 'Water and the Flow of Power', *The Ecologist*, 13, 168–74.

———. (1985), *Nature's Economy: A History of Ecological Ideas*, Cambridge University Press, Cambridge.

———. (1995), 'Nature and the Disorder of History', in M.E. Soule and G. Lease (eds), *Reinventing Nature? Responses to Postmodern Deconstruction*, Island Press, Washington (D.C.), 65–85.

———. (1997), 'The Wilderness of History', *Wild Earth*, Fall, 9–13.

Wright, J. (1990), 'Wilderness and Wasteland', *Island*, 42, 3–7.

Yearley, S. (1992), *The Green Case: A Sociology of Environmental Issues, Arguments, and Politics*, Routledge, London, 1992.

Young, I.M. (1990), *Justice and the Politics of Difference*, Princeton University Press,

Princeton (N.J.).

———. (1995), 'Communication and the Other: Beyond Deliberative Democracy', in M. Wilson and A. Yeatman (eds), *Justice and Identity: Antipodean Practices*, Bridget Williams, Wellington (NZ), 134–52.

Young, J. (1991), *Sustaining the Earth: The Past, Present and Future of the Green Revolution,* New South Wales University Press, Kensington (NSW).

Zarsky, L. (1990–91), 'Green Economics', in *21st C.: Previews of a Changing World*, Commission for the Future, Melbourne (Vic.), 22–25.

Zimmerman, M.E. (1979), 'Marx and Heidegger on the Technological Domination of Nature', *Philosophy Today*, 23, 99–112.

———. (1983), 'Toward a Heideggerian *Ethos* for Radical Environmentalism', *Environmental Ethics*, 5, 99–131.

———. (1987), 'Feminism, Deep Ecology, and Environmental Ethics', *Environmental Ethics*, 9, 21–44.

———. (1990a), 'Deep Ecology and Ecofeminism: The Emerging Dialogue', in I. Diamond and G.F. Orenstein (eds), *Reweaving the World: The Emergence of Ecofeminism*, Sierra Club, San Francisico (Calif.), 138–54.

———. (1990b), *Heidegger's Confrontation with Modernity: Technology, Politics, Art*, Indiana University Press, Bloomington (Ind.).

———. (1991), 'Deep Ecology, Ecoactivism, and Human Evolution', *ReVision*, 13, 122–28.

———. (1994), *Contesting Earth's Future: Radical Ecology and Postmodernity*, University of California Press, Berkeley (Calif.).

———. (1996), 'The Postmodern Challenge to Environmentalism', *Terra Nova*, 1(2), 131–40.

INDEX